SMALL WARS
MANUAL

SMALL WARS MANUAL

U.S. MARINE CORPS

SKYHORSE PUBLISHING

Skyhorse Publishing books may be purchased in bulk at special discounts for sales promotion, corporate gifts, fund-raising, or educational purposes. Special editions can also be created to specifications. For details, contact the Special Sales Department, Skyhorse Publishing, 555 Eighth Avenue, Suite 903, New York, NY 10018 or info@skyhorsepublishing.com.

www.skyhorsepublishing.com

10 9 8 7 6 5 4 3 2 1

Library of Congress Cataloging-in-Publication Data

United States. Marine Corps.
  Small wars manual / U.S. Marine Corps.
     p. cm.
  Originally published: Washington, D.C. : U.S.G.P.O., 1940.
  ISBN 978-1-60239-696-8
  1. United States. Marine Corps--Handbooks, manuals, etc. 2. Counterinsurgency--Handbooks, manuals, etc. 3. Low-intensity conflicts (Military science)--Handbooks, manuals, etc. 4. Guerrilla warfare--Handbooks, manuals, etc. I. Title.
  VE153.U54 2009
  355.02'18--dc22
                            2009016593

Printed in Canada

Dear Reader,

You may notice that there are some pages missing in this edition of *Small Wars Manual*. The pages were blank and were omitted to conserve paper. The omitted pages do not contain information.

—Skyhorse Publishing

DEPARTMENT OF THE NAVY
HEADQUARTERS UNITED STATES MARINE CORPS
WASHINGTON, D.C. 20380

NAVMC 2890
PL55g
1 Apr 1987

FOREWORD

1. **PURPOSE**

   To distribute a reprint of the 1940 edition of the <u>Small Wars Manual</u> as an aid
to education and training in the historical approach of Marine Corps units conducting operations in low-intensity conflicts. In the 1930's, such conflicts were
referred to as "small wars."

2. **INFORMATION**

   a. The <u>Small Wars Manual</u> is one of the best books on military operations in
peacekeeping and counterinsurgency operations published before World War II.

   b. This Manual is published for information only and is not directive in nature.
It should be read in the context of pre-World War II politics and operational
methods. Its republication does not indicate endorsement by the United States
Marine Corps of any statement in the book.

3. **RESPONSIBILITIES**

   a. Commanders should ensure that this Manual is available to Marines concerned
with the types of activities discussed in this Manual.

   b. Additional copies of this Manual may be obtained from the Marine Corps publications stock point at Marine Corps Logistics Base, Albany, Georgia, per instructions
in MCO P5600.31.

4. **RESERVE APPLICABILITY**

   This Manual is applicable to the Marine Corps Reserve.

5. **CERTIFICATION**

   Reviewed and approved this date.

JOHN PHILLIPS
Deputy Chief of Staff for
Plans, Policies and Operations

TABLE OF CONTENTS

The Small Wars Manual, United States Marine Corps, 1940, is published in 15 chapters as follows:

TABLE OF CONTENTS

TABLE OF CONTENTS

TABLE OF CONTENTS

TABLE OF CONTENTS

TABLE OF CONTENTS

TABLE OF CONTENTS

TABLE OF CONTENTS

TABLE OF CONTENTS

TABLE OF CONTENTS

TABLE OF CONTENTS

TABLE OF CONTENTS

TABLE OF CONTENTS

SMALL WARS MANUAL
UNITED STATES MARINE CORPS
1940

CHAPTER I

INTRODUCTION

RESTRICTED

UNITED STATES

GOVERNMENT PRINTING OFFICE

WASHINGTON : 1940

TABLE OF CONTENTS

The Small Wars Manual, U. S. Marine Corps, 1940, is published in
15 chapters, as follows:

SMALL WARS MANUAL

UNITED STATES MARINE CORPS

CHAPTER I

INTRODUCTION

(V)

SECTION I

GENERAL CHARACTERISTICS

1-1. **Small wars defined.**—*a.* The term "Small War" is often a vague name for any one of a great variety of military operations. As applied to the United States, small wars are operations undertaken under executive authority, wherein military force is combined with diplomatic pressure in the internal or external affairs of another state whose government is unstable, inadequate, or unsatisfactory for the preservation of life and of such interests as are determined by the foreign policy of our Nation. As herein used the term is understood in its most comprehensive sense, and all the successive steps taken in the development of a small war and the varying degrees of force applied under various situations are presented.

b. The assistance rendered in the affairs of another state may vary from a peaceful act such as the assignment of an administrative assistant, which is certainly nonmilitary and not placed under the classification of small wars, to the establishment of a complete military government supported by an active combat force. Between these extremes may be found an infinite number of forms of friendly assistance or intervention which it is almost impossible to classify under a limited number of individual types of operations.

c. Small wars vary in degrees from simple demonstrative operations to military intervention in the fullest sense, short of war. They are not limited in their size, in the extent of their theater of operations nor their cost in property, money, or lives. The essence of a small war is its purpose and the circumstances surrounding its inception and conduct, the character of either one or all of the opposing forces, and the nature of the operations themselves.

d. The ordinary expedition of the Marine Corps which does not involve a major effort in regular warfare against a first-rate power

1

GENERAL CHARACTERISTICS

may be termed a small war. It is this type of routine active foreign duty of the Marine Corps in which this manual is primarily interested. Small wars represent the normal and frequent operations of the Marine Corps. During about 85 of the last 100 years, the Marine Corps has been engaged in small wars in different parts of the world. The Marine Corps has landed troops 180 times in 37 countries from 1800 to 1934. Every year during the past 36 years since the Spanish-American War, the Marine Corps has been engaged in active operations in the field. In 1929 the Marine Corps had two-thirds of its personnel employed on expeditionary or other foreign or sea duty outside of the continental limits of the United States.

1-2. **Classes of small wars.**—*a.* Most of the small wars of the United States have resulted from the obligation of the Government under the spirit of the Monroe Doctrine and have been undertaken to suppress lawlessness or insurrection. Punitive expeditions may be resorted to in some instances, but campaigns of conquest are contrary to the policy of the Government of the United States. It is the duty of our statesmen to define a policy relative to international relationships and provide the military and naval establishments with the means to carry it into execution. With this basis, the military and naval authorities may act intelligently in the preparation of their war plans in close cooperation with the statesman. There is mutual dependence and responsibility which calls for the highest qualities of statesmanship and military leadership. The initiative devolves upon the statesmen.

b. The legal and military features of each small war present distinctive characteristics which make the segregation of all of them into fixed classifications an extremely difficult problem. There are so many combinations of conditions that a simple classification of small wars is possible only when one is limited to specific features in his study, i. e., according to their legal aspects, their military or naval features, whether active combat was engaged in or not, and many other considerations.

1-3. **Some legal aspects of small wars.**—*a.* According to international law, as recognized by the leading nations of the world, a nation may protect, or demand protection for, its citizens and their property wherever situated. The President of the United States as the Chief Executive is, under the Constitution, primarily charged with the conduct of foreign relations, including the protection of the lives and property of United States citizens abroad, save insofar as the Constitution expressly vests a part of these functions in some other branch of the Government. (For example, the participation of the

Senate in the making of treaties.)   It has been an unbroken policy of the President of the United States so to interpret their powers, beginning with the time of President Jefferson down to the present with the exception of President Buchanan.

b. The following pertinent extracts from U. S. Navy Regulations are cited:

> On occasion where injury to the United States or to citizens thereof is committed or threatened, in violation of the principles of international law or treaty right, the Commander in Chief shall consult with the diplomatic representative or consul of the United States and take such steps as the gravity of the case demands, reporting immediately to the Secretary of the Navy all the facts. The responsibility for any action taken by a naval force, however, rests wholly upon the commanding officer thereof.
>
> The use of force against a foreign and friendly state, or against anyone within the territories thereof, is illegal. The right of self-preservation, however, is a right which belongs to states as well as to individuals, and in the case of states it includes the protection of the state, its honor, and its possessions, and lives and property of its citizens against arbitrary violence, actual or impending, whereby the state or its citizens may suffer irreparable injury. The conditions calling for the application of the right of self-preservation cannot be defined beforehand, but must be left to the sound judgment of responsible officers, who are to perform their duties in this respect with all possible care and forbearance. In no case shall force be exercised in time of peace otherwise than as an application of the right of self-preservation as above defined. It must be used only as a last resort, and then only to the extent which is absolutely necessary to accomplish the end required. It can never be exercised with a view to inflicting punishment for acts already committed.
>
> Whenever, in the application of the above-mentioned principles, it shall become necessary to land an armed force in foreign territory on occasion of political disturbance where the local authorities are unable to give adequate protection to life and property, the assent of such authorities, or of some one of them, shall first be obtained, if it can be done without prejudice to the interests involved. Due to the ease with which the Navy Department can be communicated from all parts of the world, no commander in chief, flag officer, or commanding officer shall issue an ultimatum to the representative of any foreign government, or demand the performance of any service from any such representative that must be executed within a limited time, without first communicating with the Navy Department except in extreme cases where such action is necessary to save life. (U. S. Navy Regulations. NR. 722, 723, and 724.)

c. The use of the forces of the United States in foreign countries to protect the lives and property of American citizens resident in those countries does not necessarily constitute an act of war, and is, therefore, not equivalent to a declaration of war. The President, as chief executive of the nation, charged with the responsibility of the lives and property of United States citizens abroad, has the authority to use the forces of the United States to secure such protection in foreign countries.

GENERAL CHARACTERISTICS

d. The history of the United States shows that in spite of the varying trend of the foreign policy of succeeding administrations, this Government has interposed or intervened in the affairs of other states with remarkable regularity, and it may be anticipated that the same general procedure will be followed in the future. It is well that the United States may be prepared for any emergency which may occur whether it is the result of either financial or physical disaster, or social revolution at home or abroad. Insofar as these conditions can be predicted, and as these plans and preparations can be undertaken, the United States should be ready for either of these emergencies with strategical and tactical plans, preliminary preparations, organization, equipment, education, and training.

1–4. **Functions of headquarters Marine Corps.**—*a.* Small wars, generally being the execution of the responsibilities of the President in protecting American interests, life and property abroad, are therefore conducted in a manner different from major warfare. In small wars, diplomacy has not ceased to function and the State Department exercises a constant and controlling influence over the military operations. The very inception of small wars, as a rule, is an official act of the Chief Executive who personally gives instructions without action of Congress.

b. The President, who has been informed of a given situation in some foreign country through the usual agencies at his disposal, makes the decision concerning intervention. In appropriate cases this decision is communicated to the Secretary of the Navy. The senior naval officer present in the vicinity of the disturbance may then be directed to send his landing force ashore, or given authority to do so at his discretion; the Marine Corps may be ordered to have an expeditionary force ready to proceed overseas with the minimum delay. These instructions are communicated to the Marine Corps via the Secretary of the Navy or Assistant Secretary. Frequently a definite number of men is called for and not a military organization; for example, 500 men (not one battalion). It is desirable, however, that a definite military organization which approximates the required strength and characteristics for accomplishing the mission be specified, such as one infantry battalion; one infantry regiment (plus one motor transport platoon), etc. The word often comes very suddenly and calls for the immediate concentration of the forces, ready to take passage on a certain transport which will be made available at a given time and place. Generally there are no other instructions than that the force shall report to * * *, "the Commander Special Service Squadron," for example. Thereupon Headquarters Marine Corps designates the force, its personnel, organization, arms, and equipment; all necessary stores are provided and orders issued for the commanding officer of the force to report in person or by dispatch to the SOP or other authority in the disturbed area. With the present organized Fleet Marine Force ready for movement at a moment's notice, the Marine Corps now has available a highly trained and well equipped expeditionary force for use in small wars, thus eliminating in a large measure the former practice of hastily organizing and equipping such a force when the emergency arose. Accompanying these simple organization and movement orders are the monograph, maps, and other pertinent intelligence data of the disturbed area, to the extent that such information is on file and can be prepared for delivery to the Force Commander within the time limit. Thereafter Headquarters confines itself to the administrative details of the personnel replacements and the necessary supply of the force in the field.

c. The operations of the Force are directed by the Office of the Naval Operations direct or through the local naval Commander if he is senior to the Force Commander.

1–5. **Phases of small wars.**—*a.* Small wars seldom develop in accordance with any stereotyped procedure. Certain phases of those listed below may be absent in one situation; in another they may be combined and undertaken simultaneously; in still others one may find that the sequence of events or phases may be altered.

The actual operations of small wars may be arbitrarily divided into five phases as follows:

Phase 1. Initial demonstration or landing and action of vanguard.
Phase 2. The arrival of reenforcements and general military operations in the field.
Phase 3. Assumption of control of executive agencies, and cooperation with the legislative and judicial agencies.
Phase 4. Routine police functions.
Phase 5. Withdrawal from the Theater of Operations.

b. First phase.—Initial demonstration or landing and action of vanguard.

(1) One of the most common characteristics of the small wars of the United States is that its forces "dribble in" to the countries in which they intervene. This is quite natural in view of the national policy of the government. It is not at war with the neighboring state; it proposes no aggression or seizure of territory; its purpose is friendly and it wishes to accomplish its objectives with as little military display as possible with a view to gaining the lasting friendship of the inhabi-

GENERAL CHARACTERISTICS

tants of the country. Thus our Government is observed endeavoring to accomplish its end with the minimum of troops, in fact, with nothing more than a demonstration of force if that is all that is necessary and reasonably sufficient. This policy is carried on throughout the campaign and reenforcements are added by "driblets," so many companies, or a battalion, or a regiment at a time, until the force is large enough to accomplish its mission or until its its peacetime limitations in personnel have been reached. Even after landing, instructions probably will be received not to exert any physical force unless it becomes absolutely necessary, and then only to the minimum necessary to accomplish its purpose. Thus orders may be received not to fire on irregulars unless fired upon; instructions may be issued not to fire upon irregular groups if women are present with them even though it is known that armed women accompany the irregulars.

(2) During the initial phase small numbers of troops may be sent ashore to assume the initiative, as a demonstration to indicate a determination to control the situation, and to prepare the way for any troops to follow. This vanguard is generally composed of marine detachments or mixed forces of marines and sailors from ships at the critical points. Owing to its limited personnel the action of the vanguard will often be restricted to an active defense after seizing a critical area such as an important seaport or other city, the capital of a country or disturbed areas of limited extent.

c. Second phase.—The arrival of reenforcements and general military operations in the field.

During this period the theater of operations is divided into areas and forces are assigned for each. Such forces should be sufficiently strong to seize and hold the most important city in the area assigned and to be able to send combat patrols in all directions. If certain neutral zones have not been designated in the first phase, it may be done at this time if deemed advisable. During this phase the organization of a native military and police force is undertaken. In order to release ships' personnel to their normal functions afloat, such personnel are returned to their ships as soon as they can be relieved by troops of the expeditionary force.

d. Third phase.—Assumption of control of executive agencies, and cooperation with the legislative and judicial agencies.

If the measures in phase 2 do not bring decisive results, it may be necessary to resort to more thorough measures. This may involve the establishment of military government or martial law in varying degree from minor authority to complete control of the principal agen-

GENERAL CHARACTERISTICS

cies of the native government; it will involve the further strengthening of our forces by reenforcements. More detachments will be sent out to take other important localities; more active and thorough patrolling will be undertaken; measures will be taken to intercept the vital supply and support channels of the opposing factions and to break the resistance to law and order by a combination of effort of physical and moral means. During this period the marines carry the burden of most of the patrolling. Native troops, supported by marines, are increasingly employed as early as practicable in order that these native agencies may assume their proper responsibility for restoring law and order in their own country as an agency of their government.

e. Fourth phase.—Routine police functions. (1) After continued pressure of the measures in phase three, it is presumed that sooner or later regular forces will subdue the lawless elements. Military police functions and judicial authority, to the extent that they have been assumed by our military forces, are gradually returned to the native agencies to which they properly belong.

(2) Our military forces must not assume any judicial responsibility over local inhabitants beyond that expressly provided by proper authority. The judicial powers of commanders of detached posts must be clearly defined in orders from superior authority. Furthermore, as long as the judicial authority rests squarely upon the shoulders of the civil authorities, the military forces should continually impress and indoctrinate them with their responsibility while educating the people in this respect. Each situation presents certain characteristics peculiar to itself; in one instance officers were clothed with almost unlimited military authority within the law and our treaty rights; in another, less authority was exercised over the population; and in the third instance the forces of occupation had absolutely no judicial authority. The absence of such authority is often a decided handicap to forces of occupation in the discharge of their responsibilities. If the local judicial system is weak, or broken down entirely, it is better to endow the military authorities with temporary and legal judicial powers in order to avoid embarrassing situations which may result from illegal assumption.

(3) During this phase the marines act as a reserve in support of the native forces and are actively employed only in grave emergencies. The marines are successively withdrawn to the larger centers, thus affording a better means for caring for the health, comfort, and recreation of the command.

GENERAL CHARACTERISTICS

f. Fifth phase.—Withdrawal from the theater of operations. Finally, when order is restored, or when the responsible native agencies are prepared to handle the situation without other support, the troops are withdrawn upon orders from higher authority. This process is progressive from the back country or interior outward, in the reverse order to the entry into the country. After evacuation of the forces of intervention, a Legation Guard, which assumes the usual functions of such a detachment, may be left in the capital.

1-6. **Summary.**—*a.* Since the World War there has been a flood of literature dealing with the old principles illustrated and the new technique developed in that war: but there always have been and ever will be other wars of an altogether different kind, undertaken in very different theaters of operations and requiring entirely different methods from those of the World War. Such are the small wars which are described in this manual.

b. There is a sad lack of authoritative texts on the methods employed in small wars. However, there is probably no military organization of the size of the U. S. Marine Corps in the world which has had as much practical experience in this kind of combat. This experience has been gained almost entirely in small wars against poorly organized and equipped native irregulars. With all the practical advantages we enjoyed in those wars, that experience must not lead to an underestimate of the modern irregular, supplied with modern arms and equipment. If marines have become accustomed to easy victories over irregulars in the past. they must now prepare themselves for the increased effort which will be necessary to insure victory in the future. The future opponent may be as well armed as they are; he will be able to concentrate a numerical superiority against isolated detachments at the time and place he chooses; as in the past he will have a thorough knowledge of the trails. the country, and the inhabitants; and he will have the inherent ability to withstand all the natural obstacles, such as climate and disease, to a greater extent than a white man. All these natural advantages, combining primitive cunning and modern armament, will weigh heavily in the balance against the advantage of the marine forces in organization, equipment, intelligence. and discipline, if a careless audacity is permitted to warp good judgment.

c. Although small wars present a special problem requiring particular tactical and technical measures. the immutable principles of war remain the basis of these operations and require the greatest ingenuity in their application. As a regular war never takes exactly

8

GENERAL CHARACTERISTICS

the form of any of its predecessors, so, even to a greater degree is each small war somewhat different from anything which has preceded it. One must ever be on guard to prevent his views becoming fixed as to procedure or methods. Small wars demand the highest type of leadership directed by intelligence, resourcefulness, and ingenuity. Small wars are conceived in uncertainty, are conducted often with precarious responsibility and doubtful authority, under indeterminate orders lacking specific instructions.

d. Formulation of foreign policy in our form of government is not a function of the military. Relations of the United States with foreign states are controlled by the executive and legislative branches of the Government. These policies are of course binding upon the forces of intervention, and in the absence of more specific instructions, the commander in the field looks to them for guidance. For this reason all officers should familiarize themselves with current policies. A knowledge of the history of interventions, and the displays of force and other measures short of war employed by our Government in the past, are essential to thorough comprehension of our relations with foreign states insofar as these matters are concerned.

Section II

STRATEGY

1-7. The basis of the strategy.—*a.* The military strategy of small wars is more directly associated with the political strategy of the campaign than is the case in major operations. In the latter case, war is undertaken only as a last resort after all diplomatic means of adjusting differences have failed and the military commander's objective ordinarily becomes the enemy's armed forces.

b. Diplomatic agencies usually conduct negotiations with a view to arriving at a peaceful solution of the problem on a basis compatible with both national honor and treaty stipulations. Although the outcome of such negotiations often results in a friendly settlement, the military forces should be prepared for the possibility of an unfavorable termination of the proceedings. The mobilization of armed forces constitutes a highly effective weapon for forcing the opponent to accede to national demands without resort to war. When a time limit for peaceful settlement is prescribed by ultimatum the military-naval forces must be prepared to initiate operations upon expiration of the time limit.

c. In small wars, either diplomacy has not been exhausted or the party that opposes the settlement of the political question cannot be reached diplomatically. Small war situations are usually a phase of, or an operation taking place concurrently with, diplomatic effort. The political authorities do not relinquish active participation in the negotions and they ordinarily continue to exert considerable influence on the military campaign. The military leader in such operations thus finds himself limited to certain lines of action as to the strategy and even as to the tactics of the campaign. This feature has been so marked in past operations, that marines have been referred to as State Department Troops in small wars. In certain cases of this kind the State Department has even dictated the size of the force to be sent to the theater of operations. The State Department materially influences the strategy and tactics by orders and instructions which are promulgated through the Navy Department or through diplomatic representatives.

d. State Department officials represent the Government in foreign countries. The force generally nearest at hand to back up the authority of these agents is the Navy. In such operations the Navy is performing its normal function, and has, as a component part of its organization, the Fleet Marine Force, organized, equipped, and trained to perform duty of this nature. After the Force has landed, the commander afloat generally influences the operations only to the extent necessary to insure their control and direction in accordance with the policy of the instructions that he has received from higher authority. He supports and cooperates with the Force to the limit of his ability. In the latter stages of the operation the local naval commander may relinquish practically all control in order to carry out routine duties elsewhere. In such case the general operations plan is directed by, or through, the office of the Naval Operations in Washington.

e. Wars of intervention have two classifications; intervention in the internal, or intervention in the external affairs of another state. Intervention in the internal affairs of a state may be undertaken to restore order, to sustain governmental authority, to obtain redress, or to enforce the fulfilment of obligations binding between the two states. Intervention in the external affairs of a state may be the result of a treaty which authorizes one state to aid another as a matter of political expediency, to avoid more serious consequences when the interests of other states are involved, or to gain certain advantages not obtainable otherwise. It may be simply an intervention to enforce certain opinions or to propagate certain doctrines, principles, or standards. For example, in these days when pernicious propaganda is employed to spread revolutionary doctrines, it is conceivable that the United States might intervene to prevent the development of political disaffection which threatens the overthrow of a friendly state and indirectly influences our own security.

1-8. Nature of the operations.—*a.* Irregular troops may disregard, in part or entirely, International Law and the Rules of Land Warfare in their conduct of hostilities. Commanders in the field must be prepared to protect themselves against practices and methods of combat not sanctioned by the Rules of War.

b. Frequently irregulars kill and rob peaceful citizens in order to obtain supplies which are then secreted in remote strongholds. Seizure or destruction of such sources of supply is an important factor in reducing their means of resistance. Such methods of operation must be studied and adapted to the psychological reaction they will produce upon the opponents. Interventions or occupations are usually peaceful and altruistic. Accordingly, the methods of procedure must rigidly conform to this purpose; but when forced to resort to arms to carry out the object of the intervention, the operation must be pursued energetically and expeditiously in order to overcome the resistance as quickly as possible.

c. The campaign plan and strategy must be adapted to the character of the people encountered. National policy and the precepts of civilized procedure demand that our dealings with other peoples be maintained on a high-moral plan. However, the military strategy of the campaign and the tactics employed by the commander in the field must be adapted to the situation in order to accomplish the mission without delay.

d. After a study has been made of the people who will oppose the intervention, the strategical plan is evolved. The military strategical plan should include those means which will accomplish the purpose in view quickly and completely. Strategy should attempt to gain psychological ascendancy over the outlaw or insurgent element prior to hostilities. Remembering the political mission which dictates the military strategy of small wars, one or more of the following basic modes of procedure may be decided upon, depending upon the situation:

(1) Attempt to attain the aims of the intervention by a simple, clear, and forceful declaration of the position and intention of the occupying force, this without threat or promise.

(2) By a demonstration of the power which could be employed to carry out these intentions.

(3) The display of the naval or military force within the area involved.

(4) The actual application of armed force. During the transitory stage or prior to active military operations, care should be taken to avoid the commission of any acts that might precipitate a breach. Once armed force is resorted to, it should be applied with determination and to the extent required by the situation. Situations may develop so rapidly that the transition from negotiations to the use of armed force gives the commander little or no time to exert his influence through the use of the methods mentioned in subparagraphs (2) and (3) above.

e. The strategy of this type of warfare will be strongly influenced by the probable nature of the contemplated operations. In regular

STRATEGY

warfare the decision will be gained on known fronts and probably limited theaters of operations; but in small wars no defined battle front exists and the theater of the operations may be the whole length and breadth of the land. While operations are carried out in one area, other hostile elements may be causing serious havoc in another. The uncertainty of the situation may require the establishment of detached posts within small areas. Thus the regular forces may be widely dispersed and probably will be outnumbered in some areas by the hostile forces. This requires that the Force be organized with a view to mobility and flexibility, and that the troops be highly trained in the use of their special weapons as well as proper utilization of terrain.

f. Those who have participated in small wars agree that these operations find an appropriate place in the art of war. Irregular warfare between two well-armed and well-disciplined forces will open up a larger field for surprise, deception, ambuscades, etc., than is possible in regular warfare.

1–9. **National war.**—*a.* In small wars it can be expected that hostile forces in occupied territory will employ guerilla warfare as a means of gaining their end. Accounts of recent revolutionary movements, local or general, in various parts of the world indicate that young men of 18 or 20 years of age take active parts as organizers in these disturbances. Consequently, in campaigns of this nature the Force will be exposed to the action of this young and vigorous element. Rear installations and lines of communications will be threatened. Movements will be retarded by ambuscades and barred defiles, and every detachment presenting a tempting target will be harassed or attacked. In warfare of this kind, members of native forces will suddenly become innocent peasant workers when it suits their fancy and convenience. In addition, the Force will be handicapped by partisans, who constantly and accurately inform native forces of our movements. The population will be honeycombed with hostile sympathizers, making it difficult to procure reliable information. Such difficulty will result either from the deceit used by hostile sympathizers and agents, or from the intimidation of friendly natives upon whom reliance might be placed to gain information.

b. In cases of levees en masse, the problem becomes particularly difficult. This is especially true when the people are supported by a nucleus of disciplined and trained professional soldiers. This combination of soldier and armed civilian presents serious opposition

14

to every move attempted by the Force; even the noncombatants conspire for the defeat of the Force.

c. Opposition becomes more formidable when the terrain is difficult, and the resistance increases as the Force moves inland from its bases. Every native is a potential clever opponent who knows the country, its trails, resources, and obstacles, and who has friends and sympathizers on every hand. The Force may be obliged to move cautiously. Operations are based on information which is at best unreliable, while the natives enjoy continuous and accurate information. The Force after long and fatiguing marches fails to gain contact and probably finds only a deserted camp, while their opponents, still enjoying the initiative, are able to withdraw or concentrate strong forces at advantageous places for the purpose of attacking lines of communication, convoys, depots, or outposts.

d. It will be difficult and hazardous to wage war successfully under such circumstances. Undoubtedly it will require time and adequate forces. The occupying force must be strong enough to hold all the strategical points of the country, protect its communications, and at the same time furnish an operating force sufficient to overcome the opposition wherever it appears. Again a simple display of force may be sufficient to overcome resistance. While curbing the passions of the people, courtesy, friendliness, justice, and firmness should be exhibited.

e. The difficulty is sometimes of an economical, political, or social nature and not a military problem in origin. In one recent campaign the situation was an internal political problem in origin, but it had developed to such a degree that foreign national interests were affected; simple orderly processes could no longer be applied when it had outgrown the local means of control. In another instance the problem was economic and social; great tracts of the richest land were controlled and owned by foreign interests; this upset the natural order of things; the admission of cheap foreign labor with lower standards of living created a social condition among the people which should have been remedied by orderly means before it reached a crisis.

f. The application of purely military measures may not, by itself restore peace and orderly government because the fundamental causes of the condition of unrest may be economic, political, or social. These conditions may have originated years ago and in many cases have been permitted to develop freely without any attempt to apply corrective measures. An acute situation finally develops when conditions have reached a stage that is beyond control of the civil authorities and it is too late for diplomatic adjustment. The solution of such problems being basically a political adjustment, the military measures to be applied must be of secondary importance and should be applied only to such extent as to permit the continuation of peaceful corrective measures.

g. The initial problem is to restore peace. There may be many economic and social factors involved, pertaining to the administrative, executive, and judicial functions of the government. These are completely beyond military power as such unless some form of military government is included in the campaign plan. Peace and industry cannot be restored permanently without appropriate provisions for the economic welfare of the people. Moreover, productive industry cannot be fully restored until there is peace. Consequently, the remedy is found in emphasizing the corrective measures to be taken in order to permit the orderly return to normal conditions.

h. In general, the plan of action states the military measures to be applied, including the part the forces of occupation will play in the economic and social solution of the problem. The same consideration must be given to the part to be played by local government and the civil population. The efforts of the different agencies must be cooperative and coordinated to the attainment of the common end.

i. Preliminary estimates of the situation form the basis of plans to meet probable situations and should be prepared as far in advance as practicable. They should thereafter be modified and developed as new situations arise.

SECTION III

PSYCHOLOGY

1-10. **Foreword.**—*a.* While it is improbable that a knowledge of psychology will make any change in the fundamentals of the conduct of small wars, it will, however, lead to a more intelligent application of the principles which we now follow more or less unconsciously through custom established by our predecessors.

b. Psychology has always played an important part in war. This knowledge was important in ancient wars of masses; it becomes more so on the modern battlefield, with widely dispersed forces and the complexity of many local operations by small groups, or even individuals, making up the sum total of the operation. In former times the mass of enemy troops, like our own, was visible to and under the immediate control of its leaders. Now troops are dispersed in battle and not readily visible, and we must understand the psychology of the individual, who operates beyond the direct control of his superiors.

c. This difficulty of immediate control and personal influence is even more pronounced and important in small wars, on account of the decentralized nature of these operations. This fact is further emphasized because in the small wars we are dealing not only with our own forces, but also with the civil population which frequently contains elements of doubtful or antagonistic sentiments. The very nature of our own policy and attitude toward the opposing forces and normal contacts with them enable the personnel of our Force to secure material advantages through the knowledge and application of psychological principles.

d. This knowledge does not come naturally to the average individual. A study of men and human nature supplemented by a thorough knowl-

PSYCHOLOGY

edge of psychology should enable those faced with concrete situations of this type to avoid the ordinary mistakes. The application of the principles of psychology in small wars is quite different from their normal application in major warfare or even in troop leadership. The aim is not to develop a belligerent spirit in our men but rather one of caution and steadiness. Instead of employing force, one strives to accomplish the purpose by diplomacy. A Force Commander who gains his objective in a small war without firing a shot has attained far greater success than one who resorted to the use of arms. While endeavoring to avoid the infliction of physical harm to any native, there is always the necessity of preventing, as far as possible, any casualties among our own troops.

e. This is the policy with which our troops are indoctrinated; a policy which governs throughout the period of intervention and finds exception only in those situations where a resort to arms and the exercise of a belligerent spirit are necessary. This mixture of combined peaceful and warlike temperament, where adapted to any single operation, demands an application of psychology beyond the requirements of regular warfare. Our troops at the same time are dealing with a strange people whose racial origin, and whose social, political, physical and mental characteristics may be different from any before encountered.

f. The motive in small wars is not material destruction. It is usually a project dealing with the social, economic, and political development of the people. It is of primary importance that the fullest benefit be derived from the psychological aspects of the situation. That implies a serious study of the people, their racial, political, religious, and mental development. By analysis and study the reasons for the existing emergency may be deduced; the most practical method of solving the problem is to understand the possible approaches thereto and the repercussion to be expected from any actions which may be contemplated. By this study and the ability to apply correct psychological doctrine, many pitfalls may be avoided and the success of the undertaking assured.

g. The great importance of psychology in small wars must be appreciated. It is a field of unlimited extent and possibilities, to which much time and study should be devoted. It cannot be stated in rules and learned like mathematics. Human reactions cannot be reduced to an exact science, but there are certain principles which should guide our conduct. These principles are deduced by studying the history of the people and are mastered only by experience in their practical application.

PSYCHOLOGY

1-11. **Characteristics.**—The correct application of the principles of psychology to any given situation requires a knowledge of the traits peculiar to the persons with whom we are dealing. The individual characteristics as well as the national psychology are subjects for intensive study. This subject assumes increasing importance in minor operations. A failure to use tact when required or lack of firmness at a crucial moment might readily precipitate a situation that could have been avoided had the commander been familiar with the customs, religion, morals, and education of those with whom he was dealing.

1-12. **Fundamental considerations.**—The resistance to an intervention comes not only from those under arms but also from those furnishing material or moral support to the opposition. Sapping the strength of the actual or potential hostile ranks by the judicious application of psychological principles may be just as effective as battle casualties. The particular methods and extent of the application of this principle will vary widely with the situation. Some of the fundamental policies applicable to almost any situation are:

1. Social customs such as class distinctions, dress, and similar items should be recognized and receive due consideration.

2. Political affiliations or the appearance of political favoritism should be avoided; while a thorough knowledge of the political situation is essential, a strict neutrality in such matters should be observed.

3. A respect for religious customs.

Indifference in all the above matters can only be regarded as a lack of tact.

1-13. **Revolutionary tendencies.**—*a.* In the past, most of our interventions have taken place when a revolution was in full force or when the spirit of revolution was rampant. In view of these conditions (which are so often encountered in small wars) it may be well to consider briefly some of the characteristics of revolutions.

b. The knowledge of the people at any given moment of history involves an understanding of their environment, and above all, their past. The influence of racial psychology on the destiny of a people appears plainly in the history of those subject to perpetual revolutions. When composed largely of mixed races—that is to say, of individuals whose diverse heredities have dissociated their ancestral characteristics—those populations present a special problem. This class is always difficult to govern, if not ungovernable, owing to the absence of a fixed character. On the other hand, sometimes a people who have been under a rigid form of government may affect the most violent revolutions. Not having succeeded in developing progressively, or in

19

PSYCHOLOGY

adapting themselves to changes of environment, they are likely to react violently when such adaptation becomes inevitable.

c. Revolution is the term generally applied to sudden political changes, but the expression may be employed to denote any sudden transformation whether of beliefs, ideas, or doctrines. In most cases the basic causes are economic. Political revolutions ordinarily result from real or fancied grievances, existing in the minds of some few men, but many other causes may produce them. The word "discontent" sums them up. As soon as discontent becomes general a party is formed which often becomes strong enough to offer resistance to the government. The success of a revolution often depends on gaining the assistance or neutrality of the regular armed forces. However, it sometimes happens that the movement commences without the knowledge of the armed forces; but not infrequently it has its very inception within these forces. Revolutions may take place in the capital, and by contagion spread through the country. In other instances the general disaffection of the people takes concrete form in some place remote from the capital, and when it has gathered momentum moves on the capital.

d. The rapidity with which a revolution develops is made possible by modern communication facilities and publicity methods. Trivial attendant circumstances often play highly important roles in contributing to revolution and must be observed closely and given appropriate consideration. The fact is that beside the great events of which history treats there are the innumerable little facts of daily life which the casual observer may fail to see. These facts individually may be insignificant. Collectively, their volume and power may threaten the existence of the government. The study of the current history of unstable countries should include the proper evaluation of all human tendencies. Local newspapers and current periodicals are probably the most valuable sources for the study of present psychological trends of various nations. Current writings of many people of different classes comprise a history of what the people are doing and thinking and the motives for their acts. Thus, current periodicals, newspapers, etc., will more accurately portray a cross section of the character of the people. In studying the political and psychical trends of a country, one must ascertain whether or not all news organs are controlled by one political faction, in order to avoid developing an erroneous picture of the situation.

e. Governments often almost totally fail to sense the temper of their people. The inability of a government to comprehend existing condi-

PSYCHOLOGY

tions, coupled with its blind confidence in its own strength, frequently results in remarkably weak resistance to attack from within.

f. The outward events of revolutions are always a consequence of changes, often unobserved, which have gone slowly forward in men's minds. Any profound understanding of a revolution necessitates a knowledge of the mental soil in which the ideas that direct its course have to germinate. Changes in mental attitude are slow and hardly perceptible; often they can be seen only by comparing the character of the people at the beginning and at the end of a given period.

g. A revolution is rarely the result of a widespread conspiracy among the people. Usually it is not a movement which embraces a very large number of people or which calls into play deep economic or social motives. Revolutionary armies seldom reach any great size; they rarely need to in order to succeed. On the other hand, the military force of the government is generally small, ill equipped, and poorly trained; not infrequently a part, if not all of it, proves to be disloyal in a political crisis.

h. The majority of the people, especially in the rural districts, dislike and fear revolutions, which often involve forced military service for themselves and destruction of their livestock and their farm produce. However, they may be so accustomed to misgovernment and exploitation that concerted effort to check disorderly tendencies of certain leaders never occurs to them. It is this mass ignorance and indifference rather than any disposition to turbulence in the nation as a whole, which has prevented the establishment of stable government in many cases.

i. Abuses by the officials in power and their oppression of followers of the party not in power, are often the seeds of revolution. The spirit which causes the revolution arouses little enthusiasm among the poor natives at large unless they are personally affected by such oppression. The revolution, once started, naturally attracts all of the malcontents and adventurous elements in the community. The revolution may include many followers, but its spirit emanates from a few leaders. These leaders furnish the spark without which there would be no explosion. Success depends upon the enthusiastic determination of those who inspire the movement. Under effective leadership the mass will be steeped in revolutionary principles, and imbued with a submission to the will of the leader and an enthusiastic energy to perform acts in support thereof. Finally, they feel that they are the crusaders for a new deal which will regenerate the whole country. In extremely remote, isolated, and illiterate sections an educated revolutionary leader may easily lead the inhabitants to believe that they, in the act of taking up arms, are actually engaged in repelling invasion. Many such ruses are employed in the initial stages and recruiting is carried on in this manner for long periods and the inhabitants are in a state of ignorance of the actual situation.

j. How is this situation to be met? A knowledge of the laws relating to the psychology of crowds is indispensible to the interpretation of the elements of revolutionary movements, and to their conduct. Each individual of the crowd, based on the mere fact that he is one of many, senses an invincible power which at once nullifies the feeling of personal responsibility. This spirit of individual irresponsibility and loss of identity must be overcome by preventing the mobilization or concentration of revolutionary forces, and by close supervision of the actions of individuals.

k. Another element of mob sentiment is imitation. This is particularly true in people of a low order of education. Attempt should be made to prevent the development of a hero of the revolutionary movement, and no one should be permitted to become a martyr to the cause. Members of a crowd also display an exaggerated independence.

l. The method of approaching the problem should be to make revolutionary acts nonpaying or nonbeneficial and at the same time endeavor to remove or remedy the causes or conditions responsible for the revolution. One obstacle in dealing with a revolution lies in the difficulty of determining the real cause of the trouble. When found, it is often disclosed as a minor fault of the simplest nature. Then the remedies are also simple.

m. The opposing forces may employ modern weapons and technique adapted to regular organized units, but the character of the man who uses these weapons remains essentially the same as it always was. The acts of a man are determined by his character; and to understand or predict the action of a leader or a people their character must be understood. Their judgments or decisions are based upon their intelligence and experience. Unless a revolutionary leader can be discountenanced in the eyes of his followers, it may be best to admit such leadership. Through him a certain discipline may be exercised which will control the actions of a revolutionary army; for without discipline, people and armies become barbarian hordes.

n. In general, revolutionary forces are new levies, poorly trained, organized, and equipped. Yet they can often be imbued with an ardent enthusiasm and are capable of heroism to the extent of giving their lives unhesitatingly in support of their beliefs.

PSYCHOLOGY

1-14. **Basic instincts.**—*a*. It is perfectly natural that the instinct of self-preservation should be constantly at work. This powerful influence plays an important part in the attitude of the natives in small wars. It is not surprising that any indication of intervention or interposition will prompt his instinct of self-preservation to oppose this move. Every means should be employed to convince such people of the altruistic intention of our Government.

b. Fear is one of the strongest natural emotions in man. Among primitive people not far removed from an oppressed or enslaved existence, it is easy to understand the people's fear of being again enslaved; fear of political subjugation causes violent opposition to any movement which apparently threatens political or personal liberty.

c. Another basic instinct of man is self-assertion. This is a desire to be considered worthy among his fellow beings. Life for the individual centers around himself. The individual values his contacts as good or bad according to how he presumes he has been treated and how much consideration has been given to his own merits. This instinct inspires personal resentment if his effort is not recognized. Pride, which is largely self-assertion, will not tolerate contradiction. Self-respect includes also the element of self-negation which enables one to judge his own qualities and profit by the example, precept, advice, encouragement, approval, or disapproval of others. It admits capacity to do wrong, since it accepts the obligation of social standards. In dealing with foreign peoples credit should be readily accorded where merited, and undue criticism avoided.

d. There are also peoples and individuals whose instinctive reaction in contact with external influence is that of self-submission. Here is found a people who, influenced by the great power of the United States, are too willing to shirk their individual responsibility and are too ready to let others shoulder the full responsibility for restoring and, still worse, maintaining order and normalcy. In this event, if the majority of the natives are thus inclined, the initial task is quite easy, but difficulty arises in attempting to return the responsibility to those to whom it rightfully belongs. As little local responsibility as possible to accomplish the mission should be assumed, while the local government is encouraged to carry its full capacity of responsibility. Any other procedure weakens the sovereign state, complicating the relationship with the military forces and prolonging the occupation.

PSYCHOLOGY

e. States are naturally very proud of their sovereignty. National policy demands minimum interference with that sovereignty. On occasion there is clash of opinion between the military and local civil power in a given situation, and the greatest tact and diplomacy is required to bring the local political authorities to the military point of view. When the matter is important, final analysis may require resort to more vigorous methods. Before a compromise is attempted, it should be clearly understood that such action does not sacrifice all the advantages of both of the opposing opinions.

f. The natives are also proud individually. One should not award any humiliating punishments or issue orders which are unnecessarily hurtful to the pride of the inhabitants. In the all-important interest of discipline, the invention and infliction of such punishments no matter how trivial must be strictly prohibited in order to prevent the bitterness which would naturally ensue.

g. In revolutions resort may be had to sabotage. Unless the circumstances demand otherwise, the repair of damage should be done by civilian or prison labor. This will have a more unfavorable psychological effect on the revolutionists than if the occupying forces were employed to repair the damage.

h. Inhabitants of countries with a high rate of illiteracy have many childlike characteristics. In the guidance of the destinies of such people, the more that one shows a fraternal spirit, the easier will be the task and the more effective the results. It is manifestly unjust to judge such people by our standards. In listening to peasants relate a story, whether under oath or not, or give a bit of information, it may appear that they are tricky liars trying to deceive or hide the truth, because they do not tell a coherent story. It should be understood that these illiterate and uneducated people live close to nature. The fact that they are simple and highly imaginative and that their background is based on some mystic form of religion gives rise to unusual kinds of testimony. It becomes a tedious responsibility to elicit the untarnished truth. This requires patience beyond words. The same cannot be said for all the white-collar, scheming politicians of the city who are able to distinguish between right and wrong, but who flagrantly distort the truth.

i. The "underground" or "grapevine" method of communication is an effective means of transmitting information and rumors with unbelievable rapidity among the natives. When events happen in one locality which may bring objectionable repercussions in another upon receipt of this information, it is well to be prepared to expect the

PSYCHOLOGY

speedy transmission of that knowledge even in spite of every effort to keep it localized or confidential. The same means might be considered for use by intelligence units in disseminating propaganda and favorable publicity.

j. Often natives refuse to give any information and the uninitiated might immediately presume that they are members of the hostile forces or at least hostile sympathizers. While the peasant hopes for the restoration of peace and order, the constant menace and fear of guerrillas is so overpowering that he does not dare to place any confidence in an occasional visiting patrol of the occupying forces. When the patrol leader demands information, the peasant should not be misjudged for failure to comply with the request, when by so doing, he is signing his own death warrant.

k. Actual authority must not be exceeded in demanding information. A decided advantage of having military government or martial law is to give the military authorities the power to bring legal summary, and exemplary punishment to those who give false information. Another advantage of such government is the authority to require natives to carry identification cards on their persons constantly. It has been found that the average native is not only willing and anxious, but proud to carry some paper signed by a military authority to show that he is recognized. The satisfaction of this psychological peculiarity and, what is more important or practical, its exploitation to facilitate the identity of natives is a consideration of importance. This also avoids most of the humiliating and otherwise unproductive process often resorted to in attempting to identify natives or their possible relationship to the opposing forces.

l. There are people among whom the spirit of self-sacrifice does not exist to the extent found among more highly civilized peoples or among races with fanatical tendencies. This may account for the absence of the individual bravery in the attack or assault by natives even where their group has a great preponderance of numbers; among certain peoples there is not the individual combat, knifing, machete attacks by lone men which one encounters among others. This may be due to the lack of medical care provided, lack of religious fanaticism, lack of recognition for personal bravery, or lack of provision for care of dependents in case of injury or death. Psychological study of the people should take this matter into consideration and the organization, tactics, and security measures must be adapted accordingly.

m. It is customary for some people to attempt to place their officials under obligation to them by offering gifts, or gratuitous services of different kinds. This is their custom and they will expect it to prevail among others. No matter how innocent acceptance may be, and in spite of the determination that it shall never influence subsequent actions or decisions, it is best not to be a party to any such petty bribery. Another common result of such transaction is that the native resorts to this practice among his own people to indicate that he is in official favor, and ignorant individuals on the other hand believe it. Needless to say, when it is embarrassing, or practically impossible to refuse to accept a gift or gratuity, such acceptance should not influence subsequent decisions. To prevent subsequent requests for favors the following is suggested: Accept the gift with the proper and expected delight; then, before the donor has an opportunity to see you and request a favor, send your servant with a few American articles obtainable in our commissaries and which are considered delicacies by the natives. The amount should be about equal in value, locally, to the gift accepted; and usually the native will feel that he has not placed you under an obligation.

n. Sometimes the hospitality of the natives must be accepted, and it is not intended to imply that this should not be done on appropriate occasions. On the contrary, this social intercourse is often fruitful of a better mutual understanding. Great care must be exercised that such contacts are not limited to the people of any social group or political party. This often leads to the most serious charges of discrimination and favoritism which, even though untrue, will diminish the respect, confidence, and support of all who feel that they are not among the favored. If opportunities are not presented, they should be created to demonstrate clearly to all, that contacts are not discriminatory and that opinions and actions are absolutely impartial.

1–15. Attitude and bearing.—*a.* A knowledge of the character of the people and a command of their language are great assets. Political methods and motives which govern the actions of foreign people and their political parties, incomprehensible at best to the average North American, are practically beyond the understanding of persons who do not speak their language. If not already familiar with the language, all officers upon assignment to expeditionary duty should study and acquire a working knowledge of it.

b. Lack of exact information is normal in these operations, as is true in all warfare. Lack of information does not justify withholding orders when needed, nor failing to take action when the situation demands it. The extent to which the intelligence service can obtain information depends largely on the attitude adopted toward the

PSYCHOLOGY

loyal and neutral population. The natives must be made to realize the seriousness of withholding information, but at the same time they must be protected from terrorism.

c. From the very nature of the operation, it is apparent that military force cannot be applied at the stage that would be most advantageous from a tactical viewpoint. Usually turbulent situations become extremely critical before the Government feels justified in taking strong action. Therefore, it is of the utmost importance to determine the exact moment when the decision of a commander should be applied. In a gradually developing situation the "when" is often the essence of the decision. Problems which illustrate the results of too hasty or tardy decisions will be of value in developing thought along these lines. The force commander should determine his mission and inform all subordinates accordingly. Commands should be kept fully informed of any modification of the mission. The decisions of subordinate commanders should be strictly in accordance with the desires of their commanders. For the subordinate commander, the decision may be to determine when he would be justified in opening fire. For example, the patrol leader makes contact with a known camp and at the last moment finds that women camp followers are present in the camp. Shall he fire into the group? Insofar as it is practicable, subordinate military leaders should be aided in making such decisions by previously announced policies and instructions.

d. Delay in the use of force, and hesitation to accept responsibility for its employment when the situation clearly demands it, will always be interpreted as a weakness. Such indecision will encourage further disorder, and will eventually necessitate measures more severe than those which would have sufficed in the first instance. Drastic punitive measures to induce surrender, or action in the nature of reprisals, may awaken sympathy with the revolutionists. Reprisals and punitive measures may result in the destruction of lives and property of innocent people; such measures may have an adverse effect upon the discipline of our own troops. Good judgment in dealing with such problems calls for constant and careful surveillance. In extreme cases, a commanding officer may be forced to resort to some mild form of reprisal to keep men from taking more severe action on their own initiative. However, even this action is taken with the full knowledge of possible repercussions.

e. In dealing with the native population, only orders which are lawful, specific, and couched in clear, simple language should be

PSYCHOLOGY

issued. They should be firm and just, not impossible of execution nor calculated to work needless hardship upon the recipient. It is well to remember this latter injunction in formulating all orders dealing with the native population. They may be the first to sense that an order is working a needless hardship upon them, and instead of developing their support, friendship, and respect, the opposite effect may result.

f. An important consideration in dealing with the native population in small wars is the psychological approach. A study of the racial and social characteristics of the people is made to determine whether to approach them directly or indirectly, or employ both means simultaneously. Shall the approach be by means of decisions, orders, personal appeals, or admonitions, unconcealed effort, or administrative control, all of which are calculated to attain the desired end? Or shall indirect methods by subtle inspiration, propaganda through suggestion, or undermining the influential leaders of the opposition be attempted? Direct methods will naturally create some antagonism and encourage certain obstruction, but if these methods of approach are successful the result may be more speedily attained. Indirect approach, on the other hand, might require more time for accomplishment, but the result may be equally effective and probably with less regrettable bitterness.

g. Propaganda plays its part in approach to the people in small wars, since people usually will respond to indirect suggestion but may revolt against direct suggestion. The strength of suggestion is dependent upon the following factors:

(1) *Last impression*—that is, of several impressions, the last is most likely to be acted upon.

(2) *Frequency*—that is, repetitions, not one after another but intervals separated by other impressions.

(3) *Repetition*—this is distinguished from frequency by being repetitions, one after the other, without having other kinds of impressions interspersed.

h. The strongest suggestion is obtained by a combination of "frequency" and "last impressions." Propaganda at home also plays its part in the public support of small wars. An ordinary characteristic of small wars is the antagonistic propaganda against the campaign or operations in the United States press or legislature. One cannot afford to ignore the possibilities of propaganda. Many authorities believe that the Marine Force should restrict publicity to a minimum in order to prevent the spread of unfavorable and antagonistic propaganda at home. However, it is believed that when representatives of the press demand specific information, it should be given to them, if it is not of a confidential nature or such as will jeopardize the mission. Sometimes marines are pressed with the question: "Why are you here?" The best method to follow when a question of public policy is involved is to refer the individual to appropriate civil authorities.

i. There is an axiom in regular warfare to strike the hardest where the going is the easiest. In small wars also, it is well to strike most vigorously and relentlessly when the going is the easiest. When the opponents are on the run, give them no peace or rest, or time to make further plans. Try to avoid leaving a few straggling leaders in the field at the end, who with their increased mobility, easier means of evasion, and the determination to show strength, attempt to revive interest by bold strokes. At this time, public opinion shows little patience in the enterprise, and accepts with less patience any explanation for the delay necessary to bring the operation to a close.

j. In street fighting against mobs or rioters, the effect of fire is generally not due to the casualties but due to the fact that it demonstrates the determination of the authorities. Unless the use of fire is too long delayed, a single round often is all that is necessary to carry conviction. Naturally one attempts to accomplish his mission without firing but when at the critical moment all such means have failed, then one must fire. One should not make a threat without the intention to carry it out. Do not fire without giving specific warning. Fire without specific warning is only justified when the mob is actively endangering life or property. In disturbances or riots when a mob has been ordered to disperse, it must be feasible for the mob to disperse. Military interventions are actually police functions, although warlike operations often ensue. There is always the possibility of domestic disturbances getting beyond the control of local police. Hence the necessity of employing regular forces as a reserve or reenforcements for varying periods after the restoration of normal conditions.

k. The personal pride, uniform, and bearing of the marines, their dignity, courtesy, consideration, language, and personality will have an important effect on the civilian attitude toward the forces of occupation. In a country, for example, where the wearing of a coat, like wearing shoes, is the outward and unmistakable sign of a distinct social classification, it is quite unbecoming for officers who accept the hospitality of the native club for a dance, whether local ladies and

PSYCHOLOGY

gentlemen are in evening clothes or not, to appear in their khaki shirts. It appears that the United States and their representatives have lost a certain amount of prestige when they place themselves in the embarrassing position of receiving a courteous note from a people ordinarily considered backward, inviting attention to this impropriety. On the other hand, care should be exercised not to humiliate the natives. They are usually proud and humiliation will cause resentment which will have an unfavorable reaction. Nothing should be said or done which implies inferiority of the status or of the sovereignty of the native people. They should never be treated as a conquered people.

l. Often the military find themselves in the position of arbiters in differences between rival political factions. This is common in serving on electoral missions. The individual of any faction believes himself in possession of the truth and cannot refrain from affirming that anyone who does not agree with him is entirely in error. Each will attest to the dishonest intentions or stupidity of the other and will attempt by every possible means to carry his point of view irrespective of its merits. They are excitable beings and prone to express their feelings forcibly. They are influenced by personal partiality based upon family or political connections and friendship. Things go by favor. Though they may appear brusque at times they feel a slight keenly, and they know how to respect the susceptibilities of their fellows.

m. In some revolutions, particularly of economic origin, the followers may be men in want of food. A hungry man will not be inclined to listen to reason and will resort to measures more daring and desperate than under normal conditions. This should be given consideration, when tempted to burn or otherwise destroy private property or stores of the guerrillas.

n. In the interior there are natives who have never been 10 miles from their home, who seldom see strangers, and much less a white man or a foreigner. They judge the United States and the ideals and standards of its people by the conduct of its representatives. It may be no more than a passing patrol whose deportment or language is judged, or it may be fairness in the purchase of a bunch of bananas. The policy of the United States is to pay for value received, and prompt payment of a reasonable price for supplies or services rendered should be made in every instance. Although the natives of the capitals or towns may have a greater opportunity to see foreigners and the forces of occupation, the Marine Corps nevertheless represents the United States to them also, and it behooves every marine to conduct himself accordingly. There is no service which calls for greater exercise of judgment, persistency, patience, tact, and rigid military justice than in small wars, and nowhere is more of the humane and sympathetic side of a military force demanded than in this type of operation.

1–16. **Conduct of our troops.**—*a.* In addition to the strictly military plans and preparations incident to the military occupation of a foreign country, there should be formulated a method or policy for deriving the greatest benefit from psychological practices in the field. To make this effective, personnel of the command must be indoctrinated with these principles. While it is true that the command will generally reflect the attitude of the commander, this will or desire of the supreme authority should be disseminated among the subordinates of all grades. The indoctrination of all ranks with respect to the proper attitude toward the civilian population may be accomplished readily by means of a series of brief and interesting lectures prepared under the direction of the military commander and furnished all units. These lectures may set forth our mission, the purpose of our efforts, our accomplishments to date in the betterment of conditions, our objectives of future accomplishment, etc.

b. Uncertainty of the situation and the future creates a certain psychological doubt or fear in the minds of the individual concerned; if the individual is entirely unaccustomed to it, and the situation seems decidedly grave, his conduct may be abnormal or even erratic. This situation of uncertainty exists, ordinarily to a pronounced degree in small wars, particularly in the initial phases of landing and occupation. The situation itself and the form of the orders and instructions which the marine commander will receive are often indefinite. In regular warfare, clear cut orders are given, or may be expected, defining situations, missions, objectives, instructions, and the like, in more or less detail; in small wars, the initial orders may be fragmentary and lack much of the ordinary detail. However unfortunate this may be, or how difficult it may make the task, this is probably the normal situation upon landing. In order to be prepared to overcome the usual psychological reaction resulting from such uncertainty, studies and instructions in small wars should be accompanied by practice in the issuance of orders.

c. The responsibility of officers engaged in small wars and the training necessary are of a very different order from their responsibilities and training in ordinary military duties. In the latter case, they simply strive to attain a method of producing the maximum

PSYCHOLOGY

physical effect with the force at their disposal. In small wars, caution must be exercised, and instead of striving to generate the maximum power with forces available, the goal is to gain decisive results with the least application of force and the consequent minimum loss of life. This requires recourse to the principles of psychology, and is the reason why the study of psychology of the people is so important in preparation for small wars.

d. In major warfare, hatred of the enemy is developed among troops to arouse courage. In small wars, tolerance, sympathy, and kindness should be the keynote of our relationship with the mass of the population. There is nothing in this principle which should make any officer or man hesitate to act with the necessary firmness within the limitation imposed by the principles which have been laid down, whenever there is contact with armed opposition.

1–17. Summary.—*a.* Psychological errors may be committed which antagonize the population of the country occupied and all the foreign sympathizers; mistakes may have the most far-reaching effect and it may require a long period to reestablish confidence, respect, and order. Small wars involve a wide range of activities including diplomacy, contacts with the civil population and warfare of the most difficult kind. The situation is often uncertain, the orders are sometimes indefinite, and although the authority of the military commander is at time in doubt, he usually assumes full responsibility. The military individual cannot afford to be intimidated by the responsibilities of his positions, or by the fear that his actions will not be supported. He will rarely fail to receive support if he has acted with caution and reasonable moderation, coupled with the necessary firmness. On the other hand inaction and refusal to accept responsibility are likely to shake confidence in him, even though he be not directly censured.

b. The purpose should always be to restore normal government or give the people a better government than they had before, and to establish peace, order, and security on as permanent a basis as practicable. Gradually there must be instilled in the inhabitants' minds the leading ideas of civilization, the security and sanctity of life and property, and individual liberty. In so doing, one should endeavor to make self-sufficient native agencies responsible for these matters. With all this accomplished, one should be able to leave the country with the lasting friendship and respect of the native population. *The practical application of psychology is largely a matter of common sense.*

Section IV

RELATIONSHIP WITH THE STATE DEPARTMENT

1–18. **Importance of cooperation.**—*a.* One of the principal obstacles with which the naval forces are confronted in small war situations is the one that has to do with the absence of a clean-cut line of demarcation between State Department authority and military authority.

b. In a major war, "diplomatic relations" are summarily severed at the beginning of the struggle. During such a war, diplomatic intercourse proceeds through neutral channels in a manner usually not directly detrimental to the belligerents. There are numerous precedents in small wars which indicate that diplomacy does not relax its grip on the situation, except perhaps in certain of its more formal manifestations. The underlying reason for this condition is the desire to keep the war "small," to confine it within a strictly limited scope, and to deprive it, insofar as may be possible, of the more outstanding aspects of "war." The existence of this condition calls for the earnest cooperation between the State Department representatives and naval authorities.

c. There are no defined principles of "Joint Action" between the State Department and the Navy Department by which the latter is to be restricted or guided, when its representatives become involved in situations calling for such cooperation. In the absence of a clearly defined directive, the naval service has for guidance only certain general principles that have been promulgated through Navy Regulations.

1–19. **Principles prescribed by Navy Regulations.**—*a.* The principles referred to as set forth in Navy Regulations, 1920, are, for ready reference, herein quoted:

718 (1) The Commander in Chief shall preserve, so far as possible, the most cordial relations with the diplomatic and consular representatives of the United States in foreign countries and extend to them the honors, salutes, and other official courtesies to which they are entitled by these regulations.

33

RELATIONSHIP WITH THE STATE DEPARTMENT

(2) He shall carefully and duly consider any request for service or other communication from any such representatives.

(3) Although due weight should be given to the opinions and advice of such representatives, a commanding officer is solely and entirely responsible to his own immediate superior for all official acts in the administration of his command.

719. The Commander in Chief shall, as a general rule, when in foreign ports, communicate with local civil officials and foreign diplomatic and consular authorities through the diplomatic or consular representative of the United States on the spot.

b. The attitude of the Navy Department towards the relationship that should exist between the naval forces and the diplomatic branch of the Government is clearly indicated by the foregoing quotations. Experience has shown that where naval and military authorities have followed the "spirit" of these articles in their intercourse with foreign countries, whether such intercourse is incident to extended nonhostile interposition by our forces or to minor controversies, the results attained have met with the approval of our Government and have tended towards closer cooperation with the naval and military forces on the part of our diplomats.

c. It should be borne in mind that the matter of working in cooperation with the State Department officials is not restricted entirely to higher officials. In many cases very junior subordinates of the State Department and the Marine Corps may have to solve problems that might involve the United States in serious difficulties.

1-20. **Contact with State Department representatives.**—The State Department representative may be of great help to the military commander whose knowledge of the political machinery of the country may be of a general nature. It is therefore most desirable that he avail himself of the opportunity to confer immediately with the nearest State Department representative. Through the latter, the commander may become acquainted with the details of the political situation, the economic conditions, means of communication, and the strength and organization of the native military forces. He will be able to learn the names of the governmental functionaries and familiarize himself with the names of the leading officials and citizens in the area in which he is to operate. Through the diplomatic representative the military commander may readily contact the Chief Executive, become acquainted with the government's leading officials and expeditiously accomplish many details incident to the occupation of the country.

Section V

THE CHAIN OF COMMAND—NAVY AND MARINE CORPS

1-21. **Navy regulations.**—*a*. Article 575, Navy Regulations, 1920 states: "When serving on shore in cooperation with vessels of the Navy, brigade commanders or the officer commanding the detachment of marines shall be subject to the orders of the Commander in Chief, or, in his absence, to the orders of the senior officer in command of vessels specially detailed by the Commander in Chief on such combined operations so long as such senior officer is senior in rank to the officer commanding the brigade or the detachment of marines. When the brigade commander or the officer commanding the detachment is senior to the senior officer in command of the vessels specially detailed by the Commander in Chief on such combined operations, or when, in the opinion of the Commander in Chief, it is for any reason deemed inadvisable to intrust such combined command to the senior officer afloat, the Commander in Chief will constitute independent commands of the forces ashore and afloat, which forces will cooperate under the general orders of the Commander in Chief."

b. In article 576, it is provided that: "The brigade commander or other senior line officer of the Marine Corps present shall command the whole force of marines in general analogy to the duties prescribed in the Navy Regulations for the senior naval officer present when two or more naval vessels are serving in company, but the commander of each regiment, separate battalion, or detachment shall exercise the functions of command over his regiment, battalion, or detachment in like general analogy to the duties of the commander of each naval vessel."

1-22. **Control of joint operations.**—In a situation involving the utilization of a marine force in a small war campaign, the directive

THE CHAIN OF COMMAND

for the marine force commander usually requires him to report to the senior officer present in the area of anticipated operations. The Major General Commandant exercises only administrative control over the marine force; its operations are controlled by the Chief of Naval Operations directly, or through the senior naval officer present, if he be senior to the marine force commander. Consequently, no operation plans or instructions with regard to the tactical employment of the marine force originate in the office of the Major General Commandant.

1-23. **The directive.**—*a.* In situations calling for the use of naval and marine forces in operations involving protection of life and property and the preservation of law and order in unstable countries, the burden of enforcing the policies of the State Department rests with the Navy. The decisions with regard to the forces to be used in any situation are made by the Secretary of the Navy as the direct representative of the President. Through the Chief of Naval Operations, the Secretary of the Navy exercises control of these forces. The directive issued to the naval commander who is to represent the Navy Department in the theater of operations is usually very brief, but at the same time, clearly indicative of the general policies to be followed. The responsibility for errors committed by the naval commander in interpreting these policies and in carrying out the general orders of the Navy Department rests with such naval commander.

b. If, as is the usual situation, the naval commander is the senior officer present in the theater of operations, his sole directive may be in the form of a dispatch. A typical directive of this type is set forth as follows:

INTERNAL POLITICAL SITUATION IN (name of country) HAS REQUIRED SENDING OF FOLLOWING NAVAL FORCES (here follow list of forces) TO ( ) WATERS WITH ORDERS TO REPORT TO SENIOR NAVAL OFFICER FOR DUTY POLICY OF GOVERNMENT SET FORTH IN OPNAV DISPATCHES ( ) AND ( ) YOU WILL ASSUME COMMAND OF ALL NAVAL FORCES IN ( ) WATER AND AT ( ) AND IN FULLEST COOPERATION WITH AMERICAN AMBASSADOR AND CONSULAR OFFICERS WILL CARRY OUT POLICY OF US GOVERNMENT SET FORTH IN REFERENCE DISPATCHES.

c. Under the provisions of the foregoing directive, a naval commander concerned would be placed in a position of great responsibility and in accomplishing his task, he would necessarily demand the highest degree of loyalty and cooperation of all those under his command. The usual procedure, adopted by the naval commander, would be first to make a careful estimate of the situation, then arrive at a decision, draw up his plan based on this decision, and issue the necessary operation orders.

1-24. Naval officer commanding ashore.—If the force to be landed consists of naval and marine units and is placed under the direct command of a naval officer, matters with regard to the relationship between the forces ashore and the naval commander afloat will give rise to little or no concern. The naval officer afloat will, under such conditions, usually remain in the immediate vicinity of the land operations, maintain constant contact with all phases of the situation as it develops, and exercise such functions of command over both the forces ashore and those afloat as he considers conducive to the most efficient accomplishment of his task. Commanders of marine units of the landing force will bear the same relationship toward the naval officer in command of the troops ashore as it set down for subordinate units of a battalion, regiment, or brigade, as the case may be.

1-25. Marine officer commanding ashore.—*a.* When the force landed comprises a marine brigade or smaller organization under the command of a marine officer, and such forces become engaged in a type of operation that does not lend itself to the direct control by the naval commander afloat, many questions with regard to the relationship between the marine forces ashore and the naval forces afloat will present themselves. The marine force commander, in this situation, should not lose sight of, and should make every effort to indoctrinate those under his command with the idea that the task to be accomplished is a "Navy task"; that the responsibility for its accomplishment rests primarily with the immediate superior afloat; and that regardless of any apparent absence of direct supervision and control by such superior, the plans and policies of the naval commander afloat must be adhered to.

b. The vessels of the naval force may be withdrawn from the immediate theater of operations; the naval commander may assign certain vessels to routine patrol missions along the coast; while he, himself, may return to his normal station and maintain contact with the marine force and the vessels under his command by radio or other means of communication.

c. The directive issued to the marine force commander will usually provide that he keep in constant communication with the naval commander afloat in order that the latter may at all times be fully informed of the situation ashore. The extent to which the marine force commander will be required to furnish detailed information

THE CHAIN OF COMMAND

to the naval commander will depend on the policy established by the latter. As a general rule, the naval commander will allow a great deal of latitude in the strictly internal administration of the marine force and the details of the tactical employment of the various units of that force. He should, however, be informed of all matters relative to the policy governing such operations. In case the naval commander does not, through the medium of routine visits, keep himself informed of the tactical disposition of the various units of the marine force. he should be furnished with sufficient information with regard thereto as to enable him to maintain a clear picture of the general situation.

d. Usually the naval commander will be required to submit to the Navy Department. periodically, a report embracing all the existing economic, political, and tactical phases of the situation. The naval commander will, in turn, call upon the marine force commander for any reports of those matters as are within the scope of the theater in which the force is operating.

e. Estimates of this sort carefully prepared will often preclude the necessity of submitting detailed and separate reports on the matters involved and will greatly assist the naval commander in his endeavor. through the coordination of the other information at his disposal, to render to the Navy Department a more comprehensive analysis of the situation confronting him.

f. When questions of major importance arise, either involving a considerable change in the tactical disposition and employment of the marine force, or the policies outlined by the naval commander, the latter should be informed thereof in sufficient time to allow him to participate in any discussion that might be had between the political. diplomatic, and military authorities with regard thereto. It should be remembered that in making decisions in matters of importance, whether or not these decisions are made upon the advice of our diplomatic representatives, the marine-force commander is responsible to his immediate superior afloat.

g. In addition to the principles that are necessarily adhered to incident to the "chain of command," a marine-force commander on foreign shore habitually turns to the Navy for assistance in accomplishing the innumerable administrative tasks involved in the small-war situations. Matters with regard to water transportation for evacuation of personnel, matters concerning supply, matters involving intercourse with our diplomatic representatives in countries in the vicinity of the theater of operations, matters relating to assistance from the Army in supply and transportation, and any number of other phases of an administrative nature can be more expeditiously and conveniently handled through the medium of the naval commander whose prerogatives and facilities are less restricted than those of the commander in the field.

1-26. **Marine—Constabulary.**—When there is a separate marine detachment engaged in the organization and training of an armed native organization, the commanding officer of this detachment occupies a dual position. Although he is under the supervision of the Chief Executive of the country in which he is operating, he is still a member of the naval service. In order that there may be some guide for the conduct of the relationship that is to exist between the marine-force commander and the marine officer in charge of the native organization, fundamental principles should be promulgated by the Secretary of the Navy.

1-27. **Direct control by Navy Department.**—If the naval vessels that participate in the initial phases of the operation withdraw entirely from the theater of operations, the command may be vested in the marine-force commander or in the senior naval officer ashore within the theater. In such case, the officer in command on shore would be responsible directly to the Chief of Naval Operations. His relationship with the Chief of Naval Operations would then involve a combination of those principles laid down for the relationship that exists between the forces on shore and the naval commander afloat, and the relationship that the latter bears to the Navy Department as its representative.

SECTION VI

MILITARY—CIVIL RELATIONSHIP

1-28. **Importance.**—*a.* All officers of the naval establishment, whether serving with the force afloat, the forces ashore, or temporarily attached to the national forces of another country, are required by the Constitution and by Navy Regulations to observe and obey the laws of nations in their relations with foreign states and with the governments or agents thereof.

b. One of the dominating factors in the establishment of the mission in small war situations has been in the past, and will continue to be in the future, the civil contacts of the entire command. The satisfactory solution of problems involving civil authorities and civil population requires that all ranks be familiar with the language, the geography, and the political, social, and economic factors involved in the country in which they are operating. Poor judgment on the part of subordinates in the handling of situations involving the local civil authorities and the local inhabitants is certain to involve the commander of the force in unnecessary military difficulties and cause publicity adverse to the public interests of the United States.

1-29. **Contact with national government officials.**—*a.* Upon the arrival of the United States forces at the main point of entry the commander thereof should endeavor, through the medium of the United States diplomatic representative, to confer with the Chief Executive of the government, or his authorized representative and impart such information as may be required by the directive he has received. Such conference will invariably lead to acquaintance with the government's leading officials with whom the military commander may be required to deal throughout the subsequent operation.

b. Meetings with these officials frequently require considerable tact. These officials are the duly elected or appointed officials of the government, and the military commander in his association with them, represents the President of the United States. These meetings or conferences usually result in minimizing the number of officials to be

MILITARY—CIVIL RELATIONSHIP

dealt with, and the way is thereby speeded to the early formulation of plans of action by the military commander. When the mission is one of rendering assistance to the recognized government, the relationship between its officials and the military commander should be amicable. However, if animosity should be shown or cooperation be denied or withdrawn, the military commander cannot compel the foreign government officials to act according to his wishes. Ordinarily an appeal to the Chief Executive of the country concerned will effect the desired cooperation by subordinate officials. Should the military commander's appeal be unproductive, the matter should be promptly referred to the naval superior afloat or other designated superior, who will in turn transmit the information to the Navy Department and/or the State Department as the case may be.

c. In most of the theaters of operations, it will be found that the Chief Executive maintains a close grip on all phases of the national government. The executive power is vested in this official and is administered through his cabinet and various other presidential appointees. Some of these appointed officials exercise considerable power within their respective jurisdictions, both over the people and the minor local officials. Some of them exercise judicial as well as executive functions, and are directly responsible to the President as head of the National Government.

d. It follows, therefore, that in the type of situation which involves the mission of assisting a foreign government, the military commander and his subordinates, in their associations with national governmental officials, as a rule will be dealing with individuals who are adherent to the political party in power. This situation has its advantages in that it tends to generate cooperation by government officials, provided of course, the Chief Executive, himself, reflects the spirit of cooperation. At the same time, it may have the disadvantage of creating a feeling of antagonism toward our forces by the opposite political party, unless the military commander instills in all members of his command the necessity for maintaining an absolute nonpartisan attitude in all their activities.

e. Political affiliation in most countries is a paramount element in the lives of all citizens of the country. Political ties are taken very seriously and serve to influence the attitude and action of the individual in all his dealings.

f. When subordinate military commanders are assigned independent missions which bring them into contact with local and national governmental officials, they should make every effort to acquaint themselves with the political structure of the locality in which they are to be stationed. The principal guide for the conduct of their associations with the civil officials will be, of course, the regulation previously referred to which governs the relations between members of the naval service and the agents of foreign governments. The amenities of official intercourse should be observed and the conventions of society, when and where applicable, should be respected. When assuming command within a district or department, an officer should promptly pay his respects to the supreme political authority in the area, endeavor to obtain from him the desired information with regard to the economic situation in that locality and indicate by his conduct and attitude that he is desirous of cooperating to the extent of his authority with those responsible for the administration of the foreign government's affairs.

g. In giving the fullest cooperation to the civil authorities, the military commander should insist on reciprocal action on their part toward the military forces. Interference with the performance of the functions of civil officials should be avoided, while noninterference on the part of those authorities with the administration of the military forces should be demanded. In brief, a feeling of mutual respect and cooperation between members of the military forces and civil officials on a basis of mutual independence of each other should be cultivated.

1-30. **Cooperation with law-enforcement agencies.**—*a.* United States forces, other than those attached to the military establishment of the foreign country in which they are operating will not, as a rule, participate in matters concerning police and other civil functions. The military forces usually constitute a reserve which is to be made available only in extreme emergency to assist the native constabulary in the performance of its purely police mission.

b. The mission of our forces usually involves the training of native officers and men in the art of war, assisting in offensive operations against organized banditry and in such defensive measures against threatened raids of large organized bandit groups as are essential to the protection of lives and property. When the civil police functions are vested in the native military forces of the country, these forces are charged with the performance of two definite tasks—a military task involving the matters outlined above and a police task involving in general the enforcement of the civil and criminal laws. The native military forces control the traffic of arms and ammunition; they see that the police, traffic, and sanitary regulations are observed; they assume the control and administration of government prisons; and they perform numerous other duties that, by their nature, may obviously, directly or indirectly, play an important part in the accomplishment of the military mission.

c. It follows, therefore, that by cooperating to the fullest extent of his authority with the native forces in the performance of civil police functions, the military commander will, without actually participating in this phase of the picture, be rendering valuable assistance towards the accomplishment of the ultimate mission assigned to the combined military forces. Due to the fact that in most cases the individuals occupying the important positions in those native organizations performing police duties, are United States officers and enlisted men, questions arising with regard to cooperation and assistance are easy of solution. Adherence, on the part of our personnel, to the dictates of the local laws and regulations, and a thorough knowledge of the scope of authority vested in the native police force is essential to the end that we do not hamper this force in the performance of its duty, and to the end that we maintain the respect and confidence of the community as a whole.

d. With regard to the contact that is had with those connected with the judicial branch of the government, very little need be said. The magistrates and judges of the various courts are usually political appointees, or are elected to the office by the national congress. Consequently, they are affiliated politically with the party in power, national and/or local. In most situations, the civil courts will continue to function. Although this procedure is not always conducive to the best interests of the military forces, it is a situation that normally exists and must be accepted. The manner in which the judiciary performs its functions may have a profound effect on the conduct of a small war campaign. In the first place, the apprehension and delivery of criminals, including guerrillas, by the armed forces to the courts will serve no useful purpose if these courts are not in sympathy with the military authorities; and in the second place, a lack of cooperation on the part of the courts, insofar as the punishment of outlaws is concerned, may have a tendency to place the local inhabitants in fear of assisting the military forces. In view of this situation, every endeavor should be made to generate a friendly attitude on the part of these law-enforcement officials in order that their cooperation may be had.

1–31. **Contact with inhabitants.**—*a.* Whether a military commander be stationed at a headquarters in a metropolis or assigned to the smallest outpost, he must necessarily come into contact with the civilian population. By "contact" in this case is implied intercourse in daily life. The transaction of daily routine involves the association with the civilian element, even in the most tranquil territory. The purchase of fresh provisions, fuel, and other necessities of camp life involve the relationships with merchants, bankers, those in charge of public utilities, and many others. In relations with these persons, whether they be business or social, a superiority complex on the part of the military commander is unproductive of cooperation. The inhabitants are usually mindful of the fact that we are there to assist them, to cooperate with them in so doing, and while dignity in such relationship should always obtain, the conduct of the military authority should not be such as to indicate an attitude of superiority.

b. Association with civilians may be other than business or social. The same daily occurrences that take place in the United States between members of the naval forces and our own police and civilian population frequently take place on foreign soil. Damage to private property by the military forces is frequently the cause of complaints by members of the civilian population. Dealings with civilians making claims for damages incurred through the conduct of our personnel should be as equitable as the facts warrant. Even where the responsibility rests with the United States, the settlement of such claims is necessarily protracted by the required reference to the Navy Department, and the lack of facilities through which to afford prompt redress is oftentimes the cause of bad feelings. If the military commander were supplied with a fund to be used for the prompt adjustment of limited claims, the foregoing condition might be materially improved. However, under existing laws and regulations the amicable adjustment of matters involving injury and damage to the civilian population and their property calls for the highest degree of tact and sound judgment.

c. Cordial relationship between our forces and the civilian population is best maintained by engendering the spirit of good will. As previously stated, a mutual feeling of dislike and aversion to association may exist between members of rival political parties. Conservatives and liberals, or by whatever label they may be known, are frequently prone to remain "die hards" when their political candidate is unsuccessful at the polls. It is, therefore, highly important for a military commander to ascertain the party affiliation of the persons with whom he comes into contact. The homely advice: "Don't dabble in politics" is wise, and military authorities should scrupulously avoid discussing the subject.

MILITARY—CIVIL RELATIONSHIP

d. Akin to politics is the subject of religion. The people of many countries take their religion as seriously as their politics. Consequently members of the United States forces should avoid any attitude that tends to indicate criticism or lack of respect for the religious beliefs and practices observed by the native inhabitants.

e. Relations between our military forces and the civilians might easily be disturbed if the former were to get into altercations with the public press. Freedom of speech is another liberty of which the inhabitants of many countries are not only proud, but jealous. Editors of the local newspapers are not always averse to criticizing the actions of troops other than their own. Nothing can be gained by the marine commander in jumping into print and replying to such newspaper articles, other than possibly starting a controversy which may make his further retention in that locality undesirable. When a matter is so published and it is considered detrimental, the subordinate marine commander should bring it to the attention of his immediate superior for necessary action by higher authority.

f. Every endeavor should be made to assure the civilian population of the friendliness of our forces. No effort should be spared to demonstrate the advantage of law and order and to secure their friendly cooperation. All ranks should be kept mindful of the mission to be accomplished, the necessity for adhering to the policy of the United States and of observing the law of nations.

g. Foreign nationals are often the underlying cause of intervention; almost invariably they are present in the country during the occupation. Generally their concern is for the security of their lives and property; sometimes they have an exaggerated opinion of their importance and influence. Generally the condition of political unrest does not react directly against foreigners, and it often happens that the foreign resident does not consider himself in any danger until he reads of it in a foreign newspaper, whereupon his imagination becomes active. Foreign cooperation may at times be a greater obstacle to success than the foreign mercenaries in a revolutionary party, when, for equally unworthy purposes, they render aid openly or secretly to the revolutionists in order to assure themselves of the protection or favor of any new government. Any discontented faction of natives can usually secure the sympathy or support from some group of investors or speculators who think they can further their own interests or secure valuable concessions by promoting a revolution. In any event, in dealing with these corporations and in

MILITARY—CIVIL RELATIONSHIP

receiving reports from them, it may often be wise to scrutinize their actions carefully to determine if they have any ulterior motives. In interventions, the United States accords equal attention to the security of life and property of all foreign residents.

○

SMALL WARS MANUAL
UNITED STATES MARINE CORPS
1940

+

CHAPTER II

ORGANIZATION

UNITED STATES
GOVERNMENT PRINTING OFFICE
WASHINGTON : 1940

TABLE OF CONTENTS

The Small Wars Manual, U. S. Marine Corps, 1940 is published in fifteen chapters as follows:

SMALL WARS MANUAL
UNITED STATES MARINE CORPS

CHAPTER II

ORGANIZATION

▼

SECTION I

THE ESTIMATE OF THE SITUATION

2-1. **General.**—*a.* It has been stated in the previous chapter that the President, as the Chief Executive, makes the decision which initiates small war operations and that this decision is promulgated through the regular channels to the commander of the intervening force. Upon the receipt of instructions from higher authority, it is incumbent on each commander in the chain of command to make an estimate of the situation to determine the best course of action and how it is to be carried out.

b. This estimate follows the general outline of a normal "Estimate of the Situation" although certain points which are peculiar to small war operations should be emphasized. In particular decisions must be made as to: the composition of the staff; the size of the force required to accomplish the mission, or how to employ the force available most advantageously; the proportion of the infantry, supporting arms and services best suited for the situation; and the requisition and distribution of special weapons and equipment which are not included in the normal organization but which are considered necessary.

c. If sufficient information of the probable theater of operations has not been furnished, maps, monographs, and other current data concerning the country must be obtained, including information on the following: past and present political situation; economic situation; classes and distribution of the population; psychological nature of the inhabitants; military geography, both general and physical; and the military situation.

THE ESTIMATE OF THE SITUATION

2–2. **The mission.**—In a major war, the mission assigned to the armed forces is usually unequivocal—the defeat and destruction of the hostile forces. This is seldom true in small wars. More often than not, the mission will be to establish and maintain law and order by supporting or replacing the civil government in countries or areas in which the interests of the United States have been placed in jeopardy, in order to insure the safety and security of our nationals, their property and interests. If there is an organized hostile force opposing the intervention, the primary objective in small wars, as in a major war, is its early destruction. In those cases where armed opposition is encountered only from irregular forces under the leadership of malcontents or unrecognized officials, the mission is one of diplomacy rather than military. Frequently the commander of a force operating in a small wars theater of operations is not given a specific mission as such in his written orders or directive, and it then becomes necessary for him to deduce his mission from the general intent of the higher authority, or even from the foreign policy of the United States. In any event, the mission should be accomplished with a minimum loss of life and property and by methods that leave no aftermath of bitterness or render the return to peace unnecessarily difficult.

2–3. **Factors to be considered in estimating enemy strength.**— *a. Political status.*—(1) In the majority of our past small wars operations, intervention has been due to internal disorder which endangered foreign lives and property, or has been undertaken to enforce treaty obligations.

(2) In the first instance, the chaotic condition usually has been brought about as a result of the tyrannical measures adopted by the party in control of the government, by the unconstitutional usurpation of power by a political faction for the sake of gain, or because of intense hatred between rival factions which culminated in a revolt against the recognized government. As the result of such action, a state of revolution existed which was detrimental to internal and external peace and good will. The intervening power was faced usually with one of two alternatives; either to intervene between the warring factions, occupy one or more proclaimed neutral zones, and endeavor by pacific or forceful action to make the rival parties accept mediation and settlement of the controversy; or to assist, by pacific or forceful action, one side or the other, or even to support a new party, in the suppression of the disorders.

THE ESTIMATE OF THE SITUATION

(3) In the second instance, that of enforcing treaty obligations, the immediate cause of intervention usually has been the neglect and repeated refusal of the local government to carry out its obligations under the terms of a commercial or political treaty. The intervening forces sought, by show of force or by actual field operations, to enforce those obligations. If such action was unsuccessful, the intervening power in some cases deposed the party in control and established a *de facto* or *de jure* government which would carry out the provisions of the treaty. This often resulted in active opposition by the ousted party against the intervening forces who were giving aid, force, and power to the new government.

(4) It is evident from the above that the internal political organization of the country concerned, the strength of the forces which may oppose the intervention, and the external obligations of the country as a member of the family of nations, should be carefully considered in the estimate of the situation. In addition, the estimate must include the probable effect which the intervention will have upon the public opinion of the citizens of the intervening power and upon the good will of other countries. The latter, in particular, is of great importance since the friendship and trade relations of countries which are not sympathetic to the intervention may be alienated by such action.

b. Economic status and logistic support available.—The ability of a hostile force to oppose the intervening force may be limited by the availability of subsistence, natural resources, finances, arms, equipment, and ammunition. The forces opposing the intervention often live off the country by forcing contributions of money, subsistence, and other supplies from the peaceful inhabitants, or by donations from local civilians sympathetic to their cause. Even though the country concerned may be heavily indebted to their own citizens as well as to foreign powers, funds are often diverted from the state treasury or may be received from foreign sources for the purchase of modern arms and munitions of war. As a result, the intervening force usually finds the forces opposing them armed and equipped with modern weapons and capable of sustaining themselves in the field for an unlimited period. This is especially true if, as is usually the case, the hostile forces resort to guerrilla warfare.

c. Geographical features.—That part of the estimate of the situation which considers the geographical features of the theater of operations is fully as important in small wars as in a major war. It covers the general terrain features, the geographical divisions of the

THE ESTIMATE OF THE SITUATION

country as fixed by relief, suitable debarkation places, the character and suitability of routes of communication, the distribution of population, the location of principal cities, the political divisions of the state, and the strategical and tactical aspects of the frontiers. The location and extent of plain, plateau, and mountain regions, and of open, wooded, or jungle areas will affect the organization, equipment, and field operations of the intervening force. If a state of revolution is the basic cause of intervention, the political divisions within the country are particularly important and may in themselves, determine the strategic plan of operation. Of special significance, also, are those areas in which the majority of foreign citizens and interests are concentrated, since the establishment of neutral zones and similar protective operations usually will be initiated in those localities.

d. Climatic conditions.—Climatic conditions in the probable theater of operations will affect the organization, clothing, equipment, supplies, health, and especially the operations of the intervening forces. A campaign planned for the dry season may be entirely different from one planned for the rainy season. This is particularly true in countries where the road system is primitive, or where dependence is placed on river transportation for the movement of troops and supplies. Weather conditions during certain seasons of the year may increase the difficulties of combat operations in the theater of operations, but if properly evaluated, they should not be considered as insurmountable obstacles.

e. Information and security service of the enemy.—It can be stated as an accepted premise that, in small wars, the intelligence service of the opposing forces will be superior initially to that of the intervening force. From the point of view of the intervening power, the intervention is usually considered a friendly effort to assist the occupied country to reestablish peace and order within its boundaries. From the viewpoint of the majority of the citizens of the occupied country, however, this action by an alien power is an unfriendly one. Although the majority of these inhabitants will not actively oppose the intervention, many of them will indirectly assist the native forces with information relative to the movements of the intervening forces. This is especially true of those citizens who have relatives among the native forces operating in the field. To off-set this situation, recourse must be had to propaganda clearly stating the definite purpose of the intervening forces in order to show the friendly aid that is being offered to the country. Friendships should be made with the inhabitants in an honest and faithful endeavor to assist them to

4

resume their peaceful occupations and to protect them from the illegal demands made upon them by the malcontents. The liberal use of intelligence funds will be of assistance in obtaining information of hostile intentions. Secrecy and rapidity of movements, the distribution of false information regarding proposed operations, and the use of code or cipher messages will aid in preventing the hostile forces from gaining information of contemplated movements of our own forces. Routine patrols must be avoided. An effort must be made to learn the terrain and to become familiar with and utilize every road and trail in the theater of operations. Above all, an active and aggressive campaign against the hostile forces in the field is the most effective method of destroying their intelligence service. A guerrilla band which is constantly harassed and driven from place to place soon loses contact with its own sources of information; it becomes confused and its intelligence system breaks down. As the occupation continues, superiority in this respect will gradually be obtained by the intervening forces.

f. Material characteristics.—Irregulars in small wars are not encumbered with modern supply loads or other impedimenta which reduce the speed at which troops can march. Their knowledge of the terrain and their mobility permits them to move quickly and safely to avoid combat and then to launch an attack against a defenseless village or some isolated outpost. In the past, these irregulars have been armed with old types of weapons, most of which have been considered obsolete, while the intervening forces have been equipped with superior modern weapons. Due to the ease with which modern arms and equipment can be obtained from outside sources, it can be expected that, in the future, irregulars will have weapons and equipment equally as effective as those of the intervening forces. Except for aviation, therefore, the decided advantage in arms and equipment enjoyed by intervening troops in the past will seldom obtain in the future.

g. Composition, condition, and disposition of enemy forces.—When the intervening forces initially enter a small wars country, they usually find the opposing elements organized into fairly large groups controlling certain definite areas. If these large groups can be engaged and decisively defeated, armed opposition to the intervention may be brought to an end and an early peace achieved. If this fails, the larger groups either retire to more remote areas, or are dispersed into numerous small bands which remain in the same general locality, and the action becomes one of protracted guerrilla warfare.

THE ESTIMATE OF THE SITUATION

h. Racial characteristics, morale, and skill.—The leadership of the opposing forces in small wars must not be underestimated. Very often the opposition is led by men who have been trained in the United States or European military schools and who have had much experience in practical soldiering in their specialized type of warfare. Irregular forces in active operations always attract foreign soldiers of fortune of varied experience and reputation whose fighting methods influence the character of opposition encountered. (For further details, see Section III, Chapter I, "Psychology.")

2-4. **Relative strength.**—After considering the strength of your own forces, always keeping the mission of the force in mind, the factors of the *Enemy and Own Strength* are compared in order to arrive at a definite conclusion as to the relative combat efficiency of the opposing forces.

2-5. **Enemy courses of action.**—The probable intentions of the opposing forces will depend a great deal, initially, upon the causes leading up to the intervention. In the majority of cases, their purpose will be to hold the area in which they are located as a section seceding from that part of the country which is first occupied by the intervening force. Even though under the control of a single leader, they will seldom oppose the landings or engage in offensive action, initially, against the forces of occupation. As the movement inland begins, strong defensive action can be expected from the larger hostile groups. When these have been dispersed, smaller bands or groups will operate actively not only against the intervening forces but also against the towns and population then under control of the latter. Finally, the opposition usually will degenerate into guerrilla warfare. So many small bands will be in the field that a definite conclusion as to the probable intentions of any one of them will be difficult to determine. Generally their intentions will be to make surprise attacks against the intervening forces in superior numbers and against undefended local villages and towns. To offset such action, patrols must be strong enough in numbers and armament to withstand any anticipated attack or ambush, and the principal villages and towns must be given adequate protection. Further, by energetic patrolling of the area and vigorous pursuit of the hostile forces once contact is gained, the irregulars should be forced to disband completely or to move to more remote and less fertile areas. The pursuit of these small bands must be continuous.

THE ESTIMATE OF THE SITUATION

2-6. **Own courses of action.**—The intervening force commander must choose the best course of action to follow in order to accomplish his mission. This will necessarily result in a scheme of maneuver, either strategical or tactical. To accomplish this mission, it may be necessary to make a show of force in occupying the State capital, for often the history of the country will indicate that he who holds the capital holds the country. Again, he may be forced to occupy the principal cities of the country, or a certain area, the economic resources of which are such that its possessor controls the lifeblood of the country. More frequently, it will be necessary to initiate active combat operations against the large groups of opposing forces which occupy certain areas. The entire scheme of maneuver will frequently result in the occupation of the coastal area initially with a gradual coordinated movement inland, thus increasing the territory over which control and protection may be established. As this territory extends, it will be necessary to create military areas within it under the control to subordinate commanders. The area commander in turn will seek to control his area by use of small detachments to protect the towns and to conduct active operations against irregular groups until the area becomes completely pacified.

2-7. **The decision.**—When the force commander has finally selected the best course of action and determined, in general terms, how it may be executed, he makes his decision, which consists of a statement of his course of action followed by how it is to be carried out, and why. The decision indicates the commander's general plan of action as expressed in paragraph 2 of an operation order. The basic principle underlying any decision in a small-wars operation is that of initiating immediately energetic action to disband or destroy the hostile forces. This action should hasten the return of normal peace and good order to the country in the shortest possible time.

2-8. **Supporting measures.**—After the basic decision has been reached, the Force Commander must consider carefully the supporting measures which are required to put it into effect. The mission; the operations required to carry out the scheme of maneuver; the organization, armament, and leadership of the opposing forces; the terrain, geography, and climate in the theater of operations; the natural resources and routes of communication within the country to be occupied; all must be considered and all will affect the formulation of the campaign and operation plans. These factors will determine the size and composition of the commander's staff; the organiza-

7

tion of the force; the type of infantry weapons and the proper proportion of aircraft, artillery, and other supporting arms and services required; and the administrative and logistic details. When these supporting measures have been determined, the commander evolves his campaign and operation plans.

2–9. **Campaign and operation plans.**—*a.* In military operations of small wars, strategical and tactical principles are applied to attain the political objective of the government. The political objective indicates the general character of the campaign which the military leader will undertake. The campaign plan indicates the military objective and, in general terms, the nature and method of conducting the campaign. It will set forth the legal aspects of the operations and the corelated authority and responsibilities of the force. If military government or some form of political control is to be instituted, the necessary directives are included in the campaign plan. This plan also indicates the general nature of employment of the military forces. It indicates what use, if any, will be made of existing native forces or of those to be organized.

b. The operation plan prescribes the details of the tactical employment of the force employed and the important details of supply and transportation for that force. It may indicate the territorial division of the country for tactical or administrative control. It provides also for the most efficient employment, maintenance, and development of the existing signal communication system. If the campaign plan calls for the organization of a native constabulary, detailed plans must be made for its early organization and training. If the campaign plan calls for the employment of local armed civilians or guards, or if such action is considered necessary or advisable, plans must be made for the organization, training, equipment, supply, clothing, subsistence, pay, shelter, and employment of such troops. If the mission calls for the supervision of elections, this plan must include the necessary arrangements for the nonmilitary features of this duty as well as the tactical disposition of the force in the accomplishment of the task.

c. Tactical operations of regular troops against guerrillas in small wars are habitually offensive. Even though operating under a strategic defensive campaign plan, regular combatants in contact with hostile forces will emphasize the principle of the offensive to gain psychological supremacy. Isolated forces exposed to possible attack by overwhelming numbers must be well protected in positions prepared to develop the greatest possible effect of their weapons. Reverses, particularly at first, must be avoided at all costs.

d. The initiation of a campaign before adequate preparations have been made, may well be as fatal in a small war as in regular warfare. Prolonged operations are detrimental to the morale and prestige of the intervening forces. They can be avoided only by properly estimating the situation and by evolving as comprehensive, flexible, and simple a plan as possible before the campaign begins.

SECTION II

THE STAFF IN SMALL WARS

2–10. **Command and staff responsibility in small wars.**—A force engaged in small wars operations, irrespective of its size, is usually independent or semi-independent and, in such a campaign, assumes strategical, tactical, and territorial functions. Strategical decisions and territorial control are usually matters for the attention of the high command in major warfare. In small wars the Force Commander must be prepared to make or recommend decisions as to the strategy of the operation, and his staff must be able to function as a GHQ staff. In short, the force must be prepared to exercise those functions of command, supply, and territorial control which are required of the supreme command or its major subdivisions in regular warfare. More extensive planning is required than would ordinarily be expected of the same size unit that is part of a higher command. For these reasons, it is obvious that a force undertaking a small wars campaign must be adequately staffed for independent operations even if the tables of organization do not specify a full staff complement. Whether or not the executive staff is relieved of all operative functions will depend on the size and composition of the force and the situation. It is possible to visualize an independent regiment in such a situation that the demands placed upon the organization would make it inadvisable for a member of the Executive Staff to operate the various activities pertaining to his Executive Staff section. Likewise it is possible that the Executive Staff of a much larger force can operate the activities of their sections after the situation is thoroughly under control. The staff organization must be fitted to the unit after consideration of its size, composition, and the situation confronting it.

2–11. **The Force Commander.**—One of the first decisions of the force commander must make is the size and composition of his staff. He then considers the extent to which he will decentralize authority to his staff and to subordinate commanders. This decision will greatly influence his assignments of officers to specific staff and command duties. The assignment of officers according to their attainments, temperaments, and special qualifications, is one of the most important measures to insure smooth and efficient operation of the organizations or establishments. The larger the unit, the more important this becomes. The force commander must be able to issue directives only, leaving the details to his subordinates. He contents himself with seeing that the work is properly done and that the principle of the directive is not departed from, always holding himself ready to rule on doubtful points and to advise subordinates who are having difficulty.

2–12. **Staff procedure.**—*a.* The staff of a unit or organization consists of those officers specifically provided for the purpose of assisting the commander in exercising his command functions. It is divided into two groups: the Executive, or General, Staff (Chief of Staff; F–1, Personnel; F–2, Intelligence; F–3, Plans and Training; and F–4, Supply), who comprehend all the functions of command; and the Special Staff, which includes the heads of technical, supply, and administrative services, and certain technical specialists. Usually, the Executive Staff is not an operating agency; in a small force, Executive Staff officers may, or may not, actually operate one or more of the services under their sections. The organization of the staff is shown diagrammatically in Plate 1. Staff principles and functions, as defined in the "War Department Field Manual 101–5," remain fundamentally the same irrespective of the type of operation.

b. The staff, in close cooperation, works out the plans enunciated by the commander, formulates the orders and instructions for putting the plans into execution, and by observation and inspection insures proper execution. Staff officers must keep themselves informed of the situation at all times, and be able to place before the commander information in such thoroughly digested form as will enable him to come to a sound and prompt decision without having to consider an infinite number of details.

c. Staff conferences, staff visits, staff inspections, measures to insure adequate liaison, and provision for administrative details are the usual methods employed by all staff organizations to facilitate the proper performance of their specific duties. This procedure unifies the efforts of the staff in furthering the accomplishment of the will of the commander.

d. Administrative procedure and the details of the organization and routine of the various staff offices are largely dependent on the requirements of the particular situation. It is important that essential information be immediately available and that every item coming under the cognizance of the staff section or special staff officer concerned receive proper attention and be disseminated to individuals concerned. This entails the formulation of a systematic office routine and proper allocation of duties to individuals. Executive staff sections are not offices of permanent record. Each of these sections keeps a journal (Plate II) which is the daybook of the section. It contains briefs of important written and verbal messages, both received and sent, and notations of periodic reports, orders, and similar matters that pertain to the section. If an item is received or issued orally, it is entered in detail; if written, the entry may be either a reference to the file number of the document or a brief of its contents. A brief notation is also made of instructions and directions pertaining to the section which have been given by the commander or a member of the section to someone outside of the section. The journal is closed when directed by the commander, at the end of the day, a phase, or other period. These journals are the permanent records of the activities of the sections; combined, they form the record of events of the organization. For further details, see FM 101–5.

THE STAFF IN SMALL WARS

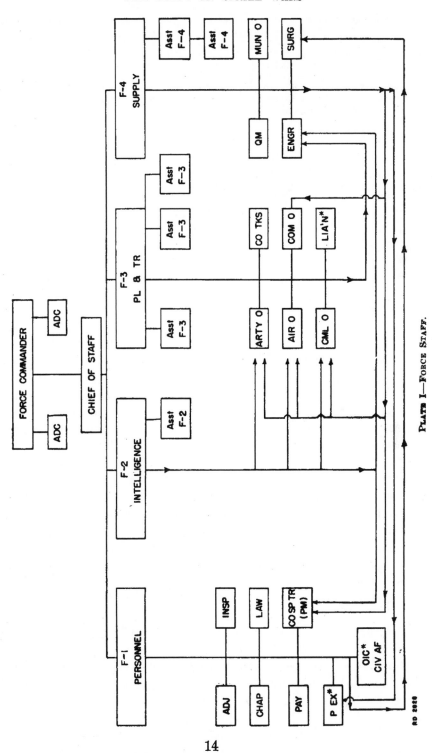

PLATE I.—FORCE STAFF.

THE STAFF IN SMALL WARS

The foregoing diagram shows the staff of a small wars force consisitng of a reinforced brigade. Special Staff Officers are here grouped under Executive Staff Sections under which they would normally perform the major portion of their functions. The diagram presumes a situation in which the function of the Executive Staff is that of direction, and the function of the Special Staff is that of operation.

Arrows indicate important functions with other staff sections.

Asterisks indicate assignments of staff officers in certain situations, although their functions may be assumed by other members of the staff if not of sufficient importance to warrant the detail of a separate officer.

JOURNAL[1]

Hq .... (Unit.)
.... (Place.)
.... (Date.)

From: .... (Date and hour.)
To: .... (Date and hour.)

Time[2]	Serial No.	Time dated[3]	Incidents, messages, orders, etc.		Dispositions	Remarks
			In	Out		

[1] Insert F-1, F-2, F-3, F-4, or "Consolidated," as appropriate.
[2] Refers to time of receipt or sending in the office keeping the journal.
[3] Refers to time the information was sent and thus calls attention to how old it is.

2-13. **The chief of staff.**—*a.* In a force no greater than a regiment or a reinforced regiment, the executive officer may perform all of the duties of chief of staff. In larger forces, the chief of staff usually will be an officer specially detailed for the purpose. His principal duties are to act as military adviser to the commander and to coordinate the activities of the staff. (See "War Department Field Manual 101-5.") He conducts all routine business in order to enable the commander to devote his time and efforts to more important matters. During the temporary absence of the commander, the chief of staff makes such decisions as the situation may demand; in each case he is guided by the policies, general instructions, or his intimate knowledge of the commander's wishes.

b. The chief of staff prescribes the internal organization of the various sections so as to fix responsibility for the initiation and supervision of work in order to secure efficiency and teamwork. He decides which members of one staff section will understudy the members of another staff section. He makes sure that the special staff is properly organized. Each chief of section will be so engrossed in his own work that, at times, one section will infringe on the duties of another. The chief of

THE STAFF IN SMALL WARS

staff must adjust this at once. His diplomacy and tact in adjusting such situations at the start will have a favorable reaction on the entire command.

c. As the organization progresses, it often develops that certain duties should be shifted from one unit to another. The chief of staff should see that such changes are made promptly. The map section of the engineers has been shifted logically, at times, from that unit to the second section. If a military government has not been established, civil relations may be shifted from the first to the second section.

d. During the concentration period, the chief of staff will be particularly interested in the plans of the staff sections and their arrangements for:

(1) Receiving incoming details and individuals.

(2) Prompt issue of equipment.

(3) Prompt completion of medical and other administrative inspections.

(4) Facilities for training.

(5) Coordination of training of all units.

(6) Organization of the Intelligence Service to meet the probable requirements of the situation.

(7) Organization of the Provost Service to meet adequately the probable demands that will be made upon it in the theater of operations.

e. The chief of staff should supervise the plans for increasing the intelligence personnel and for the establishment of provost services if it can be foreseen that the operations may result in the occupation of a country or a large section of it. The forces of occupation have four weapons with which to act: (*a*) Moral effect of the presence of troops; (*b*) intelligence service; (*c*) provost service (including Exceptional Military Courts); and finally (4) offensive action. The intelligence and provost services should be carefully considered in connection with "peaceful occupation." In the past, scant attention has been given to these services in the preparation of operation plans for small wars operations. As a rule, they have been established only when the necessities of operation forced it upon the higher command. In most cases an increase of personnel in intelligence units will be required over that allowed in organization tables when the operations include the complete occupation of a country or of large areas of it.

f. The provost service, including the exceptional military-court system, represents the military government to the mass of the people, with whom it comes in direct contact, and is the normal active instrument for the maintenance of tranquillity, freeing the natives from agitation and intimidation by their own countrymen. The provost service, more than any other element of the forces except the Intelligence Service, should understand the people, their temperament, customs, activities, and the everyday working of the average native mind. It warrants a well-founded and complete organization, including provost marshals and judges with legal knowledge, good and loyal interpreters, and sufficient clerical assistance to dispatch business with justice and celerity.

2–14. **The first section (personnel)—F–1.**—*a.* The assistant chief of staff, F–1, coordinates the activities of those agencies performing the functions that he is charged with in the "War Department Field Manual 101–5." He cooperates with the second section on matters pertaining to prisoners of war, espionage, etc., and with the third and fourth sections in regard to quartering, priorities of replacement, and allotment of time for recreational work. He is responsible for certain provisions of the administrative order, and must cooperate with the fourth section in this matter. Because he is charged with those functions which relate to the personnel of the command as individuals, he is brought into close contact with the adjutant, the inspector, the chaplain, the law officer, the surgeon, the provost marshal, the paymaster, the communications officer, the exchange officer, and the commanding officer, special troops.

b. The first section organizes the personnel of the staff section, and makes assignments of the clerical personnel, orderlies, and specialists therein.

c. Prior to leaving the United States, this section formulates a plan covering the replacements to accompany the force, numbers and classes of replacements to be dispatched later, dates that such replacements are desired, and priorities. This plan may appear as an annex to an appropriate administration order. In determining the number of replacements to be provided, the losses which may be incurred among the various classes of troops must be estimated. An ample margin should be allowed for casualties in transit and during the landing, and consideration given to the climatic and sanitary conditions en route and within the area of operations, the types of operations contemplated, the branch of service, and the time required for replacements to arrive. After arrival in the theater of operations, F–1 should insure by timely planning that complete information as to the needs of the force reaches the appropriate headquarters in the United States in sufficient time for replacements to arrive when needed. He should cooperate closely with the third section in estimating, well in advance of actual needs, changes in conditions that will require replacements, augmentation, or reduction of the Force. When replacements or reenforcements are recieved, they are distributed in accordance with priorities formulated by the third section.

d. F-1, in collaboration with the Commanding Officer, Special Troops, is charged with the allocation of space to the various headquarters' offices. Whatever the contemplated duration of the occupation, force headquarters should be so located and space so allocated thereat as to facilitate either the expansion or the reduction of its activities. In selecting and allocating office space, the first section confers with all members of the staff relative to their needs, and particularly with the fourth section, which supervises rentals and purchases.

e. Until personnel is specifically designated to take active charge of military government, the first section prepares plans as necessary for its establishment. Usually it will be advisable to organize a special staff section for this purpose. If the military government is an independent organization apart from the force, the first section acts as the liaison agent between the force commander and the staff of the military governor. For details, see Chapter XIII, "Military Government."

f. Since post exchanges are established for the welfare and convenience of the enlisted men, supervision of this activity comes under the first section. See paragraph 2-36.

g. The first section is charged with the rendition of reports concerning, and the handling of, civilian prisoners or prisoners taken from hostile forces. If a local constabulary is operating in cooperation with the intervening force, such prisoners usually are turned over to the former for trial by the constabulary courts martial or by civil courts; otherwise they are held at the disposal of the force commander.

h. The first section prepares and promulgates regulations governing the conduct of personnel in their associations with friendly natives in an effort to further cordial public relations. Social customs in countries in which small wars operations usually occur differ in many respects from those in the United States. Violation of these customs, and thoughtless disrespect to local inhabitants, tend to create animosity and distrust which makes our presence unwelcome and the task of restoring law and order more difficult.

i. The first section prepares and transmits to the fourth section such parts of the force administrative orders as affect the activities of the first section. These are principally: Replacements; military police; postal service; care and disposition of civilian prisoners and prisoners taken from the hostile forces; payment of the command; and post exchange supplies.

j. The records kept in the office of the first section should be reduced to the minimum. See paragraph 2–12, *d*. The following documents are needed in order to function efficiently:

(1) Section journal.

(2) A suspense file of orders, memoranda, and letters of instructions, which later are turned over to the adjutant.

(3) Copies of important communications which affect the force continuously. (The originals are kept in the adjutant's files.)

(4) A situation map should be kept posted, showing the status of matters pertaining to the first section at all times.

2–15. **The second section (intelligence) F–2.**—*a. General.*—(1) The assistant chief of staff F–2 constitutes the Bureau of Enemy Information. This section must keep in close touch with all other staff sections and is responsible for the dissemination of enemy information which may affect the operations of those agencies. This includes not only information of the military situation, but the political, economic, and social status of the occupied area, together with the attitude and activities of the civil population and political leaders insofar as those elements may affect the accomplishment of the mission.

(2) The duties of the intelligence officer are outlined in "War Department Field Manual 101–5." In addition, the following are of special importance in small wars operations:

(*a*) The names and descriptions of leaders, areas in which they operate, and the methods and material means which they employ in combat.

(*b*) Hostile propaganda in occupied territory, adjacent territory or countries, and our own country; and the methods, means, and agents used for its propagation.

(*c*) Liaison with government and local officials of the occupied country or areas, and with the civil representatives of our own and foreign governments therein.

(*d*) Close liaison with the commander of aviation in arranging for aerial reconnaissance.

(*e*) Maintenance of cordial relations with the local, American, and foreign press, and censoring of all press releases.

b. Duties prior to embarkation.—(1) During the concentration phase prior to embarkation, the second section will be primarily concerned with obtaining all available information relative to the country in which it is proposed to operate. Monographs, maps, and other pertinent information normally should be furnished by the Force General Staff. In no type of warfare is the latest current information more vital. For this reason the second section should immediately establish liaison with the corresponding sections of the naval and military services and with the nearest representatives of the State Department.

(2) The selection, organization, and training of the commissioned and enlisted intelligence personnel of both the headquarters and combat units should be carried on concurrently with the F–2 estimate of the situation. (See paragraph 2–13, *e*.) Every effort should be made to obtain personnel conversant with the language of the country. The force of interpreters will generally be augmented by the employment of natives. The second section, in conjunction with F–4, should compile and obtain approval of an "Allowance and Pay Table for Interpreters," based on the scale of wages of the country concerned, and funds should be allocated for payment thereunder prior to embarkation.

(3) A résumé of the available information of the theatre of operations should be completed as soon as practicable and reproduced and disseminated throughout the command. The following form is suggested for compiling this information. Some items listed therein may not be applicable in every situation, and additional items may be of great value in certain situations.

A FORM FOR A STUDY OF THE THEATER OF OPERATIONS

1. POLITICAL.—*a. History.*
   b. System of Government.
      (1) Form of government (dictatorship, republic, etc.).
      (2) Organization and method of operation.
      (3) Political subdivisions.
   c. Internal political situation.
      (1) Present government (head of state and other political leaders; personalities).
      (2) Political issues.
      (3) Analysis of parties.
      (4) Regional and social differences.
      (5) The press.

THE STAFF IN SMALL WARS

1. POLITICAL.—*a. History*—Continued.
    d. International politics.
        (1) Bearing of internal political situation on international policies.
        (2) Foreign policies.
        (3) Foreign relations.
    e. Summation (How does this affect the contemplated operations?).

2. ECONOMIC.—*a. General economic characteristics.*
        (1) Natural resources.
        (2) Degree of economic development.
        (3) Dependence on foreign trade.
    b. National productive capacity.
        (1) Agriculture.
        (2) Mining.
        (3) Manufacture.
        (4) Shipbuilding.
    c. Commerce.
        (1) Domestic trade.
        (2) Foreign trade.
    d. Transportation.
        (1) Railroads.
        (2) Highways.
        (3) Water.
        (4) Air.
    e. Communication.
        (1) Cables.
        (2) Radio.
        (3) Telegraph.
        (4) Telephone.
    f. Finance (method of financing government).
    g. Population (economic aspects; present population analysis of population, labor, and social conditions).
    h. Plans for industrial mobilization.
    i. Economic penetration by foreign interests.
    j. Influence of economic situation on foreign relations.
    k. General conclusions (reference to economic self-sufficiency, capacity for production of war supplies and food supplies, and degree of dependence on maintenance of trade routes).

3. GEOGRAPHY (PHYSICAL).[1]—*a. General topography and hydrography.*
    b. Rivers and water supply.
    c. Climatic conditions.
    d. Critical areas (areas the loss of which would seriously hamper the country under consideration).
    e. Vital areas (*areas essential to the country concerned*).
        (1) Routes of approach.
        (2) Roads, trails, and railroads.
        (3) Harbors and beaches near critical areas.

---

[1] Such geographical items as have been considered under political or economic headings should be omitted.

THE STAFF IN SMALL WARS

3. GEOGRAPHY (PHYSICAL).—*a. General topography and hydrography.*—Contd.
    (4) Communications.
    (5) General terrain considerations.
  f. Conclusions (the effect of general terrain considerations on operations. The most favorable theater of operation from a standpoint of physical geography).

4. PSYCHOLOGICAL SITUATION.[2]—*a. General racial characteristics; types, races, etc.*
  b. Education.
  c. Religion.
  d. Attitude of inhabitants toward foreigners.
  e. Susceptibility to propaganda (influence of church, press, radio, or other agency).
  f. Conclusions.

5. COMBAT ESTIMATE.—*a. Coordination of national defense.*
    (1) Military forces (government and opposition).
    (2) Supreme commander (government and opposition).
  b. Personnel.
    (1) Estimated strength of components of both government and hostile forces.
    (2) Government forces and leaders.
    (3) Hostile forces and leaders.
  c. Training, efficiency, and morale (government and hostile forces).
    (1) Individual.
    (2) Unit and combined training.
    (3) Training of reserves.
    (4) System of promotion of officers.
    (5) Efficiency.
    (6) Morale.
  d. Recruiting methods.
    (1) Government forces.
    (2) Hostile forces.
  e. Equipment and supplies available.
    (1) To government forces.
      (*a*) On hand.
      (*b*) Replacement possibilities and sources.
    (2) To hostile forces.
      (*a*) On hand.
      (*b*) Replacement possibilities and sources.
  f. Method of conducting combat.
  g. Navy.
    (1) Strength.
    (2) Organization.
    (3) Training, efficiency, and morale.
  h. Conclusions.

6. GENERAL CONCLUSION (Relative value should be given to all factors and final conclusions must be based on the study as a whole).

---

[2] Discuss only such items as are not covered fully elsewhere in the study. Refer to other paragraphs where appropriate.

THE STAFF IN SMALL WARS

(4) (*a*) Available maps are usually inaccurate and of small scale; their procurement is costly and the supply limited. They have often proved so unreliable as to detail as to be valueless except for the purpose of correction. It is often more practical and economical to obtain maps only for headquarters and executive staff sections of all units, providing means for the reproduction and distribution of corrected sections or of new maps made after arrival in the theater of operations. In small-wars operations where engineer troops have not been present, map reproduction has been made a responsibility of the second section; in other cases, the map-reproduction section of the engineers has been transferred to the force headquarters intelligence section. In any event, the second section is responsible for the procurement and distribution of maps.

(*b*) Aerial photography, in addition to its other military uses, will play an important part in the development of new maps and obtaining accurate information for the correction of old ones after reaching the theater of operations. The procurement of an initial supply of film and other materials for this purpose is essential.

(5) In order to establish favorable press relations at the start, and to avoid the publication of harmful and incorrect information, a definite policy must be adopted as to who will receive representatives of the press, what information will be furnished, and what means will be provided for obtaining it. Even though the campaign may be too insignificant to have correspondents and photographers attached for the entire operation, they will invariably be present at the beginning. In some cases, officers have been permitted to act as correspondents; if this is done, a definite agreement must be made relative to the class of information which will be furnished.

(6) If a military government is not established, civil relations with the local officials, native civilians, and foreign nationals, including citizens of the United States, become a function of the second section. Best results will be obtained if the policy for dealing with the various elements is established before the force arrives in the theater of operations. After arrival, the local representatives of the State Department should be consulted and such changes made in the policy as appear to be desirable.

(7) Organizations which are opposed to intervention in the affairs of other nations, regardless of the cause, have at times disseminated their propaganda to the force. The second section is responsible for guarding against this by locating the source and notifying, through official channels, the proper civilian officials. An early statement of

the facts relating to the situation, by the commander, will usually forestall any ill effects from such propaganda.

(8) Intelligence funds, which are not a part of the quartermaster allotment, are required for the proper functioning of the second section. F–2 is responsible for requesting the allotment of such funds prior to the embarkation of the force.

c. Duties in the theater of operations.—(1) The F–2 section is primarily an office for the consolidation of information supplied by lower units, special agents, and outside sources; and for the prompt distribution of the resulting information to other staffs, sections, and organizations concerned. If circumstances require the second section to assume the duties of the officer in charge of civil affairs or other functions, additional divisions must be organized within the section under competent assistants.

(2) The following intelligence agencies are available to F–2 for the collection of information: Secret agents, voluntary informers, aviation, intelligence agencies of lower organizations (brigades, regiments, areas, etc.), other governmental departments.

(*a*) Secret agents, hired from among the inhabitants in the theater of operations, have proved valuable collectors of information in the past. They must be carefully selected and, once employed, a close watch should be kept on their activities. Usually such agents have been politically opposed to the native forces whose activities have resulted in the intervention. If they attempt to use their position for their own aggrandizement or to embarrass personal enemies, they are useless as sources of information and handicap the intervening force in gaining the confidence of the population. However, when reliable agents have been obtained in past operations, they have provided extremely valuable information. It is often advisable to pay them low regular wages and to reward them with bonuses for timely and accurate information.

(*b*) The major portion of the information obtained from voluntary informers is often false, grossly distorted, or too late to be of value unless the informer has personal reasons for making the report. Liberal cash payments for information that proved correct and timely have sometimes brought excellent results. Hired agents and informers have been of assistance in the past in uncovering the hostile sources of supply. The source of the information must be kept inviolable in order to protect the informers and to insure an uninterrupted flow of information. The universal tendency of even reliable hired agents and voluntary informers is to protect anyone with whom they are connected by politics, business, or blood.

(*c*) It is not improbable that high officials of both the party in power and the opposition may secretly support insurrectionary activities in order to insure themselves an armed following in the field in case the intervention should be ended suddenly. Such a condition increases the difficult task of securing agents who will report impartially on all disturbing elements.

(*d*) Excellent results have been obtained through the cooperation of business establishments which maintain branches or other contacts throughout the occupied areas. For financial reasons, the central office of such concerns must have timely and impartial knowledge of actual or prospective conditions throughout the country. In many cases they are dependent upon the intervening forces for protection of their personnel and property, and it is to their advantage to restore peaceful conditions as rapidly as possible. In seeking to establish such a contact, the intelligence officer should look for a business establishment with which the force normally does business. Liaison should be maintained through members of the command who visit the business house in the routine course of duty and who are publicly known to do so. It is unfair, as well as poor intelligence technique, to risk the life or the business career of a man in his community through carelessness or loose talk. Many companies have accurate and detailed maps or surveys on file which may be obtained and reproduced to supplement the small-scale maps available to the force.

(*e*) Aerial reconnaissance is invaluable in locating large movements, encampments, and affected areas. When the opposition has been broken into small groups, the lapse of time between gaining information and the arrival of a ground patrol is usually too great to give effective results. The use of observation aviation in close support of infantry patrols operating against small hostile forces is of doubtful value. The airplane discloses the presence and location of the patrols and enables the hostile groups to avoid them or to choose the time and place for making contact. Any ambush that can be located from the air should be uncovered in ample time by the exercise of a little care on the part of the patrol leader. Aerial photographic missions often will be the best or only means for securing accurate information of the terrain in the theater of operations. For further details, see Chapter IX, "Aviation."

(*f*) Subordinate units provide the force commander with detailed information on hostile activities, the terrain and geography, and the political and economic situation in the areas in which they operate.

As combat intelligence for the purpose of gaining contact with and destroying hostile armed opposition, such information usually will be of value only to the unit first gaining it. But such information, when collected from the entire theater of operations and transformed into military intelligence, provides the commander with the information he must have to dispose his forces in accordance with the situation and to prepare for eventualities. F-2 should coordinate the activities of the intelligence sections of subordinate units. The second section of a subordinate organization, quartered in the same city or town as force headquarters, should not be used as an appendage to the force intelligence section, but should be permitted and required to function in its normal manner. However, F-2 should utilize every opportunity to develop a close understanding and personal relationship with subordinate intelligence officers.

(g) F-2 should maintain close liaison with other agencies of our government established in the theater of operations. Information from such agencies concerning the higher officials of the government of the occupied state and of the opposition party, as well as of the economic condition of the state, may be accepted as sound. But because of the limited circle within which they move, as well as for other reasons, their opinion concerning the effect of the national economy on the peace of the state, and of political and social trends to which the higher classes are unsympathetic, must be accepted with care. The same applies to the opinions of America businessmen domiciled in the country. An officer possessing a working knowledge of the language, a knowledge of the psychology of the people, good powers of observation, and who has associated with the average civilian in the outlying districts for a month, is in a position to possess a sounder knowledge of the fundamental disturbing factors at work in the country that an official or businessman who may have spent years in the capital only.

(h) Close contact should be maintained also with representataives of our government in bordering states, especially with naval and military attachés. This is particularly applicable when the affected area borders the frontier.

(3) The same agencies for securing information are available to brigade (if the force consists of more than a reenforced brigade) and regimental intelligence officers as are available to F-2, except that it will be unusual for them to contact representatives of our own or foreign governments directly. Reconnaissance aviation is usually

THE STAFF IN SMALL WARS

available on request. If a regiment is operating independently in a small wars situation, the regimental intelligence section should be strengthened to fulfill adequately the functions of the F–2 section.

(4) (*a*) Even the battalion in small wars rarely operates as a unit. Its companies often occupy the more important villages in the battalion area and, in turn, send out subdivisions to occupy strategically located settlements and outposts. The battalion intelligence officer should spend as much time as possible in the field in order that he may become thoroughly familiar with the situation throughout the area.

(*b*) As soon as it is established, every detached post or station must organize and develop its own intelligence system. Each garrison must initiate active patrolling for the purpose of becoming familiar with the routes of communication, topography and geography of the district, the inhabitants, and the economic and political forces at work in the community. Routine patrols over the same roads or trails and at regular intervals of time should be avoided; rather the objective should be to discover new trails and to explore new areas with each successive patrol and to confuse the opponents by varying the dates and hours of departure. Local garrisons must become so familiar with their subdistricts that any changes or unusual conditions will be immediately apparent. Local commanders and their noncommissioned officers should be able to proceed to any point in their subdistrict via the shortest and quickest route and without the assistance of a guide or interpreter.

(*c*) Maps furnished from the higher echelons must be supplemented by road sketches and the correction or addition of all pertinent military information. This work should be undertaken immediately upon arrival, beginning with the most important unmapped roads or trails and continuing throughout the occupation until accurate large-scale maps are available of all subdistricts. A supplementary chart should be compiled indicating the distances between all points of military importance and the time factor involved for each type of transportation available and for each season of the year.

(*d*) A record should be kept of all prominent citizens in the locality, whether friendly or hostile to the intervention. Each record should show: The full name of the individual as taken from the baptismal or birth certificate (both when these records differ); the name by which the person is customarily known; all known aliases, if any; and his reputation, character, and activities. Additional information should be entered on the record as it becomes available. Duplicates are forwarded to the next higher echelon. It is only by this means that accurate and continuous information can be maintained on the inhabitants of the occupied areas, which will prove invaluable when questioning individuals, for orienting newly arriving officers, and for preparing charges when it is desired to bring suspects to trial for their activities.

(*e*) Intelligence activities are greatly handicapped if the officers attached to battalions and smaller units in the field are not familiar with the local language. This is especially true with Bn-2. Each officer should endeavor to learn the language sufficiently well to engage in social activities and to dispense with interpreters as soon as possible.

(*f*) Outpost commanders may obtain information by:

Establishing a service of information through the local mayor or senior civil official;

Weekly reports from the senior civil official in each settlement within the subdistrict;

Questioning commercial travelers;

Interrogating persons or the relatives of persons injured or molested by the hostile forces;

Close surveillance of relatives of hostile individuals;

Examination of prisoners; and

Constant observation of the movements of all able-bodied men in the district.

(*g*) Methods of extracting information which are not countenanced by the laws of war and the customs of humanity cannot be tolerated. Such actions tend to produce only false information and are degrading to the person inflicting them.

d. Intelligence records.—(1) *Study of the theater of operations.*—A thorough knowledge of the theater of operations in small wars is highly important to all officers from the force commander to the junior patrol or outpost commander. Information compiled prior to arrival in the theater must be supplemented by reconnaissance and research on the ground. See paragraph 2-15, b.

(2) *Special studies.*—From time to time the intelligence officer may be called upon to make special studies of particular localities, situations, or other factors arising during the course of the campaign.

(3) *The intelligence annex.*—A complete intelligence annex may be issued at the beginning of the operations to accompany the campaign plan. Such an annex is not usually necessary in small wars operations unless strong, organized resistance to the intervention is anticipated.

THE STAFF IN SMALL WARS

The form for the intelligence annex given in "War Department Field Manual 101-5" may be used as a guide.

(4) *The intelligence estimate.*—(*a*) The intelligence estimate during the early phases of intervention may closely parallel the F-2 estimate of a major war. It is that part of the commander's estimate of the situation which covers the hostile forces and their probable course of action. The following outline may be used as a guide for such an estimate:

F-2 ESTIMATE

(Heading)

File No.

Maps:

1. HOSTILE FORCES:

    Dispositions; strength; physical conditions; morale; training; composition; supply and equipment; assistance to be expected from other sources.

2. ENEMY'S CAPABILITIES:

    Enemy's mission; plans open to enemy; analysis of courses open to the enemy.

3. MOST PROBABLE COURSE OF ENEMY ACTION.

(Signature.)

(*b*) As the intervention continues and the hostile forces are dispersed into small groups, purely military operations usually become subordinate to civil problems. The following form may be used as a guide for an F-2 estimate of the political, economical, and civil situation:

ESTIMATE OF THE POLITICAL, ECONOMICAL, AND CIVIL SITUATION

From: Date and hour
To: Date and hour

Unit
Place
Date and hour

File No.

Maps:

1. GENERAL STATE OF TERRITORY OCCUPIED:

    State under the appropriate number of subparagraphs, a general summary of hostile activities as it exists in each subdivision of the state or territory, allotting a subparagraph to each geographic subdivision.

2. ATTITUDE OF CIVIL POPULATION:

    Discuss attitude of the leaders, whether political or military. The general attitude of the population, whether friendly, tolerant, apathetic, or hostile. Local assistance or obstruction we may expect to our efforts.

3. ECONOMIC SITUATION:

Condition of business. Employment situation. Price of foodstuffs. Condition of crops. Influx or outflow of laborers. Conditions amongst laborers.

4. POLICE OPERATION:

Police conditions. Coöperation of native forces and native Civil Police with our own. Type of crime for which most arrests are made, whether major or minor offenses. Amount and reliability of information furnished by local force or police. Arms in use by local police, type and number. If police are subject to local political leaders for their jobs. Sources of their pay and a comparison of it with other salaried positions in the locality.

5. MILITARY OPERATION:

Either discuss or refer to B-2 Reports.

6. POLITICAL SITUATION:

A discussion of the local political situation in various sections of the state or territory, as it affects the state as a whole. A discussion of national poltics and political questions. The statements or actions of national political leaders or the national political governing body. Political situation in adjacent states which may have an immediate bearing on the local situation.

7. MISCELLANEOUS:

Such items of interest bearing on the political, economic, and civil situation as does not come belong under the proceeding paragraphs.

(s) B
Major
F-2

(5) *The Journal.*—See paragraph 2-12, d.

(6) *The Intelligence Report.*—(*a*) The information which has been collected and evaluated during a given period is disseminated by means of an intelligence report or an intelligence memorandum. The period of time to be covered by the report is prescribed by higher authority, or by the unit commander. It is issued by all combat units down to and including the battalion or corresponding command for the purpose of informing superior, adjacent, and subordinate organizations of the situation confronting the unit preparing the report. It may be supplemented by a situation map or overlay. In small wars operations, it may be advisable to prepare separate reports on the military, economic, and political situations, or, if they interlock, a combined report may be submitted. The military report is similar to that given in "War Department Field Manual 101-5." The following form may be used as a guide in preparing a combined report:

PERIODIC REPORT OF INTELLIGENCE

From: Date and hour
To: Date and hour

File No.

CONFIDENTIAL

(Heading)

1. ATTITUDE OF CIVIL POPULATION TOWARD MILITARY GOVERNMENT OR OCCUPATION:
   Hostile, neutral, or friendly; by social classes.
2. POLITICAL ACTIVITIES:
   Activity of political parties during period—deductions.
3. ECONOMIC CONDITIONS:
   Condition of crops, prices of foodstuffs, if low or high, reason therefor, pests, epidemics, disasters, labor and wages, economic conditions which may tend to produce disorder and unrest.
4. LOCAL DISTURBANCES:
   Agitation or disorder caused by rumors, secret organizations, disputes over property, criminal element.
5. PROSECUTIONS:
   Prosecution of prominent people such as newspaper men, civil officials, etc.
6. HOSTILE FORCES:
   Names of leaders, strength; number and kinds of arms, localities frequented—activity during period, normal or abnormal—deductions.
7. HOSTILE ACTIVITIES:
8. MILITARY OPERATIONS:
   Synopsis of military activity to offset hostile operations and unsettled conditions.
9. ARMS AND EQUIPMENT:
   Number of arms and equipment captured, surrendered, or taken up, with general locality.
10. MISCELLANEOUS.
11. CONCLUSIONS:
    (a) General state of territory occupied.
    (b) Possible future trend of events or courses of action open to the opposition.
    (c) Most probable future trend and course of action, based on a sound estimate only.

(Signature.)

(b) Reports submitted by organization commanders in the field should be complete and detailed. It is better to send in too much information than too little. A report which is meaningless to the commander of a small detachment may be essential to the next higher echelon when considered with the information received from other sources. On the other hand, F–2 reports to higher authority may be

THE STAFF IN SMALL WARS

in the form of brief summaries, omitting the mass of detail collected by the combat organizations. Where the immediate transmission of items of information is necessary, the most rapid means of communication available is employed.

(c) The rapid dissemination of military intelligence to all organizations concerned is fully as important as the collection of original information. The distribution of intelligence reports should include the smallest separate detachment in the field. Because of the wide dispersion of troops in usual small wars operations, intelligence reports are often the only means by which a patrol commander can be kept informed of hostile activities, or plan his operations to intercept probable enemy movements.

(d) In view of the peculiar status of our forces in small wars operations, in which they frequently become involved for the sole purpose of providing military aid to the civil power of a foreign nation in order to restore peace within the boundaries of the state, the use of the term "enemy" should be avoided in all records, reports, and other documents.

(7) *The intelligence work sheet.*—As information is received by the second section, it must be recorded in an orderly fashion preliminary to the preparation of the intelligence report. This is done by means of the intelligence work sheet. No form for this is prescribed, but a convenient method is to classify the information as it is received under the headings used in the intelligence report, starting each heading with a new sheet. This provides a satisfactory means for segregating the information, and greatly facilitates the preparation of the intelligence report.

(8) *The intelligence situation map.*—A situation map, showing the latest reported disposition of the hostile forces, is kept by the second section.

2-16. **The third section (plans and training)—F-3.**—*a.* The assistant chief of staff F-3 performs the specific duties outlined in "War Department Field Manual 101-5."

b. One of the first duties of F-3 may be to prepare letters of instruction for the immediate subordinate organization commanders as outlined by the force commander. Such instructions are secret. They indicate the successive steps to be taken if the operations progress favorably, or contemplated plans in case of reverse or other eventualities. In major warfare, letters of instruction are not common in units smaller than a corps but in small wars situations, which are usually extremely vague and which present so many possibilities, some instructions of this nature will assist the commander's immediate subordinates in the execution of his scheme of maneuver and campaign plan.

c. The third section prepares the necessary organization, movement, communications, and tactical plans. Organization of the combat units includes the priority of the assignment of replacements, and recommendations for desirable changes in armament and equipment. In conjunction with F-2, he estimates the strength, armament, equipment, and tactics of the opposing forces, and determines the necessity for the attached supporting arms with the Force such as aviation, artillery, tanks, etc., and the appropriate strength thereof. Every available means of communication must be utilized; generally additional equipment and personnel will be required as a shortage of communication material may influence the plan of campaign. The prompt preparation of an air-ground liaison code is very important.

d. In conjunction with the special staff and F-4, the third section determines the number of units of fire of normal and special ammunition to be carried with the force initially, and requests replacements from the United States as necessary.

e. F-3 prepares and issues orders for all troop movements. However, he prescribes only the general location and dispositions of the technical, supply, and administrative units and the actual movement orders for these units are issued by the staff section concerned after consultation with and approval of F-3. In considering the combat missions to be assigned to the various organizations, areas, or districts in the theater of operations, he makes appropriate redistribution of personnel or requests replacements when necessary. Because of the time factor involved in the redistribution of men or the arrival of replacements from the United States, troop movements must be planned farther in advance in small wars operations than in regular warfare.

f. In small wars, the units of the force are generally so widely distributed throughout the theater of operations that the commander may have difficulty in keeping abreast of the situations existing in the various elements. Operations orders should usually be phrased in general terms and the details of execution delegated to subordinate commanders. This necessary decentralization of authority is simplified by partition of the theater and the organization of the command into areas, districts, and subdistricts.

g. By intimate contact with other staff sections, F-3 keeps informed of all pertinent matters affecting the combat efficiency of the force. He maintains close liaison with the special staff officers concerning all matters in which their duties, technical knowledge, and functions will affect the operations. He coordinates the efforts of subordinate units or the various area organizations, the supporting arms (aviation in particular), and armed native organizations, to the end that the greatest combat effectiveness is assured.

h. In addition to situation maps, overlays, and other data permitting a ready grasp of the tactical situation, the third section keeps a suspense file of all memoranda or orders emanating therefrom, and a work sheet and a section journal.

2-17. The fourth section (supply)—F-4.—*a*. The assistant chief of staff F-4 is charged with the preparation of plans, policies, priorities, and decisions incurred in the supervision and coordination of the technical, supply, and administrative services, in matters of supply, transportation, evacuation, hospitalization, and maintenance. F-4 must so exercise his supervision of these services that the troops will not be incapacitated by the lack of sufficient clothing, food, and ammunition, and so as to relieve their commanders of the worry as to whether these articles will be furnished. The specific duties of the fourth section are outlined in "War Department Field Manual 101-5."

b. F-4, in conjunction with the third section, recommends changes in types and amounts of individual, organization, combat, supplementary, and special equipment, and the units of fire of normal and special ammunition to be carried initially. In cooperation with the first section, F-4 estimates the civilian labor needed and obtainable in the theater of operations, and the number and composition of specialists units to be attached to the force for the service of supply. hospitalization, communication, and transportation. He determines the amount of supplies that can be obtained from local sources and prepares a schedule for shipment of replacements. The amounts and types of transport to be taken will depend upon the tactical and administrative requirements, the general nature of the terrain in the theater of operations, and the availability and suitability of native transport. In many situations, a large reduction in allowances or a complete change in type from that specified in organization tables. or both, may be required. See Chapter III, "Logistics."

c. The fourth section normally coordinates, supervises, and directs the supply services without in any way operating their specialities. Ordinarily these services deal directly with F-4, who settles routine matters and refers those which involve new policies to the chief of staff for decision.

THE STAFF IN SMALL WARS

d. Since our relations with the local government in the theater of operations is usually friendly, F–4 makes the necessary arrangements with the customs officials relative to the clearance of supplies and material for the force.

2–18. **The special staff.**—*a.* The special staff consists of all officers, other than the executive staff (chief of staff, F–1, F–2, F–3, and F–4), specifically provided for the purpose of assisting the commander in exercising his command functions. This special group includes the heads of the technical, supply, and administrative services, and certain technical specialists. In the Force, the executive staff and the special staff are separate and distinct, while in lower units they usually merge into each other, one officer frequently being charged with the duties of one or more special staff officers as well as with those of a member of the executive staff. Special staff officers normally assigned to a small wars force of a reinforced brigade or larger organization are listed in the succeeding paragraphs.

b. Although the special staff sections usually function under the coordination of the executive staff sections (See Plate I, paragraph 2–12, a), such staff officers are not precluded from dealing directly with the chief of staff or the force commander when necessary. Special staff officers are not "under" any one officer of the executive staff but function with any or all of them, and with each other.

2–19. **The adjutant.**—The functions of the adjutant correspond with those prescribed for the adjutant general in "War Department Field Manual 101–5." In lower units, these functions are combined with those of F–1.

b. (1) The Force postal service is operated, under orders of the adjutant, by the postal officer, or enlisted mail clerk when no postal officer is appointed. It is advisable, however, to place an officer in charge of the post office, particularly when a large portion of the force is in the field, and cash for the purchase and payment of money orders must be handled by messenger.

(2) The postmaster at the point of concentration or port of embarkation should be consulted for information on the postal forms required.

(3) Prior to sailing, and periodically thereafter as may be necessary, an order should be published giving the correct mailing address of the command, and recommending that officers and men advise their correspondents to send money only by domestic, rather than by international money orders.

(4) If the prompt and efficient dispatch and distribution of mail cannot be effected by the authorized postal section complement, the

adjutant should not hesitate to request the temporary or permanent assignment of additional personnel. Officers and men of the command must be able to send and receive mail with facility; valuables must be secure while in transit within the Force; and the mail clerk must receive promptly the signed receipt of the addressee for registered and insured articles on the postal form provided for that purpose.

c. Combat organizations conducting operations in the field should be relieved of as much routine administrative work as possible. Company first sergeants and company clerks may be assembled at battalion or area headquarters where, under the supervision of Bn–1, they are responsible for the preparation of muster rolls, pay rolls, service record-book entries, routine correspondence, etc.

2–20. **The inspector.**—*a.* In addition to the functions prescribed in "War Department Field Manual 101–5," the inspector in small wars operations is usually required to investigate claims for damages resulting from the occupation.

b. Inspections.—(1) Inspections should not interfere with tactical operations.

(2) When patrols escort the inspector from one outpost to another, they should be of a reasonable strength; it is preferable that the inspector accompany ordinary patrols demanded by routine operations.

(3) The inspector assumes no authority while making his inspection and issues no orders unless specifically authorized to do so by the force commander.

(4) No report should be made of minor discrepancies which can be and are corrected locally.

(5) When the inspector makes recommendations or notes deficiencies in his report, he should see that proper action is taken in accordance with the policy or orders of the force commander. This is particularly true with reference to matters affecting the morale and efficiency of the troops.

c. Investigations.—One of the most important duties of the inspector in small wars is to investigate matters which involve controversies between individuals of the force and local inhabitants. These investigations should be promptly, thoroughly, and fairly made, bearing in mind the interests of the individuals concerned and those of our Government. The finding of facts should be recorded and filed for future reference to meet those charges of impropriety which so often follow our withdrawal from the theater of operations.

THE STAFF IN SMALL WARS

d. Claims and damages.—(1) Claims and damages may be a source of embarrassment to the command if they are not investigated and acted upon promptly. When a special claim board is not designated, the inspector generally acts in that capacity.

(2) In every small war, claims, involving personal injury or property damage, are presented which could be settled immediately and at great savings to the Government if funds were made available for that purpose.

(3) If an injury has been done to any individual or private property is damaged, it should be reported to the proper authority without delay. The latter should order an immediate investigation even though no claim has been presented. Damages which are the result of neglect or misconduct on the part of members of the command should be determined before the departure of the individuals concerned from the locality. The investigation should determine whether the damages are the result of a wilful act, negligence, accident, unintentional injury, or of ordinary wear and deterioration. Private or public property occupied or employed by our forces should be inspected by the local commander or his representative and the native inhabitants concerned and a record made of all deficiencies or irregularities. Such an inspection is made upon taking possession of and upon vacating the property.

(4) Prior to withdrawal from the theater of operations, the force commander may issue a proclamation indicating that all claims for damages must be submitted to the designated authority before a given date. This enables the investigation and adjustment of the claims before the evacuation of the area. It has the disadvantage of encouraging a flood of unreasonable claims.

(5) No claims should be allowed for damage to property or for personal injury which is incident to military operations or the maintenance of public safety, when no criminal intent or carelessness is in question.

(6) Records of all data affecting claims, including receipts and releases, should be retained with the files of the Force or otherwise disposed of as directed by higher authority.

2–21. **The law officer.**—In small wars operations, the law officer is the legal adviser to the force commander and his staff on questions of local civil law, in addition to the functions prescribed for the "Judge Advocate" in "War Department Field Manual 101–5."

2–22. **The officer in charge of civil affairs.**—See "War Department Field Manual 101–5."

2-23. **The chaplain.**—See "War Department Field Manual 101-5."

2-24. **The paymaster.**—*a.* The paymaster is charged with those duties prescribed for "The Finance Officer" in the "War Department Field Manual 101-5," which pertain to the payment of the command, including mileage and traveling expenses of commissioned officers. In small wars operations, he must be prepared to advise the force commander regarding the trend of foreign exchange, especially whether the command shall be paid in whole or in part in United States currency or local currency.

b. The paymaster does not pay travel expenses of enlisted men, except when travel by air is involved, nor does he handle the expenses of transportation of dependents, which payments are made by the disbursing quartermaster. In the absence of a disbursing quartermaster, the paymaster may make disbursements of funds pertaining to the Quartermaster's Department, charging such disbursements to the quartermaster's appropriation involved.

2-25. **The provost marshal.**—*a.* In addition to the normal duties prescribed for the provost marshal in "War Department Field Manual 101-5," in small wars operations he has many functions relative to the control of the local civilian population, some of which are listed below:

(1) Control of circulation of civilian population.

(2) Detention of and bringing to justice offenders against the Executive Orders and the Proclamation of Intervention.

(3) Repression of crime.

(4) Enforcement of the Executive Orders and execution of the mandates of the military authority.

(5) Execution of sentences of military courts.

(6) Arrest and detention of suspects. Investigation of reports bearing on civilian activities.

(7) Special investigation of complaints made by civilians against the members of the occupation, municipal police, etc.

(8) Observe civil officials in performance of their duties and report any official violation of this trust.

(9) Custody of certain prisons and their inmates; enforcement of prison regulations; and supervision of prison labor.

(10) Issue and cancel firearms permits in accordance with Force Orders.

(11) Control the storage and release of firearms, ammunition, and explosives imported into the country. The sale of ammunition to persons possessing arms on permits in accordance with Force Orders.

b. Native prisoners should never be confined with personnel of the intervening force; separate prisons should be used. F-2 is permitted to have free access to all native prisoners for interrogation and examination. The first section is responsible for such action as may be necessary concerning prisoners in the hands of hostile forces, and for individuals who become embroiled with the friendly civil population or are arrested by the local authorities.

2-26. The commanding officer of special troops.—The commanding officer of special troops normally performs those duties prescribed for the "Headquarters Commandant" in "War Department Field Manual 101-5." In many cases he will also be the provost marshal, and charged with the duties of that officer.

2-27. The artillery officer.—The artillery officer has the functions set forth for the "Chief of Artillery" in "War Department Field Manual 101-5," and, in addition, normally serves in the dual capacity of commander of the artillery units with the force. If a landing against opposition is anticipated, the artillery officer is responsible for the artillery annexes attached to the operations orders.

2-28. The air officer.—See "War Department Field Manual 101-5." In his dual capacity of commander of the force aviation, he is responsible for the execution of all duties and operations assigned to such aviation by the force commander.

2-29. The communications officer.—*a. General duties.*—(1) The communications officer performs those functions prescribed for the "Signal Officer" in "War Department Field Manual 101-5." In addition he:

(*a*) Coordinates communication activities with the U. S. Naval Forces, native communication agencies, and communication establishments owned by commercial concerns.

(*b*) Assumes responsibility for all naval codes and ciphers.

(*c*) Supervises all encoding and decoding of dispatches.

(2) If the headquarters of the force is so located that its communication system becomes of primary importance in the chain of Naval Communication and is the principal agency for handling dispatches for the State Department, a separate communications officer with rank corresponding to that of the chiefs of section of the executive staff should be assigned to the special staff. This officer would not necessarily have to be a communications technician. By virtue of his rank and position he would be able to advise the force commander relative to communication matters, and in general

execute the communication policy, leaving the technical details of training and operation to a technical assistant or to the commander of the force communication unit.

b. Classes of communication.—The classes of communication to be handled as wire or radio messages, and the classes to be handled by letter, should be determined prior to embarkation. Authority to handle class E (personal messages) by radio should be obtained.

e. Additional communication personnel and equipment.—Organization tables do not provide sufficient personnel or material, especially radio equipment, to meet the normal requirements of small wars operations. The communication officer is responsible for augmenting the trained personnel and obtaining the additional equipment demanded by the situation.

d. Communication policy.—(1) Irrespective of the size of the force, there are certain duties relative to policy which fall to the communications officer in small wars. The more extended the force, the more involved the policy will be. Part of the policy will be dictated by the Naval Communication Service, as defined in Naval Communication Instructions, while a part will be incident to the type of intervention.

(2) The communications officer should ascertain whether the communication facilities of the country concerned are privately or publicly owned and operated, their extent, and the communication agencies employed. He should determine what, if any, communication agencies are devoted exclusively to military activities, obtaining the call signs and frequencies of the radio establishment. He should also ascertain what communication facilities are owned and operated by foreign companies. Upon arrival in the theater of operations, he should verify this information.

2–30. **The engineer officer.**—See "War Department Field Manual 101–5."

2–31. **The surgeon.**—*a.* See "War Department Field Manual 101–5."

b. In small wars operations, when the force may be widely dispersed, the force surgeon should consider:

(1) The necessity for additional medical personnel.

(2) Extra supplies of medical materials, quinine, and similar medicaments.

(3) Portable dental outfits.

(4) The preparation of medical supplies for airplane drops.

2-32. **The quartermaster.**—In addition to the functions prescribed in "War Department Field Manual 101–5," the force quartermaster is charged with:

a. The operation of sales stores.

b. The procurement of local transportation, including riding, draft, and pack animals, either by hire or purchase.

c. Recommending changes in existing system of accountability, when required.

d. Making estimates and requests for quartermaster funds, and supervising the allotment of funds as approved by the force commander.

e. Custody and disbursement of quartermaster funds, and funds from other branches of the naval service, as authorized.[3]

f. Payment for supplies and services purchased; and for damages and claims, when authorized.[3]

g. Payment for labor and transportation hired.[3]

2-33. **The chemical officer.**—See "War Department Field Manual 101–5."

2-34. **The tank officer.**—The commanding officer of the tank unit attached to the force is the technical and tactical advisor to the force commander in all matters pertaining to the use of tanks or armored cars, and to defense against mechanized forces.

2-35. **The munitions officer.**—The munitions officer performs those functions specified for the "Ordnance Officer" and the "Munitions Officer" in "War Department Field Manual 101–5."

2-36. **The post exchange officer.**—The post exchange officer is a distinct member of the force special staff. His duties are:

a. To obtain initial funds for establishment of the exchange.

b. To procure exchange supplies by purchase or on consignment.

c. To plan for the distribution of post exchange stores to outlying garrisons.

d. To conduct the exchange in accordance with regulations.

2-37. **The amusement and welfare officer.**—*a.* An officer may be specifically designated as the amusement and welfare officer and assigned to the force special staff, or these duties may be delegated to a staff officer in addition to his regular duties.

b. His duties are:

---

[3] If an officer other the force quartermaster is designated as disbursing assistant quartermaster, the duties specified under *e*, *f*, and *g* are performed by that officer.

THE STAFF IN SMALL WARS

(1) To obtain amusement funds from proceeds of the post exchange, and from the government fund "Recreation for enlisted men."

(2) To procure and administer Red Cross and Navy relief funds.

(3) To establish libraries at the bases and hospitals.

(4) To purchase and distribute current periodicals.

(5) To obtain and distribute athletic equipment and material for other forms of recreation.

c. In the initial phases of a small wars operation, the duties of the amusement and welfare officer often may be assigned to the chaplain.

SECTION III

COMPOSITION OF THE FORCE

a. General.
b. U. S. Rifle, cal. .30, M 1903.
c. U. S. Rifle, cal. .30, M 1.
d. BAR, cal. .30, M 1917.
e. BAR, cal. .30, M 1917 (mod.).
f. TSMG, cal .45, M 1928.
g. V. B. Rifle grenade, Mk I.
h. 60-mm. Mortar.
i. Hand Grenade, frag, Mk II.
j. Auto. Pistol, cal. .45, M 1911.
k. Bayonet, M 1905.
l. BMG, cal. .30, M 1917.
m. BMG, cal. .50, M 2.
n. 81-mm. Mortar, M 1.
o. 37-mm. Gun, M 4 or M 1916.

2–38. **General.**—a. It can be assumed that the Fleet Marine Force in the Marine Corps, and the reenforced infantry or cavalry brigade in the Army, will be the basic organizations for small wars operations. Major changes in their strength, organization, armament, and equipment are neither essential nor desirable. However, some slight modifications in armament and equipment may be advisable, and the proportion of supporting arms and services attached to the force may vary from the normal.

b. A force assigned a small wars mission should be tactically and administratively a self-sustaining unit. It must be highly mobile, and tactical units, such as the battalion, must be prepared to act independently as administrative organizations. The final composition of the force will depend upon its mission, the forces available, and the character of the operations.

c. The organization and armament of the opposing force may range from small, roving, guerrilla bands, equipped only with small arms, to a completely modern force armed with the latest types of material. The lack of preponderance of any arm or weapon by the opponent will be the material factor in determining what arms and weapons will be required by the intervening force. The force must be of sufficient strength and so proportioned that it can accomplish its mission in the minimum time and with the minimum losses.

d. The terrain, climatic conditions, transportation facilities, and the availability and source of supply will influence the types of arms and equipment and especially the classes of transportation required by the force.

2–39. **Infantry.**—a. *Importance.*—Infantry, the arm of close combat, has been the most important arm in small wars because, from the very nature of such wars, it is evident that the ultimate objective will be reached only by close combat. The policy that every man, regardless of his specialty, be basically trained as an infantryman has been vindicated time and again, and any tendency to deviate from that policy must be guarded against.

b. *Training.*—Infantry units must be efficient, mobile, light infantry, composed of individuals of high morale and personal courage, thoroughly trained in the use of the rifle and of automatic weapons and capable of withstanding great fatigue on long and often fruitless patrols. As they must assume the offensive under the most difficult conditions of war, terrain, and climate, these troops must be well trained and well led.

c. *Rifle companies.*—Sooner or later, it is inevitable that small wars operations will degenerate into guerrilla warfare conducted by small hostile groups in wooded, mountainous terrain. It has generally been found that the rifle platoon of three squads is the basic unit best suited to combat such tactics. Each platoon sent on an independent combat mission should have at least one and preferably two commissioned officers attached to it. It is desirable, therefore, that the number of junior officers assigned to rifle companies be increased above the normal complement authorized in the tables of organization. The number of cooks in a rifle company should also be increased to provide one for each platoon as the company often may be divided into three separate combat patrols or outpost detachments. The attachment of a hospital corpsman to each detachment is essential.

d. Machine gun companies.—The infantry machine gun company fulfills its normal roles during the initial operations in small wars. In the later phases of guerrilla warfare and pacification, it will seldom be used as a complete organization. Squads and sections often will be attached to small combat patrols, or to detached outposts for the purpose of defense. In order to conserve personnel, some machine gun units in past small wars operations have been converted into rifle organizations, and their machine guns, minus the operating personnel, distributed among outlying stations. This is not good practice. Machine gun organizations should be maintained as such, and the smaller units detached to rifle platoons and companies as the necessity therefor arises. These remarks are also applicable to the 81 mm. mortar and antitank platoons.

2-40. **Infantry weapons.**—*a. General.*—(1) The nature of small wars operations, varying from landings against organized opposition in the initial stages to patrolling the remote areas of the country against poorly armed guerrillas in the later stages, may make some changes in the armament of the infantry desirable. Whether these changes should take place before leaving the United States, or whether they should be anticipated and effected in the theater of operations, must be determined during the estimate of the situation.

(2) The arming of the infantry for small war purposes is influenced by—

(*a*) Fighting power of the enemy, with particular reference to numerical strength, armament, leadership, and tactics.

(*b*) The short ranges of jungle warfare.

(*c*) The necessity for small units to defend themselves at close quarters when attacked by superior numbers.

(*d*) The method of transporting men, weapons, and ammunition.

(*e*) The strength of, and the offensive or defensive mission assigned to, a patrol or outpost.

(*f*) The personal opinions of the officers concerned. A company commander on an independent mission in a small war is generally given more latitude in the arming of his company than he would be permitted in a major war.

(3) Ammunition supply is a difficult problem in small wars operations. A detached post or a combat patrol operating away from its base cannot depend upon immediate, routine replacement of its ammunition expenditures. The state of training of the unit in fire dis-

COMPOSITION OF THE FORCE

cipline and fire control may be an influential factor in determining the number and type of infantry weapons assigned.

b. The U. S. rifle, caliber .30, M1903.—The bolt-action magazine fed, U. S. Rifle, caliber .30, M1903, often erroneously called the Springfield rifle, eventually will be replaced by the semiautomatic rifle as the standard arm of the infantry. Its rate of fire, accuracy, and rugged dependability in the field may influence its continued use in small wars operations. When fitted with a rifle grenade discharger, this rifle acts as the propellant for the rifle grenade.

c. The U. S. rifle, caliber .30, M1.—The U. S. Rifle, Caliber .30, M1, is a gas operated, semiautomatic, shoulder rifle. It has been adopted as the standard infantry weapon by the U. S. Army to replace the M1903 rifle. It weighs approximately a half pound more than the M1903 rifle. Its effective rate of fire is from 16 to 20 rounds per minute as compared to 10 to 20 rounds per minute for the bolt-action rifle. It is especially useful against low flying aircraft and rapidly moving terrestrial targets. It requires more care and attention than the M1903 rifle, the Browning automatic rifle, or the Thompson submachine gun. It cannot be used to propel rifle grenades of either the V. B. or rod type. Whether or not it entirely replaces the M1903 rifle, the characteristics of the M1 rifle make it definitely superior to the Browning automatic rifle M1917, and the Thompson submachine gun for small wars operations. A minimum of two U. S. rifles, caliber .30, M1, should be assigned to every rifle squad engaged in small wars operations and, in some situations, it may be desirable to issue them to every member of the squad.

d. The Browning automatic rifle, caliber .30, M1917.—With the advent of the M1 rifle and the adoption of the light machine gun as an accompanying weapon for rifle units, the Browning automatic rifle, caliber .30, M1917, with its cumbersome length, weight, and ammunition supply, should no longer be seriously considered as a suitable weapon for small-wars operations.

e. The Browning automatic rifle, caliber .30, M1917 (modified.)—The Browning automatic rifle, caliber .30, M1917 (modified), is essentially the same weapon as the BAR, fitted with a bipod mount and a reduced cyclic rate of fire which convert the weapon into an effective light machine gun capable of delivering accurate, full automatic fire. It can be carried by one man, and has the mobility of a rifle on the march and in combat. Two ammunition carriers are required, the team of three men making up a light-machine-gun group. Two groups, under a corporal, comprise a light-machine-gun

COMPOSITION OF THE FORCE

squad. Its characteristics make the Browning automatic rifle (modified) the ideal accompanying and supporting weapon for rifle units. Pending the development and adoption of some other standard light machine gun, two of these rifles should be provided for every rifle platoon of three squads in small-wars operations.

f. The Thompson submachine gun, caliber .45, M1928.—(1) Because of its light weight and short over-all length which facilitate carrying in wooded, mountainous terrain, the Thompson submachine gun has been used extensively in small-wars operations as a partial substitute for the Browning automatic rifle. It has the following disadvantages as a standard combat arm: it uses the caliber .45 cartridge which is employed in no other weapon in the rifle company except the pistol; special magazines must be carried which are difficult to reload during combat if the supply of loaded magazines is exhausted; its effective range is only 150 to 200 yards; the continuous danger space is quite limited; it is not particularly accurate. With the development of a satisfactory semiautomatic rifle, the Thompson submachine gun should no longer be considered as an organic weapon in the rifle squad in small wars.

(2) The Thompson submachine gun may be issued to messengers in place of the automatic pistol, and to a limited number of machine gun, tank, transport, aviation, and similar personnel for close-in defense in small-wars operations. In some situations it may be desirable as a military police weapon. The 20-round magazine is quieter, easier to carry and handle, and is not subject to as many malfunctions as the 50-round drum.

g. The V. B. rifle grenade, mark I.—The V. B. rifle grenade has been replaced by the 60-mm. mortar as an organic weapon of the rifle company. However, it has certain characteristics which may warrant its use in small-wars operations as a substitute for or supplementary to the mortar. The grenade weighs only 17 ounces as compared to 3.48 pounds for the mortar projectile. An M1903 rifle, a grenade discharger, and the necessary grenades may be issued to each rifle squad, thus tripling the number of grenade weapons with a rifle platoon and eliminating the necessity for a separate mortar squad. The range of the rifle grenade, using the service cartridge is from 120 to 180 yards as compared to 75 to 1,800 yards for the 60-mm. mortar. The effective bursting radius of both projectiles is approximately 20 yards.

h. The 60-mm. mortar.—Two 60-mm. mortars are organically assigned to the headquarters platoon of a rifle company. A squad

COMPOSITION OF THE FORCE

of a corporal and 4 privates is required to carry one mortar and 30 rounds of ammunition therefor. A weapon of this type has proved so valuable in previous small wars that at least one mortar should be available for every rifle platoon, or the V. B. rifle grenade should be provided as a substitute weapon.

i. The hand grenade, fragmentation, mark II.—See War Department Field Manual 23–30.

j. The automatic pistol, caliber .45 M1911.—See War Department Field Manual 23–35.

k. The bayonet, M1905.—See War Department Field Manual 23–35.

l. The Browning machine gun, caliber .30, M1917.—The employment of the Browning machine gun, caliber .30, M1917, will be normal during the initial phases of a small war. In the later phases of the operations, the machine gun will be used principally for the defense of outlying stations and the Browning automatic rifle (modified) will probably replace it as the supporting weapon for combat patrols.

m. The Browning machine gun, caliber .50, M2.—The employment of this weapon as an antiaircraft and antitank weapon will be normal.

n. The 81-mm. mortar, M1.—(1) The 81-mm. mortar is one of the most valuable weapons in small wars operations. During the landing phase and the early operations against organized forces, its application will be similar to that in a major war. In some situations in which hostile artillery is weak or lacking altogether, it may be advantageous to increase the usual complement of mortars and to employ them as infantry support in place of the heavier and more cumbersome field artillery. Because of its weight, mobility, and range, the 81-mm. mortar is the ideal supporting weapon for combat patrols operating against mountainous fortified strongholds of the enemy in the later phases of the campaign. Squads and sections often may be detached for the defense of small outposts scattered throughout the theater of operations.

(2) The mortar may be fired from boats in the initial landing or in river operations by seating the base plate in a pit of sandbags, straddling the barrel, and holding and pointing it by hand as in firing grenades from the rifle. The barrel should be wrapped with burlap and the hands should be protected by asbestos gloves.

o. The 37-mm. gun, M4 or M1916.—The tactics and employment of the 37-mm. gun do not vary in small wars from those of a major operation. Opportunities for its use probably will be limited after the completion of the initial phases of the intervention.

2–41. **Infantry individual equipment.**—*a.* Infantry units in the field in small wars operations should be lightly equipped, carrying only their weapons and essential individual equipment. Rations, packs or rolls, and extra ammunition should be carried on pack animals or other suitable transport. If the situation requires the men to carry full packs, rations, and extra ammunition, their mobility is greatly reduced and they are seriously handicapped in combat.

b. Entrenching tools are seldom required after the organized hostile forces have been dispersed. In some situations, they have been entirely dispensed with during the period of pacification, patrolling, and guerrilla warfare which follows the initial operations.

c. The amount of ammunition carried in the belt is usually sufficient for a single engagement. Even with a small combat patrol, the extra ammunition should be transported in the train, if possible. The cloth bandolier is not strong enough to stand up under hard service in the field. If the bandolier is carried, a considerable quantity of ammunition is lost which is generally salvaged by hostile troops or their sympathizers. A small leather box, suspended from the shoulder and large enough to carry one folded bandolier, has proved a satisfactory substitute for the regular bandolier.

d. If field operations continue for a considerable length of time, it may be necessary to reinforce the cartridge belts, magazine carriers, and other web equipment with leather. This has been done in the past by local artisans in the theater of operations.

e. Grenade carriers of leather or heavy canvas similar in design to the Browning automatic rifle bandolier, have been improvised in recent small wars operations. Another satisfactory carrier was made by cutting off one of the two rows of five pockets on the regular grenade apron and attaching the necessary straps. Empty .30 caliber bandoliers are not satisfactory for grenade carriers.

f. The agricultural machete is far superior to the issue bolo for cutting trails, clearing fields of fire, building shelters in bivouac, cutting forage and firewood, etc. in tropical countries. The minimum issue should be two per squad engaged in active patrolling in such terrain.

g. The horseshoe roll may replace the regulation infantry pack during field operations in small wars. It is lighter in weight and easier to assemble than the regular pack; it can be easily shifted

COMPOSITION OF THE FORCE

from place to place on the shoulders, quickly discarded at halts or in combat, and readily secured to the riding or pack saddle.

h. Mounted men should not be permitted to carry rifles or other shoulder weapons in boots nor to secure their arms or ammunition to the saddles while passing through hostile areas in which contact is imminent.

2-42. **Mounted troops.**—Infantry companies, hastily converted into mounted organizations, have played an important role in many past operations. Experience has demonstrated that local animals, accustomed to the climatic conditions and forage of the country, are more suitable for mounts than imported animals. Preparation for mounted duty will consist generally in training for this duty and the provision of necessary equipment. For further details, see Chapter VII, "Mounted Detachments."

2-43. **Engineers.**—*a.* Experience has demonstrated that the construction, improvement, and maintenance of routes of communication, including railroads, is one of the most important factors in a successful small-wars campaign. This is a function of the engineers.

b. The lack of accurate maps and the limited supply of those available has handicapped all operations in the past. A trained engineer unit supplemented by the aerial photographic facilities of aviation is indispensable . Although much of the basic ground work will be performed by combat organizations, the completion and reproduction of accurate maps must be left to skilled engineer troops.

c. With the increased use of explosives in all trades and occupations as well as in military operations, demolition materials are readily available to, and are extensively employed by, irregular forces. A demolition unit is required for our own tactical and construction needs, and for counter-demolition work.

d. Engineers are trained and equipped as light infantry. They should not be so used, except in an emergency, but they form a potential reserve for combat, and for guard duty at bases and depots.

e. The proportion of engineer troops with the force will depend largely upon the means of communication available in the theater of operations, and the condition and suitability of the road net for the contemplated campaign. In most small-wars situations, the necessary manual labor involved can be obtained locally.

2-44. **Tanks and armored cars.**—*a.* The morale effect of tanks and armored cars is probably greater in small wars operations than it is in a major war. The nature of the terrain in the theater of operations will determine whether or not they can be profitably employed.

b. When strong opposition to the initial landings is expected or encountered, the employment of tanks will be a material aid and will reduce the number of casualties. Tanks are particularly valuable in assaulting towns and villages, and in controlling the inhabitants of an occupied hostile city.

c. Armored cars can be employed to patrol the streets of occupied cities, and to maintain liaison between outlying garrisons. With suitable motorized infantry escorts, they are effective in dispersing the larger hostile forces encountered in the early phases of the occupation.

d. Except for the fact that tanks and armored cars can be used more freely in small wars due to the lack of effective opposition, their tactics will be basically the same as in a major war. As the hostile forces withdraw into the more remote parts of the country, where the terrain is generally unsuited for mechanized units, their usefulness in the field will rapidly disappear.

2–45. **Transport.**—See Chapter III, "Logistics."

2–46. **Signal troops.**—*a. General.*—Signal troops install, maintain, and operate any or all of the following communication agencies: (1) Message center; (2) messenger service, including foot messengers, mounted messengers, motorcycle messengers, and messengers using motor vehicles, boats, airplanes, and railroads as a means of transportation; (3) radio service; (4) wire service, including telephone and telegraph services operated both by military and civilian personnel; (5) visual service, including all types of flags, lights, and pyrotechnics; (6) air-ground liaison; (7) pigeon service. Detailed instructions governing the duties of signal troops will be found in "War Department Field Manual 24–5."

b. Importance.—The importance of an efficient communication system cannot be overestimated. It is only through the communication system that contact is maintained with detached garrisons and units operating independently in the field. All officers and noncommissioned officers should be familiar with the capabilities and limitations of the communication system in order that full use may be made of it. In the smaller units, the commanding officer will act as his own communications officer.

c. Commercial and Government services.—When commercial radio and wire service is available, it may be convenient to execute contracts for handling certain official dispatches, particularly in the early

COMPOSITION OF THE FORCE

stages of an operation before all the communication facilities of the force can be put into operation. However, military communication facilities should be substituted therefor as soon as practicable. If the local government operates its own radio and wire service, it is generally possible to arrange for transmission of official dispatches without charge. In some instances, the occupying force will find that an agreement or protocol, covering the establishment and operation of communication agencies by the occupying force, has been established between our own country and the country involved. Such an agreement usually contains a clause stating that limited unofficial traffic may be transmitted over the communication system of the occupying force in case of interruption in the commercial system.

d. Messengers.—The employment of military messengers, either mounted or dismounted, between detached garrisons in areas of active operations is to be considered an emergency measure only, due to the hazardous nature and the uncertainty of this method of communication. In such areas, it may be advantageous sometimes to transmit messages by civilian messengers. Persons who make regular trips between the place of origin of the message and its destination should be employed. Written messages entrusted to civilian messengers should be in code or cipher.

e. Cryptography.—Codes and ciphers are used by even the smallest units in the field. It is apparent, therefore, that all officers must be thoroughly familiar with the systems utilized. In general, the use of code is simpler and more rapid than the use of cipher, due to the ease of encoding and decoding. Codes and key words and phrases for cipher messages are issued to using units to cover definite periods of time. The necessity for changing them is dependent upon the enemy's estimated ability in cryptanalysis.

f. Wire communication.—(1) In areas where the civilian population is hostile, telephone and telegraph wires are liable to be cut and long stretches carried away. The enemy is likely to carry on such operations immediately prior to hostile activities in a definite area. Wire may be taken by a resident civilian simply because he needs it to fence a field or desires it for use in building a hut, and not because he is hostile to our forces. All wire lines are subject to being "tapped" by the enemy.

(2) If there is a commercial wire system available, each garrison telephone communication system should be connected to the commercial system through their switchboards. Provided the commercial system is connected with other towns in a large network separated

units may thus be put into communication with one another. In small-war theaters, the commercial wire system will often be found to be poorly constructed with little attention paid to insulation. Rains will cause interruption in service for hours or even days at a time, due to shorted and grounded lines. Ordinarily, the administration of the commercial telephone system is left to the civilian element normally in control of the systems, the forces of occupation co-operating to the fullest extent in the repair and maintenance of the systems. In those cases where the telephone systems are owned and operated by the government of the country concerned, the same cooperation in repair and maintenance is extended.

(3) Commercial telegraph systems will generally be found to be owned and operated by the government. Although the general condition of the equipment and facilities may not measure up to the standards of a modern system, the telegraph service usually will be found to be very good. Most of the operators are capable men and are quite willing to cooperate with the occupying forces. By judicious cooperation on the part of the military in the repair and maintenance of the telegraph system, the confidence and respect of the personnel operating the system are secured, with the result that telegraphic communication is constantly improved. Except in cases of extreme emergency, no attempt should be made to employ military personnel to operate the telegraph system. In an area of active operations, it may be advantageous to do so for a limited period of time, returning the system to civilian control and operation when the period of emergency is ended. In regions where towns are far apart but telegraph lines are readily accessible, civilian telegraph operators with small portable telegraph sets are a valuable assistance to patrols having no radio set, particularly when weather conditions preclude the operation of aircraft to maintain liaison.

g. Radio communication.—(1) The rapid development of radio as a means of communication, in even the smaller countries of the world, indicates that the forces encountered in small-wars situations may be as well equipped for radio communication as are our own forces. It is highly probable that the hostile forces will attempt the interception of radio communications. This disadvantage necessitates the habitual employment of crytograms in transmitting dispatches of importance. By gaining a knowledge of our radio organization, the enemy is enabled to estimate the organization and distribution of our forces in the field. In order to offset this disadvantage, it may be necessary to curtail the use of radio communications to some extent, particularly in an area of active operations placing temporary reliance on other means of communication.

(2) Radio furnishes the most dependable means of communication with the continental United States, with naval radio stations outside the continental limits of the United States, and with ships at sea. Commercial cable facilities and commercial radio stations may also be available for exterior communication, but are employed only in exceptional cases. Exterior communication is a function of the force headquarters.

(3) American owned commercial radio stations in the theater of operations have been utilized by agreement in the past when the radio equipment with the force was limited. This is especially true when the force has furnished military protection for the property concerned.

(4) It will often happen in small wars situations, that the best method of radio control is to establish a single net for the Force, with all outlying stations a part of the same net. This is particularly applicable when the theater of operations is limited in area. When the theater of operations necessitates the wide separation of tactical units, subordinate nets are established.

(5) There are three types of radio equipment available for forces engaged in small wars operations; semi-portable, portable, and ultra-portable.

(a) Semi-portable radio equipment is of a size and weight to permit easy handling when transported by ships, railroad, motor truck, or trailers, and is intended for the use of brigades and larger units.

Power	100 watts.
Frequency:	
Transmitter	300 to 18,000 kilocycles.
Receiver	300 to 23,000 kilocycles.
Type of transmission	Radio telegraph; radio telephone.
Range:	
Radio telegraph	1,500 miles.
Radio telephone	300 miles.

(b) Portable radio equipment is designed to permit easy handling when transported by hand or on hand-drawn carts when operating ashore. It is intended for the use of regiments and battalions.

Power	15 watts.
Frequency:	
Transmitter	2,000 to 5,000 kilocycles.
Receiver	2,000 to 20,000 kilocycles.
Type of transmission	Radio telegraph; radio telephone

COMPOSITION OF THE FORCE

Range:
Radio telegraph_____ 400 miles.
Radio telephone_____ 75 miles.
Weight_____ 86 pounds.

(*c*) Ultra-portable radio equipment consists of a carrying case, having a self-contained radio transmitter, receiver, and power supply designed for transportation by one man. It is issued to units as required and is particularly useful to mobile units, such as patrols and convoy guards.

Power_____ ½ watt.
Frequency_____ 28 to 65 megacycles.
Type of transmission_____ Radio telegraph; radio telephone.
Range:
Radio telegraph_____ 10 miles.
Radio telephone_____ 5 miles.

(6) The demand for trained personnel will normally exceed the number organically assigned to communication units. The wide separation of small units in the usual small wars will require the addition of numerous sets of radio equipment to those listed in current equipment tables. The use of the ultra-portable radio equipment will also require additional operators. See paragraph 2–29.

(7) To take care of the widely separated radio equipment, each battalion designates one man of the communication platoon as an itinerant repairman. His duties are to make repairs in the field to radio sets operated by the communication personnel of the battalion. In areas of active operations, he joins patrols whose routes will take them to the garrisons where the equipment is located. He may be transported to outlying stations by airplane to make emergency repairs. In many cases, he will find it advisable to take an extra set with him to replace a set needing major repairs. No system for making major repairs can be definitely laid down that will apply to all situations. Due to the technical nature of the equipment, it is usually more convenient to have all major repair work accomplished by the communication personnel attached to the headquarters of the Force, thus obviating the duplication of test equipment as well as the necessity for maintaining large stocks of repair parts at widely separated stations.

h. Pigeon communication.—Pigeons may be carried by patrols in active areas. Although patrols are normally equipped with portable radio sets, it may be desirable to maintain radio silence except in cases of extreme emergency. In such cases, pigeons afford a dependable means of keeping higher authority informed of the progress and actions of the patrol. Crated pigeons may be dropped to patrols in the field by aircraft, small parachutes being used to cushion the fall. This method of replenishment is used when patrols are in the field longer than 3 days.

i. Air-ground liaison.—(1) Because of the nature of the terrain usually encountered and the operation of numerous ground units employed in small-wars operations, air-ground liaison is especially important. There must be the closest cooperation between aviation and ground troops. The period of each contact is limited. Panel crews must be well trained and ground-unit commanders must confine their panel messages to items of importance only.

(2) Panels which indicate the code designation of the organization or patrol are displayed in open spots upon the approach of friendly aircraft to identify the ground unit. They also indicate to the airplane observer where he may drop messages, and where panel messages are displayed for him. Panel strips are used in conjunction with identification panels for the purpose of sending prearranged signals. Letter and number groups of the air-ground liaison code are formed from the individual panel strips, and are laid out to the right of the designation panel as determined by the direction of march. When the signal has been understood by the airplane observer, it is acknowledged by a pyrotechnic signal, wing dips, or other prearranged method.

(3) The message-dropping ground should be an open space removed from high trees, bodies of water, and weeds. If possible, it should be so located that the panels can be seen at wide angles from the vertical.

(4) The method of message pick-up employed in air-ground liaison is described in detail in "War Department Field Manual 24–5." Experience has indicated that it is preferable to make a complete loop of the pick-up cord, securing the message bag at the bottom of the loop instead of the double loose-end cord described in the abovementioned Field Manual.

(5) In small wars situations, the use of pyrotechnics for communication between ground units, other than to acknowledge lamp signals or flag signals, may be considered exceptional. Pyrotechnics are normally employed for air-ground liaison only. Position lights and signal projectors are particularly useful to ground units when heavy vegetation makes the employment of panels impracticable. Aviation employs the Very pistol for air-ground liaison when its use will speed

COMPOSITION OF THE FORCE

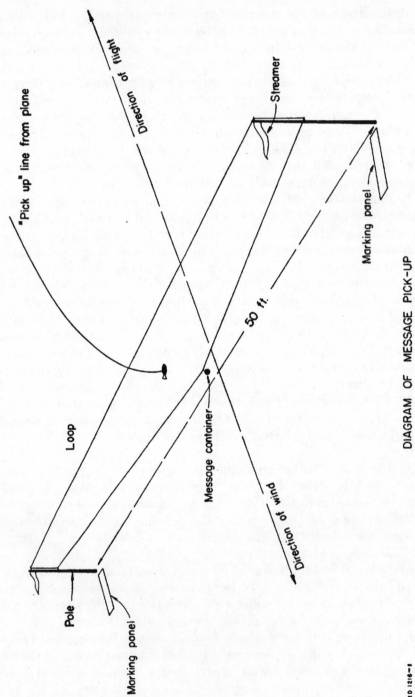

DIAGRAM OF MESSAGE PICK-UP

RD 1216–1

COMPOSITION OF THE FORCE

up the transmission of short messages by a prearranged code. This method of communication with ground units is also employed when the establishment of a message-dropping ground is prevented by heavy vegetation or other reason, or when the close approach of the airplane to the ground during a message drop would expose it to hostile rifle fire from enemy groups in the vicinity.

2–47. **Chemical troops.**—*a.* Properly employed, chemical agents should be of considerable value in small wars operations. The most effective weapons to quell civil disorders in the larger towns are the chemical hand and rifle grenades and the irritant candles. Their effectiveness has been proved so many times in civil disorders in the United States that they are now accepted weapons for such situations. Consideration should therefore be given to similar employment of these munitions in a small wars theater of operations. The burning-type hand grenade with a smoke filler may be use by patrols to indicate their location to friendly airplanes. Another use of this type of hand grenade is the development of smoke to conceal the flanking action of a large group in an attack over open ground against a strongly held and definitely located hostile position. Advantage should be taken of the prevailing wind direction and the grenades so fired that the target will be covered by the smoke cloud.

b. Chemical agents have not been employed by the United States in any small wars operations up to the present time, as their use in a foreign country is definitely against the best interests of our foreign policy. If they are employed, in some future small war, the armament, equipment, munitions, and tactics of the chemical troops will not vary from the normal doctrine. The strength of the chemical units to be included in the force will be decided by the force commander in accordance with their prospective employment as determined by the existing situation.

2–48. **Medical troops.**—*a.* The type of operation, the size of the force, the nature of the country in which operations will take place, the health conditions to be expected, and the estimated casualties from combat will determine the class or classes of field hospitals and the strength of the medical personnel that will be attached to the force. In almost every small wars operation, the number of commissioned medical and dental officers and enlisted corpsmen will be considerably in excess of that required for a corresponding force in a major war, because of the numerous small detachments of combat units scattered throughout the entire theater of operations. Special care should be taken in selecting the hospital corpsmen to accompany the force. In many cases, an enlisted corpsman will be required to make the diagnosis and administer the medication normally prescribed by a medical officer.

b. Commanding officers of all grades are responsible for sanitation and for the enforcement of sanitary regulations within their organizations and the boundaries of the areas occupied by them. They must be thoroughly conversant with the principles of military hygiene, sanitation, and first aid. Particular attention should be paid to the following:

(1) Instruction in personal hygiene of the command.

(2) The thorough washing of hands after visiting the head (latrine) and before each meal.

(3) The proper sterilization of mess gear.

(4) Vaccination against small-pox and typhoid fever.

(5) The prevention of venereal disease.

(6) The proper ventilation of quarters, and provision of adequate space therein.

(7) The carrying out of antimosquito measures.

(8) The destruction of flies, lice, and other insects.

(9) The purification of non-portable water supplies.

(10) The proper disposal of human excreta and manure.

(11) The proper disposal of garbage.

c. The medical officer, under the direction of the commanding officer, supervises the hygiene of the command and recommends such measures as he may deem necessary to prevent or diminish disease. He should investigate and make recommendations concerning the following:

(1) Training in matters of personal hygiene and military sanitation.

(2) The adequacy of the facilities for maintaining sanitary conditions.

(3) Insofar as they have a bearing upon the physical condition of the troops:

(*a*) The equipment of organizations and individuals.

(*b*) The character and condition of the buildings or other shelter occupied by the troops.

(*c*) The character and preparation of food.

(*d*) The suitability of clothing.

(*e*) The presence of rodents, vermin, and disease-bearing insects and the eradication thereof.

COMPOSITION OF THE FORCE

d. The medical personnel with the force is one of the strongest elements for gaining the confidence and friendship of the native inhabitants in the theater of operations. So long as it can be done without depleting the stock of medical supplies required for the intervening troops, they should not hesitate to care for sick and wounded civilians who have no other source of medical attention.

e. If the campaign plan contemplates the organization of armed native troops, additional medical personnel will have to be provided with the force or requested from the United States, as required.

f. See Chapters 12 and 14, Landing Force Manual, United States Navy, and Field Manuals 8-40 and 21-10, United States Army, for detailed instructions regarding military hygiene, sanitation, and first aid.

2-49. Artillery.—*a.* The amount of artillery to be included in the strength of a force assigned a small wars mission will depend upon the plan for the employment of the force, the nature of the terrain in the theater of operations, the armament and equipment of the prospective opponents, and the nature of the opposition expected. As a general rule, some artillery should accompany every expedition for possible use against towns and fortified positions, and for the defense of towns, bases, and other permanent establishments. The morale effect of artillery fire must always be considered when planning the organization and composition of the force. If the hostile forces employ modern tactics and artillery, and the terrain in the country permits, the proportion of artillery to infantry should be normal.

b. The role of artillery in small wars is fundamentally the same as in regular warfare. Its primary mission is to support the infantry. Light artillery is employed principally against personnel, accompanying weapons, tanks, and those material targets which its fire is able to destroy. Medium artillery reinforces the fire of light artillery, assists in counterbattery, and undertakes missions beyond the range of light artillery. Unless information is available that the hostile forces have heavy fortifications, or are armed with a type of artillery requiring other than light artillery for counterbattery work, the necessity for medium artillery will seldom be apparent. Antiaircraft artillery, while primarily for defense against air attack, may be used to supplement the fire of light artillery.

c. The artillery must be able to go where the infantry can go. It must be of a type that can approach the speed and mobility of foot troops. The 75-mm. gun and the 75-mm. pack howitzer fulfill

COMPOSITION OF THE FORCE

these requirements. Because the pack howitzer can be employed as pack artillery where a satisfactory road net is lacking in the theater of operations, the pack howitzer usually will be preferable to the gun in small wars situations, although the latter may be effectively employed in open country.

d. Pack artillery utilizes mules as its primary means of transport and has reasonably rapid, quiet, and dependable mobility over all kinds of terrain; however, it is incapable of increased gaits. It is especially suitable for operations in mountains and jungles. Mules required for pack purposes normally will be secured locally. The loads carried by these animals require a mule of not less than 950 pounds weight for satisfactory transportation of the equipment. If mules of this size cannot be obtained, a spare mule may be used for each load and the load shifted from one animal to the other after each 3 hours of march. One hundred horses and mules are required for pack and riding purposes with each battery. The approximate road spaces for the battery, platoon, and section, in single column, are as follows:

	Yards
Battery	400
Platoon	150
Section	52

Since there is no fifth section in the pack battery, the supply of ammunition available within the battery is limited to about 40 rounds per piece.

e. The separate artillery battalion is an administrative and tactical unit. It is responsible for the supply of ammunition to batteries so long as they remain under battalion control. When a battery is detached from the battalion, a section of the combat train and the necessary personnel from the service battery should be attached to it. In the same way, a detached platoon or section carries with it a proportional share of battery personnel and ammunition vehicles. In determining what amount of artillery, if any, should be attached to the smaller infantry units in the field, the nature of the terrain, the size and mission of infantry units, and the kind of opposition to be expected are the guiding factors. The infantry unit should be large enough to insure protection for the artillery attached to it, and the terrain and nature of the opposition should be such as to permit the attached artillery to render effective support. Also, the ammunition supply should be attached to infantry units.

COMPOSITION OF THE FORCE

No artillery should be attached to infantry units smaller than a rifle company.

A section of artillery to a rifle company.

A battery to an infantry batallion.

A battalion to an infantry regiment.

f. The employment of artillery in small wars will vary with developments and the opponent's tactics. When resistance is encountered upon landing and the advance inland is opposed, artillery will be employed in the normal manner to take under fire those targets impeding the movement. When the opponent's organization is broken and his forces widely dispersed, the role of artillery as a supporting arm for the infantry will normally pass to the 81-mm. mortar platoons. (See paragraph 2–40, n.)

g. Artillery in the march column.—(1) In marches in the presence of hostile forces tactical considerations govern the location of the artillery units in the column. Artillery should be sufficiently well forward in the column to facilitate its early entry into action, but not so far forward as to necessitate a rearward movement to take up a position for firing. It should be covered by sufficient infantry for security measures.

(2) In advance and rear guards the artillery usually marches at or near the tail of the reserve. In flank guards the artillery marches so as to best facilitate its early entry into action.

(3) The artillery with the main body, in advance, usually marches near its head. In retirement, if the enemy is aggressive, the artillery should march at or near the tail of the main body. However, when the enemy is not aggressive, it may even precede the main body, taking advantage of its mobility to relieve congestion.

(4) The difficulties to be anticipated in passing through defiles are due to the narrowness of the front and to a restricted route where the column may be subjected to concentrated infantry and artillery fire. When resistance is anticipated during the passage of a defile, the column should be organized into small groups, each composed of infantry and artillery, capable of independent action. When meeting resistance at the exit of a defile, artillery is employed to cover the debouchment. When meeting resistance at the entrance of a defile, the artillery is employed as in the attack against a defensive position.

(5) Due to the limitations in its employment at night, the entire artillery is usually placed near the tail of the main body on night marches.

COMPOSITION OF THE FORCE

h. Artillery with the outpost.—Normally the artillery which has been assigned to the advance or rear guard is attached to the outpost. The outpost commander designates the general position for the artillery, prescribes whether it shall be in position or posted in readiness, and assigns the artillery mission. Normally, the outpost artillery is placed in position. Defensive fires are prepared in advance insofar as practicable.

i. Employment of artillery on the defensive.—(1) The defense of towns, camps, etc., does not present the complex problem of ammunition supply that confronts artillery on the offensive. The ammunition available at each place usually will be ample and no question of transportation will be involved. The supply of ammunition need not affect the assignment of artillery for defensive purposes. The presence of a single piece in a defended town will often have a deterring effect on hostile forces.

(2) After the initial stages of the operation, if it appears that artillery will be required for special limited missions only, it can be used to advantage in the defense of stabilized bases, and permanent stations and garrisons. The troops not needed with the artillery can be used to relieve rifle units on special guard duty, such as at headquarters, fixed bases, and on lines of communication. The conversion of artillery into infantry units should be considered only as an emergency measure. However, artillery units of the force carry with them (boxed) the necessary rifles, other infantry weapons, and equipment required to convert them into infantry when the situation develops a need for this action.

j. Antiaircraft and base defense artillery.—(1) It can be assumed that, in the future, some hostile aviation will be encountered in small wars operations, and the inclusion of antiaircraft artillery in the force will have to be considered. To depend upon aviation alone for antiaircraft protection presupposes that friendly air forces can annihilate all hostile aircraft and all facilities for replacements. Even one hostile operating plane will be a potential threat to vital areas such as the beachhead, supply bases, and routes of communication.

(2) The comparative mobility of the .50 caliber AA machine gun makes it particularly suitable for employment in small wars operations. However, its limited range renders it impotent against any hostile aircraft other than low-flying planes. If it becomes apparent that antiaircraft machine guns, as such, are not needed, this weapon can be profitably employed for the defense of the more important bases and outlying garrisons against ground targets.

COMPOSITION OF THE FORCE

(3) Whether the 3″ AA gun will be included in the force will depend largely upon the opposition expected. This weapon may be used with restricted mobility on defensive missions against land targets in the same manner as the 75 mm. gun.

(4) It is difficult to conceive of any small wars situation in which base defense weapons of 5″ caliber would be required with the force. If the opponent can muster sufficient armament to make the inclusion of such artillery necessary in the force, the campaign will probably take on all the aspects of a major war, at least during the initial stages of the operation.

2-50. **Aviation.**—For the employment of aviation in small wars operations, see Chapter IX, "Aviation."

○

SMALL WARS MANUAL
UNITED STATES MARINE CORPS
1940

✦

CHAPTER III

LOGISTICS

RESTRICTED

UNITED STATES
GOVERNMENT PRINTING OFFICE
WASHINGTON : 1940

TABLE OF CONTENTS

The Small Wars Manual, U. S. Marine Corps, 1940, is published in 15 chapters as follows:

SMALL WARS MANUAL

UNITED STATES MARINE CORPS

———— ————

Chapter III

LOGISTICS

SECTION I

INTRODUCTION

3–1. Logistics is that branch of the military art which embraces the details of transportation and supplies.

"The Tables of Equipment, Supplies, and Tonnage, U. S. Marine Corps," set forth the equipment and supplies that are prescribed for Marine Corps expeditionary forces to take the field. These tables are a guide to the fourth section of the commander's executive staff in making a decision as to the type and amount of transportation and supplies required. However, the supply on hand at the port of embarkation, the time allowed for preparations, the ship's storage space available, the supplies in the theater of operations, the distance from home ports, when replacements can be expected on the foreign shore, and the condition of the roads and the road net within the anticipated field of operations will all be essential and controlling factors in arriving at the final decision.

SECTION II

SUPPLY

3-2. **Influence of Supply on a column.**—The "big three" of supply are Ammunition, Food, and Water. Combat troops can operate in the field for a very limited time in actual combat with only AMMUNITION, but their continued existence requires the other two, FOOD and WATER. Therefore, in order to conduct the advance inland, one of the first considerations in such a movement must be the means of supply.

Supplies may be obtained as follows:

(1) From the country en route, by requisition or other authorized method.

(2) By continuous resupply via convoys despatched from the base.

(3) By taking sufficient supplies with the column for its maintenance from the base to its destination; resupply to begin after arrival at destination.

(4) By the establishment of fortified advanced bases along the route. These advanced bases are established by detachments from the column initially and supplies built up at them by convoys dispatched from the rear or main supply base; thereafter, the column draws its supplies from these advanced bases direct.

(5) By airplane, either in plane drops or landing of transport planes on favorable terrain at the camp site of the column. (*See Chapter IX, Aviation.*)

3

SUPPLY

(6) In most small wars operations, a combination of all these methods will be used.

3–3. **Supply officers.**—Officers charged with supply have a dual mission. They must first get the supplies, then supply them to the troops. In order to carry out these duties it is essential that the officer responsible for supply has the following essential information at all times:

(1) The supplies and equipment required by the force.

(2) The supplies and equipment the force has on hand.

(3) Where the required items may be procured, from whom, and when.

(4) When, where, and in what quantities replacements will be needed.

3–4. **Storage.**—*a.* The matter of storage is very closely connected with the problem of supply and starts at the port of debarkation. Prior to or upon arrival of the expeditionary force at the port of debarkation, a decision must be made as to the location of the main-supply depot. The following factors are of importance in reaching this decision:

(1) Mission of the intervening force.

(2) Docking or lighter facilities.

(3) Availability of suitable shelter for stores.

(4) Railroads, highways, water routes available for supply purposes and types of carriers.

(5) Availability of civilian labor.

(6) Security.

(7) Location of troops; distance from supply base.

(8) Location of possible landing fields.

b. It is always desirable to have the supply base near the point of debarkation in order to facilitate unloading and segregation of stores. However, for various reasons, this is not always practicable. It will then be necessary to establish at the debarkation point a forwarding depot, and place the main depot or base at an intermediate point, between the forwarding depot and the area to be supplied. From the main supply depot, the flow of supply would ordinarily be to and through advanced supply bases, and forward to organizations in combat zones. The usual route would be via railroad, where it exists, or highway, using motor transportation to advanced supply bases in organization areas. It will usually be found advantageous to build up small stocks of essential supplies, at these advanced bases, or even farther forward at the advanced distributing points, in order to

4

SUPPLY

insure a continuous supply. This is especially necessary when operating in a theater that has a rainy season.

c. The available transportation facilities will also be an important consideration in determining the location of distributing points, and the levels at which they are to be kept.

d. Quartermaster department personnel will be kept at the depots. These units will ordinarily be organized to handle the main subdivisions with warrant officers or staff noncommissioned officers of the department as assistants or section chiefs. At these points the enlisted force should be augmented by civilian labor if available.

e. Routine replacements of depot stocks will ordinarily be maintained by timely requisitions submitted by officers in charge to the proper supply depot in the United States or, in the case of articles not normally carried by these depots, by requisitions submitted direct to the Quartermaster, Headquarters Marine Corps.

f. The foregoing replenishment should be augmented by local purchases of items available locally at reasonable prices.

g. It will be necessary to inspect existing local facilities regarding shelter for depot stocks and service units in order that proper recommendations may be made to Force Headquarters relative to preparation of formal agreement for rental. Failing this, it would be proper, in the event a long stay is anticipated, to recommend construction of suitable buildings for this purpose. Ordinarily, in tropical countries, service units may be quartered in tents.

h. The location of transportation units employed in the depot supply plan will usually be controlled by the location of the depot or bases. Such units should be reasonably close to the depots and subject to depot control.

3–5. Distribution.—*a.* Ordinarily, depots with force transportation will supply as far forward as consistent with existing conditions. Organization transportation, whether motor or pack, will carry forward from this point either directly to troops or to positions from which troops may be supplied by carrying parties. Force Headquarters units and rear echelons of all organizations will normally be supplied directly by supply depots or bases.

b. If fortified advanced bases are to be established, the decision relative to their location will be influenced by the suitability of the sites as camps. The type of shelter utilized will depend on the availability of buildings or construction material in the vicinity or the feasibility of transporting shelter material to these sites from the main or intermediate base. In the latter case, the decision will be

SUPPLY

SUPPLY PROCUREMENT AND DISTRIBUTION CHART FOR SMALL WARS MANUAL

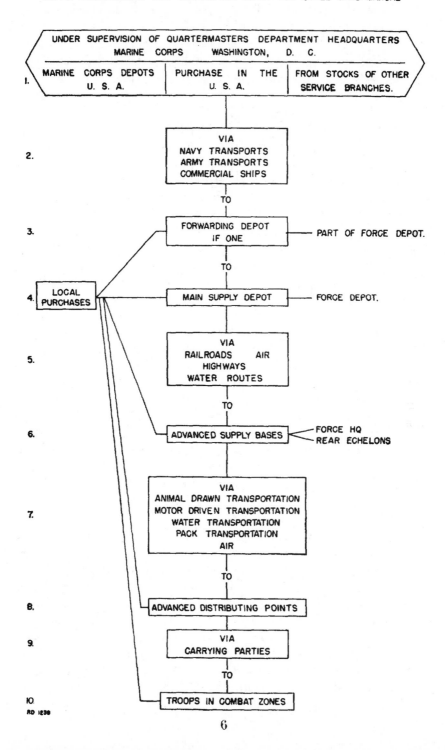

SUPPLY

influenced by the amount of transportation available at the time the bases are being established. If local shelter or transportation for such construction material is not available, the vicinity of the advanced bases should at least be cleared and developed as a camp site. An adequate water and fuel supply should be available.

c. The accompanying chart shows how procurement and supply will normally exist in small wars.

d. Description of chart.—Step (1)—Procurement here and (2) transportation to depot or forwarding depot is of course continuous, based on requisitions from the expeditionary force. Requisitions are varied, consisting of periodical requirements submitted on usual forms together with letter and, in emergencies, radio, telegraphic, or cable dispatches. Decisions as to quantities for, and places of, storage depend upon the particular situation and mission. In some instances the port of debarkation might be selected as the site of the force depot. If the operation necessitates the presence of the bulk of the force far inland, it is probable that only a forwarding depot or segregation point would be maintained at the port of debarkation, and the main depot established further inland along the line of communications. There can be no set rule regarding this arrangement. From the depot or main base, field distribution begins. Those nearest the main base would probably be supplied through the medium of advanced supply bases at which small stocks would be maintained. If possible, a daily distribution would be made to points beyond. Failing this, a periodical system of distribution would be made, carrying forward to combatant units sufficient supplies and ammunition to meet their needs for stated periods. This would entail the establishment of additional advanced dumps from which troops could be supplied either by means of their own transportation, or in some instances, by pack trains. Carrying parties might be employed at this point.

e. It is doctrine that supplies are echeloned in depth to the rear, and that some system be decided upon that results in a proper distribution forward. In most small wars situations almost every accepted principle of warfare on a large scale is subject to modification due to the irregularity of the operation. It is this characteristic that sets the "small war" in a class by itself. It is obvious then, that a successful supply plan in any small war theater must be ready to meet these irregular conditions. Here, the means offered by the specific local country and used extensively by it, should most certainly be exploited, modified, improved, where necessary, and adopted to the

7

use of our forces. This is particularly true of methods of transport. Supply officers of a small war operation should never overlook the fact that it is always possible to learn something from close observance of local facilities and customs. They may need modification and improvement in order to meet our requirements, but basically there will almost always be found something of value that can and should be used.

3–6. **Supply steps.**—From a study of the chart above, it will be apparent that some of these steps may, in certain situations, be eliminated, such as the Forwarding Depot and carrying parties where step No. 7 supplies directly to step No. 10.

3–7. **Local purchases.**—*a.* Local purchases may be made at any of the five places shown along the chain of supply, and sent to troops in combat areas.

b. Where local purchases are made by other than a regularly detailed purchasing officer, prior authorization for such practice must be secured from Force Headquarters.

3–8. **Requisitions.**—*a* Requisitions for replacements of equipment, supplies, ammunition, etc., are submitted to the nearest accountable or supply officer by the officer responsible, usually company commanders, to and through Bn–4's. Sufficient forethought must be employed to permit procurement and distribution by the required time.

b. Close teamwork should exist between the Quartermaster Department and the field commanders. It is essential that the Quartermaster know what supplies can be procured by the field commander, and likewise the field commander should know what supplies can be furnished by the Quartermaster.

3–9. **Depots, dumps, and distributing points.**—*a.* The Advanced Distributing Points may be at Area Headquarters or merely at a selected site close to combatant troops. In countries where the condition of roads in forward areas will not permit a daily delivery routine, and such occasions will be common, it will be necessary to maintain small stocks of essential supplies at these Advanced Distributing Points. In most systems of supply operating in the field, there exists the necessity for establishing permanent and temporary points of storage and points where distribution takes place. The terms commonly used to designate such points are "depot," "dump," and "distributing point." The word "depot" is used to designate a place where supplies in bulk are stored permanently and from which the first step in field distribution takes place. Such a point requires shelter, security, and close proximity to some good means of transporting supplies. This point is usually established by the organization carrying the bulk of replacement supplies.

b. The word "distributing point" signifies a position or site selected for the transfer and distribution of supplies to consuming units. It is most often used in connection with the daily distribution of automatic supplies used by troops at a fairly uniform rate, such as rations, oil, fuel, forage, etc. It simply means a spot or area to which supplies are brought by one means and turned over to another for purpose of interorganizational distribution.

c. Advanced supply bases are in reality subsidiary depots established inland to facilitate the forwarding of supplies to the distributing points.

3–10. **Chain of responsibility.**—*a.* The usual chain of responsibility of individuals connected with procurement and distribution of equipment and supplies in the field is:

```
Force Headquarters_____ _____F–4.
                                       Force QM.
                                       Force Depot QM.
Brigades_____B–4.
Regiments_____R–4.
Battalions_____Bn–4.
Companies_____Company commander.
Platoons or detachments_____Platoon or detachment commander.
```

b. In each company is a company supply sergeant, whose duties include the preparation of company requisitions and through whom requests for replacements of any kind emanating from squads, sections, and platoons should be sent to the company commander. When these requisitions are filled, the company supply sergeant is in charge of proper distribution of the new material to the lower units and individuals. This man holds the rank of sergeant and is entrusted with matters of company supply.

c. Company and detachment commanders should exercise close supervision over requisitions and the issuing of supplies. This is particularly true of rations.

3–11. **Accountability.**—*a.* Ordinarily, accountability, when it exists, extends down to the battalion in field organizations where the battalions are administrative units. From there on down to the individual, responsibility obtains.

b. There is no set rule by which decisions may be reached relative to recommending the discontinuance of all or part of accountability. In any event, such discontinuance will have to be authorized

SUPPLY

by the Quartermaster, Headquarters Marine Corps, and approved by the Major General Commandant.

c. There will be occasions when some modifications of this system will be desirable and necessary, but normally the administrative units of the force will be able to establish and conduct the routine of their rear echelons so as to permit and justify the continuance of accountability and proper records involving responsibility.

d. The absence of accountability promotes carelessness and waste and presents a serious obstacle to intelligent and economical supply. Loose handling in the responsibility for weapons and ammunition makes it easier for these articles to get into unauthorized hands and even into the hands of the opposing force.

e. The exigencies of field conditions are recognized by everyone connected with our service of supply and consideration is always given to such conditions. Headquarters, U. S. Marine Corps, is fully cognizant of hazards engendered by field conditions. Under justifiable circumstances, certificates of adjustment to accountable officers' accounts will be acceptable. The Quartermaster's Department recognizes this fact and acts accordingly but the point is, that in continuing accountability, there must be a certified record of all such unusual occurrences.

f. It may be entirely impossible for an administrative unit to obtain proper receipts for its issues, but a record for such issues can and should be kept in order that requests for replacements within the unit can be intelligently supervised by the unit supply officer. If the entire administrative unit has taken the field actively, such record should be kept by the accountable officer in the last step of the supply chain before it reaches the unit.

3–12. **Public funds.**—*a*. Public funds for procurement of such material and services as the force may find desirable and economical are usually entrusted, through official channels, to an officer designated as a disbursing assistant quartermaster.

b. These officers, when authorized by competent authority, may advance public funds to officers in outlying stations for certain local purchases. When such purchases are made, standard forms of vouchers are either prepared by the officer making the purchase, or ordinary receipts are taken by him and furnished the disbursing assistant quartermaster concerned. In order that such transaction may have proper basic authentication it has been the usual practice to write into the orders for such officers, when detailed for duty at outlying points, a specific designation as agent for the disbursing assistant quartermaster concerned which becomes the authority for advancement of public funds.

c. An officer receiving such designation as agent should, before entering on his new duties, confer with the disbursing assistant quartermaster in order that there will be complete understanding of how the money in the possession of the agent is to be accounted for when expended. If such a procedure is impracticable, the matter should be made the subject of immediate correspondence between these two officers. There exists such a multitude of regulations and decisions governing the expenditure of Government funds that no one should undertake disbursing even to the extent of a very small sum, without first learning the proper method to pursue. Such procedure will avoid explanation and correspondence later, and may be the means of saving the one concerned the necessity of making good from personal funds an amount of public funds spent in error, solely because of lack of sufficient and proper advance information. It is desired to stress this point most emphatically.

3–13. **Objective.**—The objective is the one common to all military operations, i. e., success in battle. The well trained and supplied fighter needs but proper leadership to win; therefore the task of the supply officer becomes one of considerable importance from the commander's point of view.

3–14. **Supervision of requisitions.**—*a.* The most important function of a supply officer is the supervision of requisitions. To know what, when, where, and how to get what the command needs, and then get it and distribute it, is perhaps the whole story of supply insofar as it affects the one to be supplied. The remainder consists of proper recording of what has been done; this is known as accountability.

b. The requisition is the starting point of the whole process. If it be wrong, everything else can't help but be wrong also. Never pad a requisition on the assumption that it will be cut down. Sooner or later this will become known and your requisitions will be worthless to the one who reviews them. If your real needs are cut by someone, find out why and, if you can, insist on what you ask for. But be sure you know what you want, and why. On the other hand, a requisition should never be cut without a thorough investigation.

c. Place explanations on the face of requisitions covering items that are exceptional from previous requests.

SUPPLY

3–15. Accumulation of stores.—*a.* There is a delicate balance between overstocking and understocking. Overstocking means forced issues, while understocking means privation and possibly failure.

b. Do not permit the accumulation of slow-moving stores, particularly clothing in extreme sizes. If it fails to move, report its presence and ask for disposition. Someone, elsewhere, may want the very sizes that are in excess of your needs. Arrange to turn over subsistence stores of a staple nature at least once every 90 days. Report your excess quantities to your nearest senior supply officer through official channels.

3–16. General.—*a.* The following general rules may be of assistance to persons responsible for the handling and storing of supplies:

(1) As a rule, provide an air space under all stored articles. It prevents deterioration.

(2) In the absence of buildings for storage, request that necessary security measures be taken to safeguard your stores.

(3) Visit the units that you supply.

(4) Find out how your system works and adjust it where necessary.

(5) Watch your stock of subsistence stores.

(6) Become familiar with the data contained under "Minimum safekeeping period" for subsistence stores under article 14–54, Marine Corps Manual. (Note particularly the remarks in this table.)

(7) Ask for an audience from time to time with your commander. Keep him apprised of the supply situation. Give him your picture, clearly and briefly, and then recommend desirable changes, if any. Above all, make your supply system fit into his plans.

(8) Keep in close touch with your source of supply. Know what is there and how long it will take you to get it.

(9) Get a receipt for everything that leaves your control. If field conditions are such that this is, in part, impracticable, then keep a record of all such transactions, and set down the reasons for not being able to obtain proper receipts.

(10) Keep your own supply records up-to-date and render necessary reports regarding them.

(11) When you need help, ask for it and *state facts*. Camouflage, or any attempt at it in the supply game, is fatal. If your best judgment has failed, admit it. It is a human characteristic and can rarely be cloaked by a garment of excuses.

3–17. Importance of supply.—The importance of the question of supply upon small wars is well set forth in the following extract taken from Small Wars by Callwell:

SUPPLY

The fact that small wars are, generally speaking, campaigns rather against nature than against hostile armies has been already referred to. It constitutes one of the most distinctive characteristics of this class of warfare. It effects the course of operations to an extent varying greatly according to circumstances, but so vitally at times as to govern the whole course of the campaign from start to finish. It arises almost entirely out of the difficulties as regards supply which the theaters of small wars generally present. Climate effects the health of troops, absence of communication retards the movement of soldiers, the jungle and the bush embarrass a commander; but if it were not for the difficulty as regards food for man and beast which roadless and inhospitable tracts oppose to the operations of a regular army, good troops well led would make light of such obstacles in their path. It is not the question of pushing forward the man or the horse or the gun, that has to be taken into account so much as that of the provision of the necessities of life for the troops when they have been pushed forward.

SECTION III

TRANSPORTATION

3–18. **General.**—*a.* The types of transportation used in small wars operations will vary widely, depending upon local conditions such as roads, terrain, and distances to be covered. In some cases the seasons of the year will be a controlling factor.

b. During small wars in the past every possible type of transportation known to mankind has been used, from railroad, aviation, and motor transportation to dogs, elephant, camel, and porter service.

c. It is safe to say that the type of transportation most suitable to any specific country is being utilized there. A study of these local methods, together with the local conditions, will aid the commander in determining the type of transportation to be used by the intervening forces.

d. In countries where small wars usually take place, the roads are generally bad and exist in only a few localities. When there is a season of heavy rain, it is most probable that practically all roads and trails will become impassable for trucks and tractor-trailer transportation. For that reason other means of transportation must be utilized. This may mean that railroads and air transport, where

TRANSPORTATION

they are available, will have to be used for very short hauls. Animal, cart, boat, or porter transportation will have to be used where there are no passable roads, trails, or railroads.

3-19. **Railroad transportation.**—*a*. Normal principles of loading and transporting troops and supplies will apply as they do in similar movements elsewhere, making use of whatever rail facilities the country has to offer.

b. For the use of railroads for movement inland see chapter 5, paragraph 3, Movement by Rail.

3-20. **Motor transportation.**—*a*. This type of transportation should be under the direction of officers specially qualified in its uses. It is not always known exactly what road conditions can be found in the field, and the motor transportation officer, knowing the capabilities and limitations of this type of transportation, considering the conditions of the roads, the road net, and the seasons of the year, will have to use ingenuity in carrying out the task assigned to him.

b. Trucks should be of uniform type generally, but sturdy enough to stand heavy usage. The U. S. Marine Corps equipment tables provide for ½- and 2-ton trucks; these seem to be best for our purposes.

c. Motor transport assignment varies according to the situation. Motor transportation is attached to the force by sections, platoons, or companies, as the case may be. In the case of an independent regiment, a section or more of motor transportation is usually attached.

d. Motorcycles, with or without sidecars, are of very little value in small wars. They require good roads and have some value for messenger service.

e. When needed, native-owned transportation can be used to great advantage. Native chauffeurs, mechanics, and laborers are used when practical. Sudden demands made on the native type of transportation will usually exceed the supply, resulting in very high costs for transportation; but this cannot be avoided.

3-21. **Tractor-trailer transportation.**—*a*. In certain localities it is likely that where the roads stop, there will be trails and terrain that are passable for tractors with trailers, where motortrucks will be unable to go.

b. Tractors may be available in four sizes. The lightest will weigh approximately 2 tons and run on wheels, using "Jumbo" tires, with small wheels in front and large ones in rear. The other three sizes

TRANSPORTATION

will weigh approximately 3 tons, 5 tons, and either 7 or 8 tons. All of these are to be the track-laying types.

3–22. **Transportation pools.**—*a.* Certain organizations habitually requiring transportation have vehicles along with their operators and supplies attached to them as a part of their organic organization. Other organizations request transportation as it is needed.

b. In some instances it will be more economical to operate a transportation pool. This is done by placing all transportation in the force under the Force Motor Transport Officer, who will assign the different vehicles to the different tasks as they are required.

3–23. **Aviation transport.**—For transportation of supplies and troops by aircraft, see chapter IX ("Aviation").

3–24. **Water transportation.**—*a.* In some instances river boats and lighters can be used to transport troops, animals, and supplies from the port of debarkation inland.

b. Where lakes or other inland waterways exist within the theater of operations, a most valuable method of transportation may be open to the force, and every effort should be made to utilize all water-transportation facilities available.

c. Boats for this purpose and outboard motors should be carried if it is expected that they will be needed. (See ch. **X**, "River Operations.")

3–25. **Animal transportation.**—*a.* The use of animals for the purpose of transporting supplies has been one of the most generally used methods of transportation in small-wars operations.

b. Without the pack animal, operations far into the interior of a mountainous and unsettled area, devoid of roads, are impracticable if not impossible. However, the use of pack animals is not a simple or always a satisfactory solution of a transportation problem. Crude or improvised pack equipment, unconditioned animals, and the general lack of knowledge in the elementary principles of animal management and pack transportation will tend to make the use of pack transportation difficult, costly, and possibly unsatisfactory.

c. The efficiency with which the pack train is handled has a direct and material effect on the mobility of the column which it accommodates. With an inefficient pack train the hour of starting, the route of march, and the amount of distance covered are noticeably affected. On the other hand, with conditioned animals, good modern equipment, and personnel with a modicum of training in handling packs, the pack train can accommodate itself to the march of the column and not materially hamper its mobility.

d. If time permits it is highly important to have the animals that are to be used for transporting supplies accustomed to the firing of rifles and automatic weapons, so that they will not be frightened and try to run away if a contact is made. This can be done by firing these weapons while the animals are in a place with which they are familiar and preferably while they are feeding. The firing should be done at some distance first and gradually moved closer as the animals get accustomed to the noise. In a short time the animals will pay no attention to the reports when they find that it does not hurt them. If this is impossible, and an animal carrying important cargo, such as a machine gun or ammunition, is frightened and tries to bolt, the animal should be shot to prevent the loss of these supplies and to prevent them from falling into the hands of the opposing forces.

e. Pack animals must be conditioned before being taken on an extended march or heavy losses of animals will result.

f. The march should begin immediately after the last animal is packed.

3-26. Important points in packing.—*a*. Loads and distances traveled must be adjusted to the condition of the animals. Pack animals must not be overloaded.

b. In packing up, the time interval between placing the loads on the first and the last animal should be reduced to an absolute minimum. This time interval should never exceed 30 minutes.

c. All equipment should be assembled neatly and arranged the night before a march is to begin. Every single item should be checked, otherwise needless delays will result in the morning.

d. All cargoes should be weighed, balanced, and lashed up the night before a march is to begin.

e. A standard system should be established for stowing all pack gear and cargo loads at each halt for the night. This facilitates the checking of equipment after the halt and greatly reduces the number of lost pieces. A satisfactory system is to place the pack saddles on the ground in a row just in rear of the picket line or, if the animals are pastured at night, place them on a line in a space suitable for packing up in the morning. The harness, lash ropes, and all other gear that belong to that particular saddle and its load should be placed on top of each saddle. The loads should be placed in a row parallel to the saddles; each load in rear of the saddle on which it is to be packed. Only by careful planning and by systematic arrangement can delays in packing up be averted.

TRANSPORTATION

3–27. Pack mules.—*a.* The mule is the ideal pack animal for supply trains, pack trains with foot patrols, and pack trains with detachments mounted on mules. The mule has certain advantages over the horse which fit him for this work, namely:

(1) The mule withstands hot weather better, and is less susceptible to colic and founder than the horse.

(2) A mule takes better care of himself, in the hands of an incompetent driver, than the horse.

(3) The foot of the mule is less subject to disorders.

(4) The mule is invariably a good walker.

(5) Age and infirmity count less against a mule than a horse.

b. Pack mules are habitually driven and not led. However, pack mules carrying weapons and ammunition will, for purposes of safety, be led in column by having the leader of each mule drive the mule that precedes him. His mule will, in turn, be driven by the man next in rear of it.

3–28. Pack horses.—*a.* Any good riding horse of normal conformation, good disposition, and normal gaits can be used as a pack horse. The pack animals of a detachment mounted on horses should always be horses. This is necessary in order to maintain the mobility of the mounted detachment. Each pack horse is led alongside a ridden horse. On very narrow trails and at any time when it is impossible for two horses to travel abreast, the pack horse is led behind the ridden horse.

b. Horses properly packed can march at the same gaits as the ridden horse.

3–29. Pack bulls.—*a.* Under certain conditions, bulls can be used to good advantage as pack animals. A pack bull with its wide spreading hoof can negotiate mud in which a mule with its small hoof will bog down. While slower than mules, bulls can carry heavier cargoes than the mules usually found in most small war theaters. Good pack bulls can carry from two hundred (200) to two hundred and fifty (250) pounds of cargo. They can make about fifteen (15) miles a day loaded but, after 5 days march, they will require a rest of from five (5) to seven (7) days if they are to be kept in condition. In employing pack bulls it is advisable to hire native bull keepers to handle them.

b. Mixed pack trains of bulls and horses do not operate smoothly due to their different characteristics.

TRANSPORTATION

3–30. **Phillips pack saddle.**—*a.* The Phillips pack saddle was developed to supply the need for a military pack saddle of simple but scientific design—a saddle that could be handled by newly organized troops with only a short period of training. The characteristics of this saddle make it ideally suited for small wars operations. It is manufactured in one design in four sizes, and all sizes are suitable for either horses or mules.

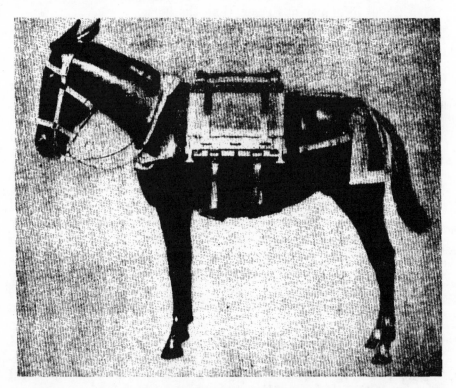

PHILLIPS PACK SADDLE, PONY SIZE.
Correctly positioned and harness properly adjusted.

(1) *Cargo-artillery type.*—75-mm. pack howitzer units are equipped with this size. It is designed for the large American pack mule.

(2) *Cavalry type.*—A size designated for the average American cavalry horse.

(3) *Pony type.*—A size designed for the Philippine and Chinese pony.

(4) *Caribbean type.*—A size designed for the Central American mules.

b. This saddle is designed for either hanger or lash loads. Hang-

TRANSPORTATION

ers for all standard equipment such as the Browning machine gun, the 37-mm. gun, ammunition in machine-gun boxes, some radio sets, and the pack kitchen can be obtained with these saddles. These hangers consist of attachments which can be quickly and easily attached to the saddle. The loads for which they are designed are simply placed in these hangers and held firmly and rigidly in place

BROWNING MACHINE GUN LOAD ON PHILLIPS PACK SADDLE, PONY SIZE.

with gooseneck clamps which can be instantly secured or released.

3–31. **McClellan saddle.**—In addition to the regular pack saddles, McClellan saddles may be used in emergency for packing. The tree of a McClellan saddle has most of the characteristics of a pack-saddle tree, and fair results may be obtained by tying the two side loads together across it and running the lashings under the quarter straps or through the cinch strap rings, spider rings (at lower part of

TRANSPORTATION

quarter straps), or the quarter strap D-rings to hold the load down.

3-32. **Pack equipment.**—*a.* The types of pack equipment in common use by the inhabitants of countries where pack transportation forms a basic part of the transportation system vary in different countries, and sometimes within a country in different areas. This native equipment, though crude, can be converted to military pur-

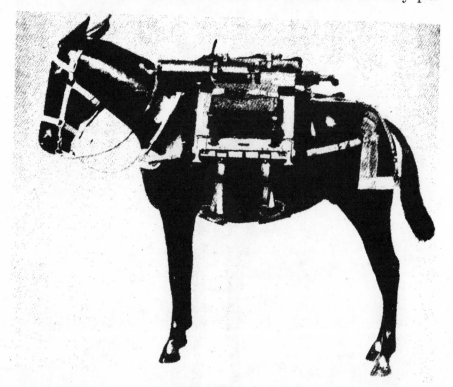

BROWNING MACHINE GUN LOAD ON PHILLIPS PACK SADDLE, PONY SIZE.
A complete fighting unit of gun, tripod, ammunition, and spare parts roll. Quick release devices on each item of load.

poses and, when no other equipment is available, must be used. Such native equipment invariably has one or more of the following defects:

(1) Highly skilled specialists are required to use it satisfactorily.

(2) Due to its crude construction it is very injurious to animals.

(3) It cannot be adjusted easily on the trails.

(4) Many military loads are extremely difficult to pack on this equipment.

TRANSPORTATION

(5) The pads, cinches, and other attachments wear out rapidly under constant usage.

(6) Packing and unpacking require a comparatively great length of time.

b. The advantages of Native Equipment are:

(1) Generally available in quantities in or near the zone of operation.

(2) Relatively cheap.

(3) Light in weight.

BROWNING MACHINE GUN LOAD.

Tripod side.

c. These advantages are the only reasons which might justify the use of native pack equipment in preference to the Phillips pack equipment. However, the cheapness of native equipment is overbalanced by the high percentage of animals incapacitated by its use; its light weight is not necessarily an advantage as an equal or greater pay load can be carried on heavier modern equipment with considerably less damage to the animal.

TRANSPORTATION

d. The aparejo, or primitive pack saddle, has many shapes, being made of leather with sometimes a wooden tree or back pieces to stiffen it and padding placed either in the leather skirts or between the leather and the animal's back, or both. This type is rather hard to pack, as it requires a complicated hitch around the load and saddle.

e. Another form in general use by civilians is the sawbuck type.

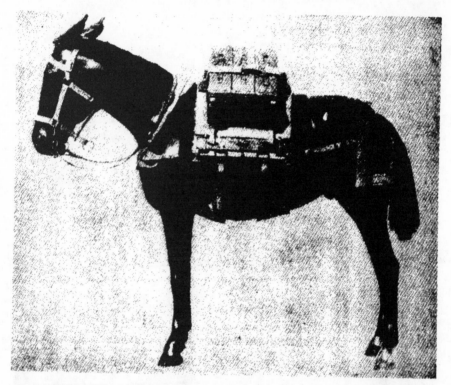

MACHINE GUN AMMUNITION LOAD ON PHILLIPS PACK SADDLE, PONY SIZE.
Seven hundred and fifty rounds on each side with space on top of saddle for additional equipment.

It consists of a wooden tree formed of two bars fitting the saddle place (bearing surfaces) and four straight wooden pieces which form two crosses, one at the pommel and one at the cantle, fastened to these bars. This type may be used with flat straw pads or blankets or both. It has the advantage of absolute rigidity in the frame or tree, requires little skill to construct of materials easily available, and less skill to pack than the aparejo described above.

3-33. **Native packers.**—Native packers have been used to good advantage. Two natives experienced in packing are generally hired

24

TRANSPORTATION

for every 10 animals, since two men are required to pack each animal and hence work in pairs. A good system is to hire a competent chief packer and allow him to hire the necessary number of packers. With such an arrangement, all orders and instructions should be issued through the chief packer and he should be held responsible for the handling of the cargoes of the animals.

3–34. **Marines as packers.**—*a.* The average marine can be trained in a fairly short time to pack mules more securely and more rapidly than the average native mule driver, and in regions where pack transportation is used, every marine should be taught to pack. The use of marines as packers has the effect of decreasing to some extent the combat strength of the column, but it has many advantages.

MACHINE GUN AMMUNITION LOAD.
Showing space on top of saddle for additional equipment.

b. In some cases it may be undesirable or impracticable to include native packers in a combat patrol. The hiring of native packers always gives the populace warning that the column is about to move out.

c. The train is more efficiently handled by marines, thus obviating the necessity of issuing orders to the train in a foreign language. Ammunition and weapon loads should always be led by marines, rather than herded or turned over to natives. The adoption of the Phillips pack saddle, coupled with the ease and rapidity with which

TRANSPORTATION

marines can be taught to use it, will warrant a greater use of marines as packers in future operations.

3–35. **Bullcarts.**—*a.* In some localities the bull-drawn cart is the principal means of transporting bulky articles, and when large quantities of supplies are required, the bullcart may be the best means of transportation available. It is a suitable means of transport when

THE NEW CAVALRY PACK COOKING OUTFIT ON THE PHILLIPS SADDLE.
This outfit is made up of many standard utensils nested to form two side loads. Each troop of cavalry is to have one pack cooking outfit.

motortrucks or tractors are impracticable and when the time element does not equire supply by the faster methods. Supplies shipped in bullcarts will ordinarily arrive in good condition, if properly loaded and protected. Weapons and munitions so transported should be constantly under special guard.

b. If it can possibly be avoided, bulls should not be purchased for Government ownership. Private ownership is more feasible and less

TRANSPORTATION

expensive. Furthermore, it is unlikely that good animals can be purchased at a reasonable price; natives are willing to part with their aged and disabled animals, but rarely sell their good ones.

c. Whenever possible, a chief bullcart driver should be secured or appointed. He should be a man in whom the other native drivers have confidence, and through whom general instructions can be issued.

d. A definite contract should be drawn up with the native owners

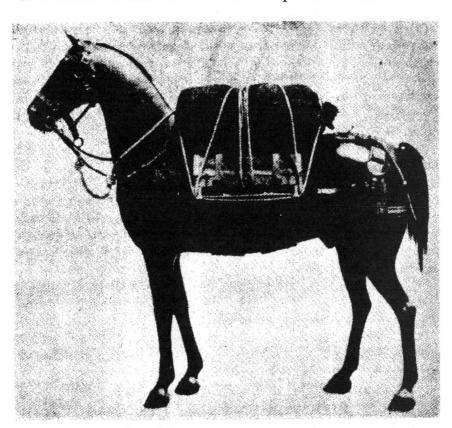

DIAMOND HITCH LOAD ON THE PHILLIPS PACK SADDLE.
The usual lash cinch is not required on this saddle.

before the movement begins. All details of pay, rationing of native drivers and animals, breakage, and damage should be clearly set forth; it is necessary to be assured that the native contractors thoroughly understand the terms of the contract. Contracts should be made on the basis of weight or bulk delivered at the destination, and the natives should not be paid until the service is completed. Deductions can be made for losses or damage to supplies en route. In some cases

TRANSPORTATION

it may be necessary to advance small sums for the feeding of the animals en route.

e. Much that is associated with the handling of bullcarts must be learned from experience. The following information, if followed by the inexperienced bullcart commander, will greatly lessen his difficulties:

(1) The bullcart is a simple outfit, but it requires an experienced bullwhacker to guide and man it.

(2) Two thousand pounds is a maximum load for a cart drawn by two yoke of bulls. If the going is bad, from one thousand (1,000) to one thousand six hundred (1,600) pounds is a sufficient load. A load of over two thousand (2,000) pounds is dangerous, regardless of road conditions or number of bulls per cart, as it is too great a strain on the cart and will cause break-downs which are almost impossible to repair on the trail.

(3) When such break-downs occur, new carts should be secured in the immediate vicinity of the break-down, or the load of the broken cart should be distributed among the remaining carts. If neither of these makeshifts is possible, sufficient of the least valuable cargo should be discarded and the loads of carts rearranged to carry all important or valuable cargo. This rearrangement of loads should be made by the chief bullcart driver under the supervision of the escort commander, if there is one.

(4) It is difficult to tell whether one pair of bulls is stronger or weaker than another. Some carts will have to be loaded lighter than others, and only an expert will be able to decide this.

(5) It is better to arrive safely with all carts, cargo, and bulls in good condition than to gamble on overloads with their resultant delays, broken cargo, and injured bulls.

(6) The weight of all military ammunition and supplies can be estimated, and ration containers are usually accurately marked with the gross weight. Thus proper loads can be assigned to all carts.

(7) When streams are to be crossed, carts should be loaded so that the top layer contains perishable cargo, such as sacked flour and sugar, thus preventing or lessening losses by wetting.

(8) In loading carts the native cart driver should be permitted to distribute and lash his load as he sees fit, insofar as is practicable. However, the driver should not be permitted to say when he has a sufficiently large load or he will start off with as light a load as possible. He should be given his share of the cargo and such assistance as he needs in loading it. He will balance his load with a slight

TRANSPORTATION

excess weight to the front to prevent the tongue from riding upwards when under way. He will test the loading by lifting the tongue before the bulls are hitched to it, to estimate the strain on the bulls when they are attached to the cart.

(9) On the first day's march, the best cart drivers should be noted. This can be done by personal observation and careful spotting of the carts that are slow, and those that cause most delays. On the morning of the second day, or sooner if it can be done without undue delay or confusion, poor carts should be placed at the head of the train, leaving the best carts in the rear. This will assist in keeping the column closed up, thus making supervision, protection, and control of the train much less difficult. When the train consists of so many carts that a mental list of the drivers is difficult, the carts should be numbered with painted numerals before departure, and a written list made of each cart by number, driver, and owner.

(10) By having a few officers or noncommissioned officers mounted, much time can be saved in checking up and clearing delays on the trail. If all trouble has to be cleared on foot, needless delays will result.

(11) Train guards must keep a careful watch on cargo to prevent drivers from breaking containers and consuming unauthorized rations en route and in camp. In camp, carts should be arranged in a park convenient for guarding and for the next day's departure.

(12) Extra bulls should be provided for a train, especially in hot weather, to facilitate getting carts out of difficulties, advancing carts up steep grades, and replacing casualties among the bulls.

(13) Any interference on the part of marines with the function of the native drivers, other than that absolutely necessary, will probably work out disadvantageously.

(14) Cargoes, especially those of rations and ammunition, should have a protective covering—such as ponchos or canvas.

(15) During hot weather, bulls cannot be worked in the heat of the day. A good schedule to follow at such times is to start the day at about 3 a. m., and travel until about 9 a. m., then give the bulls a rest until 3 p. m., when travel can be continued again until 9 p. m. In this way the carts can cover from 15 to 20 miles per day, depending upon the conditions of the roads.

(16) A marine officer in charge of a train should cooperate to the fullest extent with the native chief of the bull-cart train in allowing him to set his own schedule. The trip can be materially speeded if this is possible, and the schedule of the marines made to conform to that of the bull-cart train.

(17) In traveling through barren country, it may be necessary to carry food for the animals and, if this is the case, the pay load must be lessened in proportion. As soon as responsible natives can be found and when the route along which the supplies have to be transported is safe, it is wise to allow the train to proceed without escort. The natives, if held strictly responsible for losses, will probably not proceed if there is danger that the train will be captured, as they will have been warned of this danger before the marines. Escorting supplies by such a slow method is very tedious and costly in men. However, ammunition and weapons must be escorted.

3–36. **Trains with combat columns.**—*a.* Pack trains which carry the supplies, baggage, ammunition, and weapons of combat columns should be made as mobile as possible. Both the number of animals and the cargo loads should be as small as is consistent with the absolute needs of the column. If there is a choice, it is better to increase the number of animals than to increase the individual cargo loads.

b. In general, the pack loads accompanying a combat column should not exceed twenty-five percent (25%) of the weight of the pack animal which, for small mules and horses, would mean a maximum pay load of about one hundred and thirty (130) pounds. One hundred pounds is considered an average load. This is a general rule and the load must be varied to meet the condition of trails and the condition of the individual animal. Some combat loads will exceed this percentage, and it will be necessary to select the strongest and best conditioned animals to carry these special loads.

⊙

SMALL WARS MANUAL
UNITED STATES MARINE CORPS
1940

+

CHAPTER IV
TRAINING

RESTRICTED

UNITED STATES
GOVERNMENT PRINTING OFFICE
WASHINGTON : 1940

TABLE OF CONTENTS

The Small Wars Manual, United States Marine Corps, 1940, is published in 15 chapters as follows:

SMALL WARS MANUAL

UNITED STATES MARINE CORPS

CHAPTER IV

TRAINING

Section I

CHARACTER AND PURPOSE OF SMALL WARS TRAINING

4–1. **Relation to other training.**—*a.* Training for small wars missions is carried on simultaneously with training for naval operations overseas and training for major warfare on land. Training for naval overseas operations and major warfare on land is often applicable, in many of its phases, to small wars operations. Training that is associated particularly with small wars operations is of value in the execution of guerilla operations on the fringes of the principal front in major warfare.

b. In small wars, the normal separation of units, both in garrison as well as in the field, requires that all military qualities be well developed in both the individual and the unit. Particular attention should be paid to the development of initiative, adaptability, leadership, teamwork, and tactical proficiency of individuals composing the various units. These qualities, while important in no small degree in major warfare, are exceedingly important in small wars operations.

c. Training for small wars operations places particular emphasis upon the following subjects:

(1) Composition, armament, and equipment of infantry patrols.

(2) Formations and tactics of infantry patrols.

(3) Mounted detachments.

(4) Transportation of wounded.

(5) Planned schemes of maneuver when enemy is encountered by patrols.

(6) Security on the march.

(7) Security during halts and in camp.

(8) Organization of the ground for all-around defense.

(9) Night operations, both offensive and defensive.

(10) Employment of weapons.

3

(11) Messing. To include the feeding of troops on the trail and in small groups in garrison.

(12) Laying ambushes.

(13) Attacking a house.

(14) Street fighting.

(15) Riot duty.

(16) Defense of garrisons.

(17) Surprise attacks on enemy encampments.

(18) Stratagems and ruses.

(19) Scouting and patrolling, including tracking.

(20) Combat practice firing.

(21) Sketching and aerial photograph map reading.

(22) Marching, with particular attention paid to marching over rough, wooded trails, both dirt and rocky, under varied weather conditions. Trail cutting through dense underbrush and conservation of drinking water to be included.

(23) Bivouacs and camps.

(24) Sanitation, first aid, and hygiene.

(25) Handling of small boats on inland waterways.

(26) Air-ground liaison.

(27) Training of officers as aviation observers.

(28) Rules of land warfare.

4–2. **Tactical training.**—The current training manuals describe the combat principles for the various arms and are the basis of tactical instruction for units preparing for or participating in a small war. These combat principles may be supplemented or modified to conform with the requirements of anticipated or existing conditions. The usual enemy tactics encountered in small wars are those associated with the ambush of patrols and convoys, river fighting, and surprise attacks against garrisons and towns. These operations are described in chapters V to X, inclusive.

4–3. **Rifle company.**—*a.* The rifle company and its subdivisions are often called upon to perform independent mission. Such missions include the establishment of small garrisons in isolated communities and at strategic points along lines of communication and supply, patrol operations coordinated with the operations of aviation and other patrol units, and independent operations that may carry the rifle company and its subdivisions beyond supporting distance of their bases or friendly patrol units. Training for small wars operations, therefore, must be conducted with a view to the probable assignment of the rifle company and its subdivisions to independent missions. In the larger patrols, the patrol leader will usually find it impracticable, if not impossible, to direct the actions of each subdivision of his patrol during action against the enemy. In such cases, the leaders of the several subdivisions of the patrol must control their units in such manner as will best promote the known plan of the patrol leader. Upon contact with the enemy in the field, there will often be no opportunity for the leaders of the several subdivisions of the patrol to consult with and receive orders from the patrol leader prior to committing their units to action. They must know, in advance, his plan of action in case contact is made with the enemy and must be prepared to act independently without the slightest hesitation. In the training of patrols, the independent control of subdivisions should always be stressed.

b. The principal weapon of the combat organizations is the rifle. The man so armed must have complete confidence in his ability to hit battlefield targets and must be thoroughly imbued with the "spirit of the bayonet"—the desire to close with the enemy in personal combat and destroy him. The fact that small wars operations may be conducted in localities where the terrain and vegetation will often prevent engaging the enemy in hand-to-hand combat does not remove the necessity for training in the use of the bayonet. It is only through such training that each individual of the combat team is imbued with the "will to win." Every man attached to a combat organization must be trained in the use of the rifle grenade and hand grenade, both of which are important weapons in small wars operations. The rifleman should be given a course of training in the other infantry weapons in order that he may know their employment and functioning. Machine guns, mortars, and 37 mm. guns may, at times, be issued the infantry company to augment the fire power of its rifles. Since additional trained personnel will often not be available to man the added weapons it becomes the duty of the infantry company to organize squads for the operation of such weapons.

c. The rifle is an extremely accurate shoulder weapon. In the hands of an expert rifle shot (sniper) it is the most important weapon of the combat units. Other infantry weapons cannot replace the rifle. The rifle is exceedingly effective in the type of fire fight connected with small wars operations. A course in sniper firing is of great value in the development of individuals as snipers. Such a course may be readily improvised by placing vegetation before the line of targets on any rifle range or by using growing vegetation, provided its location makes the method practicable. Silhouette targets are shown for several seconds at irregular intervals and at different locations within the vegetation by the manipulation of ropes or wires from a pit or other shelter in the vicinity of the targets. This type of training develops fast, accurate shooters.

d. While the development of expert individual rifle shots is highly desirable, it is even more important that combat units receive a course of training in the application of musketry principles to the conditions of combat ordinarily encountered in small wars operations. Whenever facilities are available, the training program should devote considerable time to combat range firing. Every phase of actual combat should be included in this training. To make the practice realistic will require much ingenuity and skillful planning but there is no other method of training that will develop effective combat teams. Combat practice firing presents the nearest approach to actual battle conditions that is encountered in the whole scheme of military training. Exercises should be so designed that leaders are required to make an estimate of the situation, arrive at a decision, issue orders to put the decision into effect, and actually supervise the execution of orders they may issue. The degree of skill and teamwork of the unit is shown by the manner in which the orders of the leaders are executed. The conservation of ammunition should be stressed in all combat practices.

4-4. Machine gun company.—The machine gun company is organized as a unit for administrative purposes to effect uniformity in instruction, and to promote efficiency in training. During active operations in the field, however, it will often be found necessary to assign platoons, sections, or even single guns to either permanent or temporary duty with garrisons, patrols, or other units. In some cases it may be necessary to arm the personnel as riflemen to augment the number of men available for patrol duty. Machine-gun personnel are, therefore, given the course of training with the rifle as outlined in paragraph 4-3, *c* and *d.*

4-5. Mortars and 37 mm. guns.—*a.* These weapons are employed to augment the fire power of other weapons. They are of particular value in the organization of the defensive fires of small garrisons. Because of their bulk and the difficulty of effective employment in heavy vegetation, they are not normally carried by small, highly mobile patrols in the field. In an attack on an organized position, the need for both weapons is apparent.

b. The 37 mm. gun is employed against definitely located automatic weapons and for the destruction of light field works. It delivers fire from a masked position by use of the quadrant sight. When time is an important element, direct laying is used or fire may be conducted from a masked position having sight defilade only. Since its tactical employment in small wars does not vary from its normal use in major warfare, there is no need for special training applicable only to small wars situations.

c. The ability of the mortar to fire from well-concealed positions against targets on reverse slopes and under cover makes it a valuable weapon for small wars operations. Because of its mobility it will often be used as a substitute for light artillery. It can be used against targets that can not be reached by other infantry weapons. No special training is required for small wars operations.

4–6. **Troop schools.**—*a.* The troop school is an important agency of the unit commander for the training of his own personnel to meet the requirements of the training program. Troop schools may take any form that produces effective results, including informal conferences or lectures, demonstrations, sand table or squadroom instruction, as well as the formal organized school with its staff of instructors, a definite course, and fixed periods of instruction.

b. The object of the troop school is to train personnel for combat and to coordinate such training. It insures uniformity in the training of the entire command. Certain technical subjects, in which a comparatively small number of men from each organization are to be qualified, can frequently be taught more economically and thoroughly in classes or schools conducted by a higher echelon.

c. Instruction in centralized classes, whether they be company classes or those of a higher unit, does not relieve the subordinate commander from further training of troops under his command. It is his duty and responsibility to so organize his unit that each individual is placed where he may contribute most to the efficient working of the combat team. Thus, a scout may receive instruction in scouting and patrolling in a centralized class, returning to his organization upon the completion of the course. Upon his return, his training is continued under his squad leader and officers of his own unit in order that the unit may gain the advantage of the training he has received while attending the centralized class for scouts.

d. A course in a troop school is planned with one of the two following objectives:

(1) A course conducted for the purpose of developing instructors in a particular subject. As a rule, these classes are conducted by the

battalion or higher echelon. Graduates of such classes are particularly valuable as instructors in newly organized units.

(2) A course conducted for the purpose of teaching troops the mechanics and technique of their work and equipment. It does not concern itself with the development of qualified instructors. As a rule, these classes will be conducted by companies, the course being somewhat shorter than the course designed for developing instructors.

e. The group method of instruction may be used in the training of any group, regardless of its size or organization. It provides careful systematic instruction under the direct supervision of an instructor, and centralizes control within the group for the purpose of teaching the mechanics of any subject. The group method of instruction is preferable for introductory training and is especially adapted to instruction in basic military subjects. It consists of five distinct steps, as follows:

(1) Explanation of the subject or action by the instructor.

(2) Demonstration of the subject or action by the instructor and assistants.

(3) Imitation (application) by all undergoing instruction.

(4) Explanation and demonstration of common errors by the instructor and his assistants.

(5) Correction of errors by the instructor and his assistants.

Instruction should be clear and precise. Every error made by the student during the applicatory step should be corrected immediately in order to prevent the formation of faulty habits and wrong impressions. It is often easier to instruct a new recruit than to change the faulty habits of a man who has been longer in the service.

SECTION II

TRAINING DURING CONCENTRATION

4-7. Training objective.—*a.* The character of the training conducted during concentration depends upon the time available, the state of training of the individual units concentrated, the nature of the country in which operations are to be conducted, the character and armament of the forces likely to be encountered, and the type of operations that may be necessary.

b. Training during concentration is primarily concerned with preparation for the following operations:

(1) Ship-to-shore movement, against organized opposition and without opposition.

(2) Reorganization preliminary to movement inland.

(3) Movement inland, including the seizure of defended cities and towns, and operations against guerrilla forces whose tactics include surprise attacks and ambushes.

4-8. Scope of training.—*a.* During concentration it is necessary to verify the readiness of troops for the conduct of small wars operations. Deficiencies in training must be corrected, particularly if the deficiency is such as to hazard the successful prosecution of contemplated operations.

b. For subjects to be stressed during training for small wars operations, see paragraph 4-1, *c.*

4-9. Disciplinary training.—Where time is short, all training in ceremonies and close order drill should be reduced to a minimum. The disciplinary value of close order drills may be achieved through the efficient conduct and close supervision of field exercises, during instruction in bayonet fighting, and in training in the use, functioning, and care of weapons and equipment. Smartness, prompt obedience, and orderly execution can be exacted of troops during such exercises, thus increasing the value of the instruction as well as developing a higher degree of battle efficiency in the individual.

SECTION III

TRAINING EN ROUTE ON BOARD SHIP

4–10. **General.**—*a.* The relative value of training conducted aboard ship depends on the necessity for the training. The more an organization is in need of training, the more it will profit from every hour devoted to such training. The more advanced an organization is in its training, the more difficult it is to prepare a profitable schedule that can be carried into effect on board. It must be remembered that one of the main features of a system of instruction is the prevention of idleness and resultant discontent. A schedule that allows practically no time for relaxation, however, is always to be avoided.

b. The total time available for instruction is a factor to be considered when formulating the training schedule. Some organizations will be on board only during the period spent enroute to the scene of operations and will disembark immediately upon arrival thereat. Other organizations (sometimes called "floating" battalions) may be quartered on board for varying periods of time, possibly for several months.

c. The thoroughness of the instruction will be dependent upon the skillful planning of schedules, the ability of the instructors, the time allotted for each subject, and the facilities available.

4–11. **Ship routine.**—*a.* Any training to be conducted on board ship must be fitted into the ship's routine. The troop commander is in command of the troops on board, but the commanding officer of the ship is responsible for all the activities on board. The troop activities must not interfere with the normal routine of the ship, without specific permission of the commanding officer of the ship. Usually, the ship's routine will include breakfast at 0730, inspection of quarters at 0830, quarters at 0900, dinner at 1200, and supper at 1700. Friday is normally given over to field day, with Saturday morning reserved for inspection of living spaces and personnel by the commanding officer. As a result, training is limited to 4 full days per week.

TRAINING EN ROUTE ON BOARD SHIP

b. Mess facilities on board ship are usually limited. Troops will probably eat cafeteria style, using their individual mess equipment. Normally 1 hour will be ample time for the troops to be served and to complete any meal. This includes sufficient time for them to procure their mess gear, be served, wash their mess gear, and stow it.

c. Working parties will be required for serving the food, work in the galley, and handling stores. In order that interference with training may be reduced to a minimum, it is desirable that a complete unit, such as a platoon or company, be detailed daily for such duty. The duty should be assigned to troop units in rotation.

d. Emergency drills will also interfere with the schedule of training. These drills are an important part of the ship's routine. They include abandon ship, collision, fire, and fire and rescue drills. Everyone on board will participate in these drills.

4–12. **Time available for troop training.**—The time available for which definite schedules for troop training may be made up is limited to two daily periods, 0900 to 1130 and 1300 to 1600, a total of 5½ hours. Since only 4 full days per week can be definitely scheduled, the weekly schedule is limited to 22 hours of instruction. If Friday may be used for training, another 5½ hours will be available.

4–13. **Troop schools on board ship.**—*a.* Classes are organized to cover instruction in such subjects as may best prepare each member of a command to become a more proficient member of his combat team. Due to lack of space and facilities, the establishment of troop schools, employing the group method of instruction, is the accepted method for shipboard training. Classes covering essential subjects are organized for officers, noncommissioned officers (including selected privates) and privates.

b. Formations are usually limited to assemblies for quarters and inspections. At such formations, it is often possible to carry out exercises such as the manual of arms, setting-up exercises, and physical drill under arms.

4–14. **Size of classes.**—Training on board ship is generally attended by a number of distracting and annoying features such as seasickness, wet paint, scrubbing of decks, heat, etc. It is, therefore, desirable that classes be organized in small groups. Groups of 20 are the largest that one able instructor can be expected to handle efficiently. In the instruction of groups in the mechanics of the several types of weapons, care should be taken to avoid assigning too many individuals to a single weapon. Not more than two men should be assigned to one automatic rifle and not more than three to a machine gun. A man learns very little about the mechanics of a weapon by watching someone else assemble and disassemble the weapon. He must have the weapon in his own hands and perform the work himself as it is only through this method that he attains proficiency.

4-15. **Assignment to classes.**—*a.* An example of the assignment of the personnel of a rifle company to the several classes of a troop school on board ship is as follows:

Class	*Supervision Attendance*
Automatic rifle	Company, 2 per squad (18), plus instructors.
Machine gun	Do.
Grenades	Do.
Scout	Do.
Signal	Company, 3 from co. hdqtrs. and 3 per platoon hdqtrs. (12), plus instructors.
Communication	Battalion, 2 per company (cp. "Signal" and pvt. "Agent") (8).

b. Classes organized as shown above are of a convenient size. Qualified instructors are assigned to each group, the number of assistants depending upon the type of instruction and the availability of qualified personnel. The name of the class indicates the subject in which that class receives the major part of its instruction. However, each class receives instruction in such other subjects as may be considered necessary.

c. An example of a day's schedule for the automatic rifle class is as follows:

0930–1030 Functioning of automatic rifle. Lieutenant, first platoon, senior instructor.

1045–1130 Stoppages of automatic rifle. Lieutenant, first platoon, senior instructor.

1300–1330 Bayonet training. Lieutenant, bayonet instructor, a rifle company officer designated by the battalion commander, senior instructor. He coordinates all bayonet instruction within the battalion.

1345–1430 Tactics, street fighting. Company commander, instructor. Scout class joins for this period.

1445–1530 First aid, application of tourniquets. Battalion surgeon, senior instructor.

1545–1600 Talk, racial characteristics of country of destination. Company commander, senior instructor. Entire company assembles for this period.

d. Division of personnel of machine gun and howitzer units into groups for class instruction is effected similarly to the outline shown for the rifle company in paragraph 4–15, a. The daily schedules for the different classes are made up in a manner similar to the example shown for the automatic rifle class in paragraph 4–15, c.

TRAINING EN ROUTE ON BOARD SHIP

4–16. Subjects covered.—*a.* Paragraph 4–1, c, lists a number of subjects that are suitable for shipboard instruction. Deficiencies in training of the troops on board, as influenced by the tactical situation likely to be encountered, will govern the selection of subjects that are to be stressed. Having determined the training needs of the several units, the subjects to be stressed may be selected and schedules prepared accordingly.

b. In addition to the subjects listed in paragraph 4–1, c, the following subjects are particularly important and should be emphasized enroute to the theater of operations.

(1) Information of the country of destination; its people, language, topography, political and military situation.

(2) Enemy tactics likely to be encountered. Tactics to be adopted by our own troops.

(3) Relations with inhabitants of the country of destination.

4–17. Essential training.—*a.* Newly organized units will often include men who are only partially trained in handling their weapons. After formation of the unit, there may be only a short period for instruction prior to embarkation. In some cases, there will be no time for any instruction whatsoever. While enroute to the country of destination, troop schools should aim to acquaint every man with the mechanics, technique, firing, and technical employment of the weapon with which he is armed, thus increasing his value to his organization as a member of the combat team. Permission may be readily secured from the commanding officer of the ship to fire the various infantry weapons from the deck while the ship is under way. Targets may consist of articles floating at sea or articles thrown overboard (tins and boxes from the galley). For safety, shooting is conducted only from the stern of the ship. If there are no articles available to be used as targets, "white-caps" may be used as aiming points.

b. Instruction in tactics should be sufficiently adequate to give all enlisted personnel a knowledge of scouting, patrolling, security measures, and troop leading problems, appropriate to their rank. Methods of Instruction include sketches on blackboards (the best method), chalk sketches on the deck, and matches laid out on deck. The instructor explains the situation (diagram or sketch) and asks different men for their decisions and reasons for their decisions. Initiative and discussion should be encouraged. In small wars situations, the noncommissioned officer and private are often faced with problems requiring decision and subsequent immediate execution.

Section IV

TRAINING IN THE THEATER OF OPERATIONS

4–18. **System of training.**—*a.* Upon arrival in the theater of operations, immediate steps are taken to continue the training along methodical and progressive lines. The training is goverened by training programs and schedules prepared by the various organizations.

b. For each training subject functional units (squads, sections, and platoons) are employed. This places the responsibility for training progress upon the unit leader. Unfortunately, all training subjects cannot be so handled. In many instances, subjects must be taught by classes composed of individuals from several subdivisions of a unit.

4–19. **Facilities.**—As early as possible after the force is established on shore, organization commanders of higher echelons should provide their respective commands with the facilities necessary for the conduct of training. Whenever practicable, these facilities should include the establishment of training centers, troop schools, ranges for practice and record firing of infantry weapons, ranges for combat practice firing, and terrain suitable for the conduct of field exercises.

4–20. **Subjects covered.**—*a.* Paragraphs 4–1, c, and 4–16, b, list subjects suitable for training conducted in the theater of operations.

b. All training should include field exercises involving the tactical employment of troops in military situations peculiar to the terrain and enemy resistance likely to be encountered in different sections of the country.

4–21. **Training centers.**—*a.* Weapons are constantly improving and minor powers are progressively arming themselves with a greater number of improved weapons. This indicates the necessity for trained troops if our operations are to succeed without excessive casualties to personnel. When partially trained troops compose a large part of the units of the force, the establishment of a training center is highly important.

TRAINING IN THE THEATER OF OPERATIONS

b. The establishment of a training center offers the following advantages:

(1) It provides for methodical, progressive, and coordinated training.

(2) It is the central agency for the receipt and dissemination of information with respect to the unusual features of the campaign as they develop during operations in the field.

(3) It may be made sufficiently extensive to include terrain for field exercises and ranges for combat practice firing, thus providing facilities that might otherwise be denied to detached companies and battalions.

(4) It is an ideal agency for the training of replacements. All replacements, both officers and men, should be put through an intensive course of training before they are assigned to active units in the field.

(5) It supplies a location for troop schools.

(6) It provides the ranges necessary for the record firing of all infantry weapons.

c. A training center includes the following activities:

(1) Ranges for record practice: These include the ranges, courses, and courts necessary to conduct record practice with all weapons.

(2) Ranges for combat practice firing: These ranges should be sufficiently extensive to permit the maneuvering of units and the firing of all weapons under conditions similar to those encountered in the type of combat peculiar to the country in which operating.

(3) Troop schools: The unit in charge of the training center will be better able to conduct classes in special subjects than will other units of the force. Units of the force are thus enabled to send selected personnel to the training center for an intensive course of training in a particular specialty.

4–22. Troop schools.—Each theater of operations will present different problems that will require a knowledge of special subjects. A troop school is the ideal agency for such instruction. The following are a few of the subjects that may have special application:

Scouting and patrolling. (To include tracking.)

Sniping.

Handling small boats. (Launches, native canoes, etc.)

Language of the country.

Transportation. (Ox carts, small boats, animals.)

Care of animals, riding and draft.

Packing. (Pack animals, pack saddles, and their cargoes.)

First aid, hygiene, field sanitation. (An advanced course.)

Horseshoeing.

Saddlery. (Leather working.)

Cooks and bakers. (To include butchering and cooking for small units on the march and in garrison.)

Aviation observers. (For all officers.)

4–23. Organization of troop schools and training centers.—*a.* Instructors for troop schools that are conducted by the various garrison units are supplied by the units themselves. The students for such troop schools are the members of the unit and duties are so arranged that the troop school does not interfere with the normal routine of the garrison. At times, the unit will be called upon to perform some emergency type of duty that may necessitate the temporary suspension of the troop school. Instructors for a training center come from the unit in charge of the training center and from the unit or units undergoing instruction. Ordinarily, units such as complete companies are assigned to training centers for instruction. In addition, replacements are organized into casual units in the order in which they arrive for duty from the continental United States. At times, it will be advantageous to assign certain qualified individuals among the replacements to receive special instruction in one of the troop school classes conducted at training centers. Troop school instructors are members of the unit in charge of the training center.

b. The training unit is the company. Instruction may be by platoons, sections, or squads. Companies undergoing training at a training center furnish many of their own instructors. Special instructors are furnished by the unit operating the training center. The supervision and coordination of training is a function of the staff of the training center.

c. A list of subjects suitable for the troop school method of instruction is found in paragraph 4–22. Classes are organized from among selected personnel sent to the training center from the various units in the field and from among qualified replacement personnel who have just arrived. Upon completion of the assigned courses, men are sent to active units in the field. Provided existing conditions do not require otherwise, men who have been sent to the training center for specialized training are ordinarily returned to the organizations from which they were originally detailed. Replacement personnel who have completed a special course are sent to those organizations where their specialized training will be most valuable.

Section V

TRAINING PROGRAMS AND SCHEDULES

4–24. **Training instructions.**—Training programs and training schedules are the means generally used to outline the training for the various units, thus providing uniformity in training. Training memoranda may supplement training programs and training schedules.

4–25. **Training programs.**—*a.* Training programs are issued by all commands down to and including the company. They express the general plan of training of the command over a considerable period of time, usually a training cycle of 1 year, but may be issued to cover periods of 6 months, 3 months, or 1 month.

b. The essential elements of the training program include the training objective or objectives, the time available in which to accomplish the mission or missions, and such special instructions relating to the conduct of the training as may be necessary. A feature of the training program with which unit commanders are primarily concerned is the total amount of time allotted for the training of their own units. The authority issuing a training program should indicate clearly the time available, whether the training period covered by the program is 1 year, 6 months, 3 months, or 1 month.

c. Prior to the preparation of a training program, a careful estimate is made of the entire training situation. The following factors must be taken into consideration: (1) Analysis of order from higher authority; (2) mission (training objective); (3) essential subjects; (4) time available; (5) equipment and facilities available; (6) personnel; (7) local conditions (climate and terrain); (8) existing state of training; (9) organization for training; and (10) obstacles to be overcome.

d. The amount of information that should appear in the training program (order) depends upon the size of the unit and the particular situation. A small unit requires a training program in more detailed form than does a larger unit. A situation pertaining to a mobilization will demand more centralized control than will a situation normal to peacetime training. During peacetime training, brief orders containing only essential information may be considered satisfactory, since the various units will usually contain a number of experienced officers.

e. It is assumed that the First Battalion, Fifth Marines, has received the regimental training program (order), with annex showing regimental "losses" for the training period, October 1, 19____, to March 31, 19____. The battalion training program is then prepared and is sent to the various companies of the battalion and such other units as may be concerned. An annex showing battalion "losses" accompanies the battalion training program. The following is an example of such a battalion training program (order):

GENERAL ORDER ⎱
No._____10 ⎰

HEADQUARTERS 1ST BN 5TH MARINES,
Marine Barracks, Quantico, Va.,
August 20, 19____

1. The following training program governing the training of the 1st Bn 5th Marines during the period, 1 October, 19____, to 31 March, 19____, is published for the information and guidance of all concerned.

2. A conference, at which all officers of the battalion will be present, will be held at battalion headquarters at 0930, 25 August, 19____, to discuss this training program. All officers will make a careful study of the training program prior to the conference.

A_____ B. C_____,
Lieutenant Colonel, U. S. Marine Corps, Commanding.

Official:
D_____ E. F_____,
First Lieutenant, U. S. Marine Corps, Bn–3.
Distribution: A, B, X.

TRAINING PROGRAM

HEADQUARTERS 1ST BN 5TH MARINES,
Marine Barracks, Quantico, Va.,
20 August, 19__

1. TRAINING MISSIONS.—The training missions of this battalion are:

a. To secure in this command a maximum of efficiency for the march, camp, and battlefield, with a view to possible active service at any time.

b. To prepare organizations for expansion to war strength.

c. To develop instructors for training recruits in case an emergency should arise.

d. To provide personnel (individuals as well as groups) for increasing units to war strength and to provide personnel for newly organized units.

e. To develop the science and art of war.

2. TIME AVAILABLE.—*a. Training period.*—Six months, 1 October, 19____, to 31 March, 19____ (both dates inclusive).

b. Training week.—Six days, except when shortened by holidays, guard duty, police details, and working details.

TRAINING PROGRAMS AND SCHEDULES

c. Training day.—Normally, 4 hours, 0730 to 1130. During periods devoted to marksmanship, the training will be 7 hours, 0700 to 1200 and 1300 to 1500. No limiting hours are prescribed for field exercises. Calculations for field exercises should be based upon a 7-hour day. No instructional periods will be scheduled for mornings that follow night operations. As a general rule, afternoons will be available for administrative work, additional training for deficient men, athletics, troop schools, and ceremonies. Rifle companies will devote one afternoon each week to instruction in rifle marksmanship. This may take the form of gallery practice and competitions. No training will be scheduled for Wednesday or Saturday afternoons, except that during marksmanship and field exercise periods, training will be scheduled for Wednesday afternoons. Saturday mornings will be set aside for inspection except during marksmanship periods or when other instruction has been ordered by the battalion commander.

d. Training losses.—(1) One company will be detailed each day to perform the necessary guard duty and furnish police and working parties for the regiment. The Regimental Headquarters Company and Regimental Service Company will not be so detailed. Guard schedules will be issued every 2 weeks. The normal order of detail will be A, B, C, D, E, F, G, H, I, K, L, and M Company. During regimental and battalion field exercises the necessary guard duty will be performed by the regimental band. Guard mounting will be held at 1145, daily, commencing on 30 September, when Company A will take over the duty as guard company. Further details relative to guard mounting will be issued later. Instruction in interior guard duty will be carried on during the days each company is detailed to perform guard duty.

(2) The following holidays are announced: 23 November (Thanksgiving Day); 24 December to 1 January (both inclusive); 22 February (Washington's Birthday).

(3) Regimental losses are shown in annex A.

(4) Battalion losses are shown in annex B.

3. SCOPE OF INSTRUCTION.—*a. Training subjects and references.*—The applicatory system of training will be employed. Training of units and individuals will be conducted as prescribed in the following orders and publications, as applicable:

> Marine Corps Order No. 146.
> Landing Operations Manual, U. S. Navy.
> Landing Force Manual, U. S. Navy.
> Field Manuals, U. S. Army.
> Small Wars Manual, U. S. Marine Corps.

b. Use of schools.—Schools will be established and conducted, as follows:

(1) Officers' school.
    (*a*) Advanced course.
    (*b*) Orientation course (current training).
(2) Enlisted men's schools.
    (*a*) Basic course.
    (*b*) Specialist courses.
    (*c*) Drills and tactical training.
    (*d*) Field training.

TRAINING PROGRAMS AND SCHEDULES

A schedule of instruction for the officers' school will be issued by this headquarters every 2 weeks. The names of officers who will act as instructors in the several subjects will be included in the schedule. Specialist courses for enlisted personnel will be conducted by the heads of their respective sections; that is, communication officers will conduct the instruction of communication personnel, pioneer officers will conduct the instruction of pioneer personnel, etc. With the exception of the specialist courses, all other courses of instruction for enlisted personnel will be conducted under the direction and supervision of company commanders.

c. Standards of proficiency.—All material to be inspected shall be complete, immaculately clean, and serviceable. In all training covered by published regulations, the standard for all ranks is accuracy as to knowledge and precision as to execution. In tactical training, the objective is the development of the tactical judgment of all leaders and their replacements by the application of accepted tactical principles and methods to a variety of tactical situations.

d. Inspection.—A proficiency test will be held at the conclusion of each phase of training. It is to be expected that the required standard of proficiency will have been developed on the last day of training in any given subject and it is on that day that the final test will be conducted. However, instruction in a given subject may be discontinued at any time that it becomes apparent that the desired standard has been reached. The time thus saved may be utilized for other instruction. Unit progress charts will be kept by each company commander. Inspections to test proficiency in a subject will be practical and informal in nature and will not interfere with the training.

4. MISCELLANEOUS.—*a. Programs and schedules.*—(1) *Training programs.*—Company commanders will prepare company training programs for the period indicated and will submit them to this headquarters prior to 15 September. The company training program is not to be regarded as a rigid schedule of execution. It is merely the plan of the company commander, showing the approximate allotment of time and the general scheme for using that time. It is intended to be flexible and must be so considered.

(2) *Training schedules.*—Weekly training schedules will be submitted to this headquarters before noon on the Wednesday preceding the training week covered by the schedule. Alternate instruction for 1 day will be added to weekly schedules to provide for possible interruption due to inclement weather.

b. Attendance.—Men detailed to special duty will receive not less than 8 hours instruction weekly, except during weeks shortened by holidays or guard duty. Company commanders will submit requests to this headquarters, 1 week in advance, when the attendance of special duty men is desired for instructional periods. Administrative details will be so arranged that every man performing special duty will receive at least 4 hours training each week.

c. Ceremonies.—Weather permitting, there will be one regimental ceremony and one ceremony for each battalion weekly, except during the marksmanship period. All units will normally participate in the regimental ceremony. Days for ceremonies are assigned as follows: Monday, First Battalion; Tuesday, Second Battalion; Thursday, Third Battalion; Friday, Fifth Marines.

d. Athletics.—Participation in athletics is voluntary. Company commanders will encourage intercompany sports and company competition. The bat-

TRAINING PROGRAM
COMPANY B, 5th MARINES
1 OCTOBER 19__ — 31 MARCH 19__

MONTH	OCTOBER				NOVEMBER				DECEMBER					JANUARY				FEBRUARY				MARCH				
WEEK ENDING	6	13	20	27	3	10	17	24	1	8	15	22	29	5	12	19	26	2	9	16	23	2	9	16	23	30
LOSSES *																										
HOLIDAYS																										
GUARD, POLICE AND WORKING PARTIES																										
MARKSMANSHIP																										
SICK IN QRTRS																										
REGIMENTAL TRAINING																										
BATTALION TRAINING																										
BATTALION FIELD EXERCISES																										
REGIMENTAL FIELD EXERCISES																										
TACTICAL INSPECTIONS																										
TOTALS																										

HOURS AVAILABLE

SUBJECTS — DISCIPLINARY (TOTAL HOURS AVAILABLE)

1. NAVY REGULATIONS
2. CUSTOMS OF THE SERVICE
3. MILITARY COURTESY
4. HYGIENE AND SANITATION
5. INTERIOR GUARD DUTY
6. DRILL, CLOSE-ORDER
7. DRILL, EXTENDED-ORDER
8. DRILL AND COMBAT SIGNALS
9. INSPECTIONS

TECHNICAL (TOTAL HOURS AVAILABLE)

11. RIFLE AND BAYONET
12. BAYONET FIGHTING
13. PISTOL
14. AUTOMATIC RIFLE
16. THOMPSON SUB-MACHINE GUN
17. FIRST AID
18. GAS DEFENSE
19. TENT DRILL
20. MARKING
21. FIELD FORTIFICATIONS
22. THE PACK
23. SCOUTING AND PATROLLING

TACTICAL (TOTAL HOURS AVAILABLE)

24. COMMAND POST
25. PLATOON
26. COMPANY
27. MARCHING AND CAMPING
28. DEFENSE AGAINST AIRCRAFT
29. NIGHT OPERATIONS
30. SMALL MAPS
31. LANDING OPERATIONS

MISCELLANEOUS

33. OPEN TIME

* THIS TABULATION IS MADE BY EACH ORGANIZATION COMMANDER FROM INFORMATION CONTAINED IN REGIMENTAL AND BATTALION PROGRAMS. GUARD, ETC., IS COMPUTED ON THE BASIS OF A TOUR EVERY 15 DAYS.

† WHERE NO ALLOTMENT OF HOURS IS MADE TO SUBJECTS IT IS EXPECTED THAT TRAINING IN SUCH SUBJECTS WILL BE CONCURRENT WITH TRAINING IN OTHER SUBJECTS.

253960—40   (Face p. 23)

tallon athletic officer will coordinate the use of the various athletic facilities.

c. Junior officers.—Except as otherwise prescribed in regulations, each lieutenant will be given a permanent assignment to a clearly defined duty pertaining to the daily command training and administrative activities of his organization.

f. Exercises in leaving post.—Organizations will be prepared to leave the post at any time upon order of the regimental commander. The order for the exercise will include instructions relative to the amount of equipment to be carried and whether preparations will be made for prolonged field service or for only a short period of time.

g. Uniform.—Post regulations prescribe the uniform of the day for different seasons of the year. During training, the uniform of the day may be modified at the discretion of company commanders, depending upon the nature of the training; i. e., dungarees may be worn by Company D during gun drills and by all companies during instruction in scouting and patrolling; shooting coats may be worn during marksmanship training, etc.

h. Instructional methods.—Instructions in oral orders, messages, range estimation, target designation, hasty sketches, care and display of equipment, and similar subjects, will be carried on concurrently with other training.

A_____ B. C_____,
Lieutenant Colonel, U. S. Marine Corps. Commanding.

Official:

D_____ E. F_____,
First lieutenant, U. S. Marine Corps,
Bn–3.

Annexes:
A—Regimental losses.
B—Battalion losses.

Distribution: A, B, X.

f. Company training programs are ordinarily made up in tabular form and list the training subjects, the estimated number of hours to be devoted to each subject, and the allocation of these hours by weeks. These forms are convenient and useful, provided they are regarded as flexible. At best, they only estimate the time factor and indicate a proposed scheme for employing that time. They are understood to be tentative programs and should be so considered. It is assumed that Company B, Fifth Marines, has received a copy of the training program of the First Battalion. Using the battalion training program as a guide, the following is an example of the company training program prepared to cover the period October 1, 19____, to March 31, 19____:

4–26. **Training schedules.**—Training schedules are issued by a commander for that part of the training of his unit that is to be accomplished under his direct command. They are based upon the training programs and orders of higher commanders. The amount of time devoted to the several subjects is dependent upon the state of

TRAINING PROGRAMS AND SCHEDULES

proficiency of the unit for which the training schedule is prepared, more time being allotted to instruction in those subjects in which the unit is deficient. Training schedules are generally made up in tabular form and include the name of the subject, hours, place, uniform and equipment, references, and name of the instructor. When properly prepared, no additional information is required for training during the period covered by the schedule. Training schedules are issued to cover relatively short periods of time. The usual period covered is 1 week, since schedules covering a longer period are likely to be interrupted. In addition to imparting the information necessary to conduct the training, schedules are so arranged that the required standards of proficiency are reached in a minimum of time. When approved by higher authority, training schedules become instruments of execution. This does not mean, however, that a training schedule is to be followed blindly. Should it become apparent that the instruction is not accomplishing the desired result, the schedule should be varied immediately. It is assumed that Company B, Fifth Marines, has received a copy of the battalion training program of the First Battalion and that the company training program has been prepared. Using these programs as a guide, the following is an example of the weekly training schedule prepared to cover the week ending October 6, 19_____.

WEEKLY TRAINING SCHEDULE
COMPANY B, 5th MARINES
FOR THE WEEK ENDING 6, OCTOBER 19—

1. DATE DAY OF WEEK	2. HOUR	3. ASSEMBLY POINT	4. AREA OF INSTRUCTION	5. INSTRUCTOR	6. SUBJECT AND NATURE OF INSTRUCTION	7. STUDY REFERENCES	8. UNIFORM AND EQUIPMENT	9. REMARKS
MON OCT 1	0730-0830	COMPANY PARADE	DRILL FIELD	COMPANY COMMANDER	DRILL, CLOSE-ORDER	TR 420-50, SECT I-III	DAY, WITH ARMS	
	0830-0930	BARRACKS	BARRACKS	PLATOON COMMANDER	NAVY REGULATIONS	USN REGS. 1920	DAY	
	0930-1030	BARRACKS	BARRACKS	PLATOON COMMANDER	THE RIFLE AND BAYONET	TR 320-10, SECT I-II	DAY, RIFLE, BAYONET	
	1030-1130	COMPANY PARADE	DRILL FIELD	PLATOON COMMANDER	DRILL AND COMBAT SIGNALS	TR 420-40, SECT I-II	DAY, WITH CARTRIDGE BELTS	
TUES OCT 2					COMPANY ON GUARD			
WED OCT 3	0730-0830	COMPANY PARADE	DRILL FIELD	PLATOON COMMANDER	DRILL, EXTENDED-ORDER	TR 420-85	DAY, WITH ARMS	
	0830-0930	BARRACKS	BARRACKS	PLATOON COMMANDER	THE RIFLE AND BAYONET	TR 320-10, SECT I-II	DAY, RIFLE, BAYONET	
	0930-1130	COMPANY PARADE	DRILL FIELD	PLATOON COMMANDER	COMBAT PRINCIPLES, SQUAD	TR 420-105, SECT I	DAY, WITH ARMS	
THUR OCT 4	0730-0830	BARRACKS	BARRACKS	PLATOON COMMANDER	THE PACK	TR 50-80	DAY	
	0830-0930	BARRACKS	BARRACKS	PLATOON COMMANDER	CUSTOMS OF THE SERVICE	NONE	DAY	
	0930-1130	COMPANY PARADE	DRILL FIELD	PLATOON COMMANDER	SCOUTING AND PATROLLING	TR 200-5, SECT I	DAY, WITH ARMS	
FRI OCT 5	0730-0830	COMPANY PARADE	DRILL FIELD	COMPANY COMMANDER	DRILL, CLOSE-ORDER	TR 420-50, SECT I-III	DAY, WITH ARMS	
	0830-0930	DRILL FIELD	DO	PLATOON COMMANDER	MILITARY COURTESY	NONE	DO	
	0930-1130	DO	DO	DO	COMBAT PRINCIPLES, SQUAD	TR 420-105, SECT I	DO	
SAT OCT 6	0730-0900	BARRACKS	BARRACKS	COMPANY COMMANDER	PREPARATION FOR INSPECTION	NONE	DAY, WITH ARMS	
	0900-1000	COMPANY PARADE	COMPANY PARADE	COMPANY COMMANDER	INSPECTION	LFM USN, 1927, CHAP 5, SECT II DO	DO	
	1000-1130	COMPANY PARADE	BATTALION PARADE	BATTALION COMMANDER	INSPECTION OF PERSONNEL, BARRACKS AND GROUNDS	DO	DO	
INCLEMENT WEATHER SCHEDULE					RIFLE MARKSMANSHIP	SFM, VOL II, PART I, CHAP I		

RD 1034-C-3

SMALL WARS MANUAL
UNITED STATES MARINE CORPS
1940

✦

CHAPTER V

INITIAL OPERATIONS

RESTRICTED

UNITED STATES
GOVERNMENT PRINTING OFFICE
WASHINGTON : 1940

TABLE OF CONTENTS

The Small Wars Manual, U. S. Marine Corps, 1940, is published in 15 chapters as follows:

III

SMALL WARS MANUAL

UNITED STATES MARINE CORPS

Chapter V

INITIAL OPERATIONS

v

SECTION I

NEUTRAL ZONES

5–1. General.—*a*. A neutral zone is an area in which no hostilities are permitted. The establishment of neutral zones is not of recent origin; the system has been employed not only by civilized nations but also by early American Indians and by African tribes. The procedure at the beginning of a small war operation often follows a sequence that is more or less a matter of routine. First, one or more of our cruisers arrive off a foreign port in consequence of actual or potential danger to our nationals and their property. Then if the situation requires it, a ship's landing force is sent ashore at this port to suppress disorder, provide a guard for our nationals and their property in the port, including our legation or consular buildings, and, in addition, certain local government buildings, such as custom houses. If there is a prospect of fighting between the local factions, the cruiser's commander (or senior naval officer in command locally) forbids combat in areas where the lives and property of our nationals might be endangered. This is done by the establishment of neutral zones; and this procedure frequently results in the cessation of hostilities; the mutually destructive strife may become so severe that absolute chaos is imminent, and neither faction is capable of guaranteeing the security of life and property. Then the neutral forces may be forced to enlarge their sphere of action by a movement inland.

b. The foreign policy of the United States relative to domestic disorders in unstable countries is one of nonintervention. However, as a measure to safeguard our nationals and, incidental thereto, other foreign nationals, havens of refuge will no doubt be established at certain seaports of an unstable country whenever the domestic disorder threatens the lives of these nationals. To provide protection

1

en route to the haven of refuge, certain routes of evacuation, such as railroads, highways, and rivers leading to the seaports may also be designated as a part of the neutral zones. In such cases, a definite time limit may be set for refugees to clear the routes. Situations undoubtedly will arise where our individual nationals will not seek safety within the neutral zone established at the seaport, but will elect to remain with their property and goods in the interior. In such cases the responsibility of the commander of the United States forces at the seaport neutral zone should be considered to be at an end with regard to any protection to be afforded these nationals. Should any harm come to these nationals who elect to remain with their property and goods in the interior of the country, recourse must be had later to diplomatic action for redress, and recompense for loss of goods and property must be made in the case of those nationals who seek safety in the neutral zone seaports and abandon their property in the interior.

c. The establishment of a neutral zone may not necessarily be followed by further military operations; however the prolongation of the unsettled condition in the country may necessitate such action, involving a movement inland from those zones. Accordingly, Neutral Zones and Movement Inland are presented in that order in this chapter.

5–2. **Purpose, occasion, and circumstances.**—a. *Purpose.*—(1) Protect treaty rights.

(2) Assist in maintaining the existence of, or the independence of, a government in accordance with treaty provisions.

(3) Protect lives and property of our nationals located in disturbed areas and unfortified cities.

(4) Further the provisions of our national policy.

(5) Protect and prevent depredations on neutral territory of adjacent countries.

b. *Occasion.*—(1) In time of revolution, during riots, or when the local government has ceased to function.

(2) In time of war between two nations.

c. *Circumstances.*—(1) At the request of a recognized government, or at the insistence of regular local officials.

(2) At the request of the opposing factions.

(3) By forces of another power, or group of powers, without the invitation of any faction.

(4) By agreement between contending states or forces.

NEUTRAL ZONES

5-3. Basic orders.—The orders directing the establishment of a neutral zone should be brief and concise, and should contain the following information:

(1) Designation of the military force to be employed in the establishment and maintenance of the zone, and the zone force commander.

(2) The mission of the force.

(3) Information relative to the purpose, occasion and circumstances necessitating the establishment of the neutral zone.

(4) The exact time after which an area shall be considered as a neutral zone, relative to movements by land, water and air.

(5) The limits of the neutral zone.

(6) Logistic provisions, including those pertaining to the requirements of refugees.

(7) Reference to the communication plan and notification of the location of the zone force commander.

5-4. Instructions.—Additional information required should accompany the Basic Order in the form of an annex, or if there are existing general instructions relative to the establishment of neutral zones, reference should be made to them in the order. These instructions should contain, when applicable, stipulations covering the following matters:

(1) Control to be exercised by the zone force commander and the local civil authorities.

(2) Restrictions placed on opposing force(s) within limits of neutral zones at the time of establishment.

(3) Instructions relative to local authorities and civilians bearing arms within the zone.

(4) Acts to be prohibited, such as the delivery from, or passage through the zone, of supplies destined for the contending forces who are prohibited the use of the zone.

(5) Type of vessels and also land and air transportation carriers prohibited entrance to or passage through the zone.

(6) Restrictions upon the communication facilities.

5-5. Zone force commander's order.—The operation orders of the zone force commander should contain so much of the information furnished him in his orders from higher authority as will be of value to his subordinates, and also any additional information that may be pertinent. The order should contain detailed instructions for each task group of his force. If general instructions for the establishment of neutral zones have been issued by higher authority, those parts that are applicable to the immediate situation should be promulgated to the task groups of the local zone force either in the zone force commander's order or as an annex thereto. Logistic provisions, communication plan and location of the zone force commander should complete the order.

5-6. **Proclamation.**—The civilian population of the neutral zone and its vicinity, as well as the factions to be prohibited the use of the zone, should be informed of its establishment as early as practicable. This may be accomplished by the delivery of a written memorandum to the local authorities and to the heads of the contending factions, or by the publication of a proclamation in the local newspapers with a delivery of same to the local authorities and to the heads of the contending factions. Such memorandum or proclamation should be published both in English and in the local language. The delivery of the memorandum or proclamation may be made direct or through the diplomatic agent of the country represented by the zone force commander. Regardless of the method of transmission or its form, the proclamation should contain stipulations regarding the following matters:

(1) Precise date and hour at which the establishment of the neutral zone becomes effective.

(2) Area included in the neutral zone, with the boundaries or limits clearly defined by terrain features.

(3) Relationship of armed forces of contending factions with the neutral zone.

(4) Relationship of the zone force with the civilian population and local authorities within the zone.

(5) Acts to be prohibited in or over the zone.

(6) Transportation restrictions on routes of communication through the zone.

(7) Communication restrictions within the zone.

(8) Conduct or status of armed vessels within the zone.

(9) Such other information as may be necessary for a clear understanding of the exact circumstances upon which the establishment of the zone is based, the purpose to be accomplished, and the means to be used.

Section II

MOVEMENT INLAND

5–7. **Point of departure.**—*a.* As in all forms of warfare, logistic requirements must be given careful consideration in preparing strategic and tactical plans; in fact such requirements are frequently the determining factor. Before a movement inland is undertaken an analysis and estimate of the local transportation and supply facilities must be made in order to insure a reasonable rate of advance with replacement of supplies.

b. The movement inland will not always be a movement from a seaport to the interior. Frequently the movement will be made from the capital or principal city, located at the terminus of a railroad at the head of navigation on the upper part of a large river, or on a well-developed highway, with well-defined lines of communication connecting it with the seacoast. In any case the point of departure becomes a base of operations as well as a base of supply until other bases more advanced are established. Should the small-war operations be initiated by the establishment of neutral zones, one or more of them may later become a base for extended operations.

c. If the point of departure for the movement inland is to be other than a seaport, the movement to the point is made by the most convenient means. The movement will be of the same general nature as an advance in major warfare in the presence of the enemy. The special features of a movement by inland waterways are presented in chapter XII.

5–8. **Mobile columns and flying columns.**—*a.* When the successful prosecution of the campaign requires the execution of measures beyond and/or supplementary to the establishment of neutral zones, the control of seaports, or key cities along lines of communication in the affected areas, mobile columns must be projected inland from the

5

points of departure, for the purpose of pursuing, rounding-up, capturing, or dispersing any existing irregular forces; of covering productive areas; or of establishing chains of protected advanced bases in the interior.

b. Mobile columns as such differ from the so-called flying columns in one great essential—supply. A flying column is defined as a detachment, usually of all arms, operating at a distance from, and independent of, a main body or supporting troops, lightly equipped to insure mobility and sufficiently strong to exempt it from being tied to a base of supplies through a fixed line of communications. A mobile column is of the same description as the flying column with the exception that it is self-supporting to a lesser degree and is dependent for its existence on its base of supplies.

c. The movement may be made by a large force operating along a well-defined route, but will usually be made by several mobile columns operating either along separate lines of advance or following each other independently along the same route of advance at an interval of about 1 day. In some situations, columns may start from different points of departure and converge on a city or productive area. The columns may vary in size from a reinforced company to a reinforced regiment, but the size best adapted to such operations has been found to be a reinforced battalion.

d. When fortified posts with permanent garrisons are established, flying columns should operate therefrom. This is the most arduous of all operations; the idea being to combat the native guerilla at his own game on his own ground. At the beginning of such operations, the column may be of considerable strength—a company of infantry accompanied by a machine gun and howitzer detachment preceded by a mounted detachment. As the guerilla forces are dispersed, combat patrols (mounted or dismounted) consisting of two or more squads may suffice. The mission of the flying column will be to seek out the hostile groups, attack them energetically, and then pursue them to the limit. Therefore, there should be nothing in its composition or armament that would tend to reduce its mobility or independence of action beyond that absolutely necessary for combat and subsistence. Except for supplies which can be carried by the men, the column as a rule will depend upon the permanent garrisons. These posts must be established in sufficient numbers to permit of such supply—a post always being within 1 or 2 days' march of another post.

e. A flying column should never be dispatched to any area unless it is amply supplied with CASH. With available funds, not only may subsistence be purchased, but often information of the hostile forces and the terrain (guides and interpreters). The money supplied the flying column should be in SMALL denominations, principally silver; it is difficult, frequently impossible, to change bills in rural communities.

5-9. **Strength and composition of columns.**—*a.* The strength and composition of mobile columns will depend upon the probable resistance to be encountered, the terrain to be traversed, the type and condition of existing transportation, and the means of communication. Normally, the addition of mounted detachments, armored cars, and aircraft is desirable in such columns. If a march through an extensive area of undeveloped country is contemplated, an engineer unit should be included. The use of light field pieces has been limited in the past, but with the increase of armament by all classes of powers and the improvement of defensive means, they cannot be dispensed with unless there is every assurance that they will not be needed. However, as a general rule, nothing should be added to the mobile column that would tend to decrease its mobility and which is not absolutely necessary.

b. The column should be of sufficient strength to enable it to cope with the largest force likely to be encountered. While weakness in the strength of a column is dangerous, yet excessive strength should be avoided. The supply requirements of a large column necessitate considerable transportation, and results in a proportionately larger train guard as the length of the column increases. A larger train also decreases the mobility of the column.

c. If the movement is made over broken country with poor roads and trails, the column often will be forced to move in single file. A column of excessive strength for its mission will march irregularly due to the elongation of the column. Such a column will arrive at its destination in a more exhausted condition than a smaller force which is able to maintain a regular rate of march. In case an operation necessitates a large column with the corresponding large train, the train may be broken up into two columns in addition to separation of the combat force. This will prevent elongation of the column and allow a regular rate of march.

d. The numerical strength of a column may be decreased by the inclusion of an increase of automatic weapons and supporting infantry weapons above the normal allowance. The increase of ammunition necessitated thereby will not be proportionate to the decrease in the amount of subsistence. Such a decrease will also decrease the amount of transportation required.

e. By means of the modern portable light radio sets (one of which at least should be assigned to the column) and contact planes, a column can be readily reinforced when necessary. Columns moving in the same general area are better able, due to these means of communication, to keep in close touch and to render mutual support than in the past. This, with the offensive support available from aviation, must be considered in determining the composition and strength of the column.

f. Radio and contact planes may be the only reliable means of communication at the beginning of a movement. However, all means of communication must be considered, not only in deciding upon the strength of the column but also the route to be followed. Telegraph and telephone lines may be destroyed, and in the early stages of the operations it may not be worth while to repair and maintain them. If not interfered with, or when control is established and repairs effected, these land lines should be utilized. Dispatch riders (runners, foot or mounted) may not be of much value until conditions become fairly settled, but at times they may be the only means available, or they may be used to supplement other means. Where the country lends itself to the employment of armored cars, they may be used for courier service. Any courier service on a regular time schedule and via restricted routes is dangerous.

5–10. **Protective measures covering movement.**—*a.* When a column starts its movement, it is immediately concerned with the general means of insuring its uninterrupted advance through hostile territory. Usually all parts of a column are vulnerable to attack. In major warfare an army usually has such an extent of front that its rear and base are reasonably secure, and attacks are launched by the enemy at the flanks and front. In small wars, however, the frontage of the regular force is relatively narrow and the column of regular troops is liable to attack by encircling detachments of the irregular forces. Therefore the column must insure itself from an attack from every direction.

b. In major warfare, this security is effected by outposts, by advance, flank, and rear guards, by scouts, by combat patrols and connecting groups, by deployment in depth, and by means of air reconnaissance. In small wars, the principles of security are the same but their application varies with the hostile tactics, armament, and the terrain over which the forces operate. The guiding principle of security is to prevent the hostile fire from being effective against the main body of a march column. The enemy should be denied all terrain from which he may inflict losses upon the column, and the advantage of superior armament and accuracy of fire maintained to prevent the opponent from closing in to effective range of his own weapons even though he be superior numerically.

c. The nature of the terrain has a marked influence on security measures. Often in the theaters of operations, thick low brush interspersed with cactus extends along the main trails and roads making an almost impenetrable jungle, too thick for the movements therein of even small combat groups. In such cases the use of flank guards for a marching column is practically impossible, the lack of which at times permits the hostile force to establish favorable ambushes along such a route.

d. An active hostile force bent on small depredations and armed with rifles and automatic weapons will have ample opportunity to ambush the main body of a column after the advance guard has passed unless patrols are kept continually moving through the underbrush on both sides of the road at a distance from which the ambush position would be effective (normally about 20 to 40 yards). The progress of such flank patrols, however, will be slower than that of the main body with the result that these patrols will be continually falling behind. This necessitates sending out frequent patrols from the head of each organization. To prevent uncovering the head of each organization by these detachments therefrom, the patrols should be started out well ahead of the organization when opposite the rear.

e. On mountain trails with heavy growth of brush and timber which restricts or prohibits the use of flank patrols, a column may be obliged to march in single file. Its only security in this case will depend upon a prompt return of a heavy volume of fire from the part of the column attacked. When the column is restricted in its march formation, it should be divided into a number of small combat teams, each being capable of independent action.

5-11. **Establishment of advanced bases inland.**—*a.* After the mobile columns have successfully dispersed the larger groups of the hostile forces in any area, the next step is the establishment of advanced bases and fortified posts inland for the prosecution of the next phase—the operation of flying columns into the interior.

b. The particular functions of a fortified post are as follows:

(1) To cover productive areas and their lines of communication with their markets.

MOVEMENT INLAND

(2) To afford protection to the local population in that area.

(3) To form a base of supply, rest, replacement, and information for flying columns.

c. As a general rule, these posts should be located at the heads of valleys on main roads or waterways leading from seaports, and at the apexes of valley and intervalley roads and trails leading to the more difficult wooded and mountain regions—the final theater of operations.

d. The site of the post should if possible have the following characteristics:

(1) Be capable of defense by a relatively small detachment.

(2) Be of sufficient extent to permit the bivouac of a flying column of not less than 100 men with a mounted detachment.

(3) Be so situated as to control any town in the vicinity and all approaches thereto, especially roads and ravines.

(4) Be located on commanding ground overlooking the surrounding country.

(5) Be accessible to water supply and main roads.

(6) Be located near terrain suitable for a landing field.

e. In many cases, old forts, redoubts, or isolated masonry buildings with compounds can be organized for defense. Often however it will be found that conditions will warrant the construction of an entirely new fortified post from the material available in the vicinity.

f. The main requirements of a fortified post, garrisoned as it will be by only a few men is that is must not be vulnerable to a sudden attack or rush. This requirement can be met by the construction of a double line of defense; an outer line of defense (occupied only when the flying column is present) to inclose the bivouac area, and an inner line of defense to inclose the depot facilities and permanent garrison, provision being made in both lines for free use of automatic weapons and grenades. (For further details concerning the defense of towns, etc., see ch. VI.)

g. Communication with fortified posts should primarily depend upon radio and aviation. All such posts should be equipped with a radio set capable of communicating not only with its headquarters and other nearby posts, but also with the air service. A landing field at times may not be available in the vicinity of the post so recourse must be had to the use of the pick-up and drop message method of communication.

5-12. **Movement by rail.**—a. If the movement to the point of departure is opposed, or the adjacent territory not under complete control, a movement by rail will involve many tactical features not encountered in a simple rail movement. Even after the railroad is functioning and the hostile forces dispersed, raids and other operations by guerrillas may require the use of armored trains with train guards. Guards may be necessary at stations, bridges, junction points, and other critical points along the railroad.

b. In case a country, or an extensive part of it, containing railroads is to be occupied as a part of the campaign plan, then the operation order for the seizure of the seaport terminus of the railroad should include instructions directing the seizure of the rolling stock and the terminal and shop facilities. This action may prevent their destruction or their removal from the seaport area. Rolling stock having been seized in accordance with the aforesaid instructions, measures must be taken to continue the operation of the railroad service, provided the strategical plan involves the establishment of a point of departure at some place along the railroad line or at some inland terminus thereof. Opposition to such use of the railroad may be encountered in the form of organized military resistance, or by sabotage.

c. The first step taken to operate a railway train over the line where opposition may be expected, is to provide a pilot train. The engine of this train should be protected by placing armor, usually improvised, over the vital parts, supplemented by additional protection of sandbags or similar material. Several cars loaded to full-weight capacity, preferably flat cars or gondolas, that do not obstruct the view from the engine and rear cars, should be placed ahead of the engine to serve as a buffer. These flat cars will then serve as a test-load element, over mines laid in the road bed, or over bridges and viaducts that have been weakened through sabotage. The car immediately in rear of the engine should be a box or cattle car from the top of which rifle and machine gun fire may be directed over the engine to the front. The remaining cars in the pilot train should be flat cars, gondolas or cattle cars, from which troops protected by sandbags or similar material may deliver all-around fire. Some of the personnel accompanying the pilot train should consist of engineer troops to be employed in counter-demolition work and in inspecting the roadbed for mines and the bridges and viaducts for structural weakness. Where such mines are found, these engineer troops should accomplish their destruction, and in the case of weakened bridges, etc., should make the necessary repairs. The main body of the troops embarked on the pilot train should consist of sufficient personnel to protect the train and the working

MOVEMENT INLAND

parties of engineers and laborers. A number of volunteer local civilian laborers may be added to the complement in order to obviate the necessity of using the combat troops as working parties with the engineers. The combat troops should be armed with a large proportion of automatic weapons, light mortars and 37 mm. guns. Some fire-fighting equipment should also be carried with this pilot train. A few light chemical tanks, water barrels and tools for beating out a fire, should be placed in one of the cars. Irregular forces not provided with demolition equipment will probably resort to burning the wooden bridges and railroad trestles usually found in the theater of operations in small wars. Material available for putting out a fire of this nature in its initial stage will gain many hours of valuable time in the advance inland. A troop train should follow the pilot train within close supporting distance; it should contain sufficient troops, properly armed, as to be capable of dispersing any hostile forces until the arrival of additional troop trains. If the use of artillery is contemplated later in the combat operations, some of it should be carried on this troop train.

d. This troop train should have some flat cars or gondolas ahead of its engine and should also be equipped with improvised protective material for the troops. The troops on the forward flat cars or gondolas should be armed with machine guns and howitzer platoon weapons. The remainder of the train should be composed of railroad cars readily adapted to all-around defense and of such type as to permit the rapid debarkation therefrom of the troops. Depending on the capacity of the trains available, detachments of troops from the first troop train or another closely following it should be debarked at critical points along the railroad line for its protection. These protective detachments should institute a system of patrols along the line to prevent sabotage and interruption of the railroad line at points intermediary between the critical points. Aviation may render most valuable aid to these trains in the initial movement inland as well as during the period of operation of the line. On the approach to a city, defile, or other critical points, the troop train should close up on the pilot train and a reconnaissance should be made by ground troops to supplement the information furnished by the aviation. Positive information from the aviation can usually be acted upon; however, negative data from the aviation may be misleading and if acted upon, may lead to fatal results.

e. Where a good road parallels closely a railroad, a flank covering detachment in trucks may expedite the train movement.

SECTION III

MILITARY TERRITORIAL ORGANIZATION

5–13. **Purpose.**—*a.* In all warfare, territorial organization is necessary to facilitate the performance of strategical, tactical, and administrative functions by allocating appropriate tasks to various units.

b. Nearly all independent states include internal territorial subdivisions that are utilized to facilitate the execution of numerous governmental functions. In many cases the limits of these subdivisions were predetermined by the necessities of government. Usually one or more of the following factors have fixed the geographical limits of the internal territorial subdivisions of a country:

> Density of population.
> Routes of communication.
> Economic conditions.
> Geographic features.
> Racial extraction.
> Military requirements.

c. The larger subdivisions of a country, regardless of name (Department, Province, State, etc.,) are usually the political, electoral, administrative, judicial, and military districts of the country.

5–14. **Influence of the mission on territorial organization.**—*a.* The mission of the intervening force will usually come under one of the following headings:

(1) Restoration of law and order, (either by furnishing aid to the recognized government or by establishing a temporary military government).

(2) Supervision of elections.

(3) Establishment of neutral zones.

b. If the mission is to aid the local government in restoring law and order, or to establish military government until a new govern-

MILITARY TERRITORIAL ORGANIZATION

ment is organized and functioning. it is advisable to recognize the political subdivisions of the country. When the military situation requires that an arbitrary division of the country into areas be made, easily recognized topographical features should be used as boundary lines. In areas of military activity. boundaries should not split a locality such as dense forests and rugged terrain, that favors hostile operations, but should include such features in a single command area when practicable.

c. In supervising elections, political subdivisions should be recognized and followed in the assignment of personnel.

d. Neutral zones are generally as limited as the accomplishment of the mission will permit. Thus the boundaries of such zones will often be arbitrary and at other times will follow some distinctive terrain feature.

5–15. **Assignment of troops to areas.**—a. Major territorial divisions, such as areas, should have complete tactical and administrative control within their limits subject to such coordinating instructions as are issued by higher authority. This necessitates the assignment of sufficient executive and special staff personnel to enable the unit to perform all of its functions efficiently. Thus the assignment of a regiment, independent battalion, or other tactical and administrative unit to an appropriate area is advantageous. Small tactical units must have the necessary administrative staff assigned to them.

b. Large areas are usually subdivided for the reasons stated in Paragraph 5–13. Such minor divisions are usually called departments, districts, or subdistricts, depending on the size and importance of the area. Command and staff appropriate to the task are allocated to these subdivisions.

5–16. **Size and limits of areas.**—a. It is not necessary that areas be equal in military strength, population, or extent, but for reasons of organization and command previously discussed, more or less similarity in these features is desirable. Some of the considerations that should be borne in mind when defining the size and limits of specific areas are:

(1) Available troops in the theater.

(2) Location and strength of hostile force(s).

(3) Present boundaries of subdivisions of the country.

(4) Political affiliations of the inhabitants.

(5) Geographic-topographic features.

(6) Supply.

MILITARY TERRITORIAL ORGANIZATION

(7) Communication.

(8) Transportation.

(9) Distribution of population.

(10) Economic conditions.

The more important of these are discussed in order in the sub-paragraphs below.

b. In considering the troops available, it will usually be found that a considerable number are required for garrisoning areas where unrest exists, and for the protection of bases, lines of communication and the like. Often this duty can be utilized as rest periods for troops that have been engaged in active operations. A decision as to the strength of forces required in various localities will determine the location of the administrative and tactical units of the Force. This in turn should be considered in determining the size of military areas.

c. If active opposition is localized, it may be desirable to form an area of the turbulent zone in order to centralize the command so far as combat activities are concerned. The nature of the opposition has considerable influence on the composition of the force assigned to an area. A large area with varied terrain and considerable resistance to overcome might have a force of all arms for the task. This force, in turn, may have a section particularly adapted to the operation of a particular arm (mounted units, mechanized unit, or special river patrol), in which case the particular arm, if available, might well compose a district garrison.

d. Other considerations being equal, retention of existing boundaries when defining the limits of command areas is desirable for several reasons, among which are:

(1) Political, judicial, and administrative functions (insofar as the civil population is concerned) are better coordinated.

(2) The routine of the people is less disturbed; thus better information and less antagonism may result.

(3) Often such boundaries coincide with those dictated by strategy and tactics.

e. When political or other antagonisms among the inhabitants contribute materially to the difficulties of the situation, formerly established subdivisions may be divided or combined in a manner best calculated to accomplish the desired end. In cases where a step-by-step occupation of the country is necessary, territorial organization may conform to the geographical features which control the successive objectives.

MILITARY TERRITORIAL ORGANIZATION

f. Each area should have its own base(s) of supply. A landing field should be in each area. If the supply channels of one area pass through another area, positive steps must be taken by the supreme commander to insure that the flow of supplies to the far area is uninterrupted. It is highly desirable that areas have transversal as well as longitudinal lines of communication. Ordinarily an undeveloped section with poor roads or trails that might serve as a hideout or stronghold for irregulars should be incorporated in a command area in such a way that the commander controls the routes thereto.

g. Consideration of the existing wire communication installations is of importance when defining the limits of an area. Area commanders should not be forced to rely on radio and airplane communications alone, if there are other means of communication available.

h. In countries which are not well developed, maps are not usually up to the normal standard as to variety or accuracy. When defining areas, the use of a particular map designated as "official" by all units facilitates coordination and partly eliminates the confusion as to names of localities, distances, boundaries, and other matters that result from the use of erroneous maps of different origin.

SECTION IV

METHODS OF PACIFICATION

5-17. **The nature of the problem.**—The regular forces in this type of warfare usually are of inadequate numerical strength from the viewpoint of extent of terrain to be controlled. Thus the decision as to the amount of dispersion of regular forces that may be resorted to is an important problem. Detachments with offensive missions should be maintained at sufficient strength to insure their ability to overcome the largest armed bands likely to be encountered. Detachments with security missions, such as the garrison of a town or the escort of a convoy, should be of the strength essential to the accomplishment of the task.

5-18. **Methods of operations.**—Among the various methods that have been used for the pacification of an area infested with irregulars are:

(1) Occupation of an area.

(2) Patrols.

(3) Roving patrols.

(4) Zones of refuge.

(5) Cordon system.

(6) Blockhouse system.

(7) Special methods.

Each of these will be discussed in the succeeding paragraphs.

5-19. **Occupation of an area.**—*a*. This consists of dispersing the force in as many small towns and important localities as the security and patrolling required of each garrison will permit. It partakes of the nature of an active defense. When communications are good,

METHODS OF PACIFICATION

a coordinated counter-offensive may be taken up rapidly, because patrols from various garrisons will receive prompt operation orders.

b. Sometimes the requirements that certain localities be defended necessitates the application of this method; at other times pressure from the outside sources to secure protection for communities and individuals makes this method mandatory. In establishing numerous fixed posts, consideration should be given to the fact that withdrawal therefrom during active operations involves protests from those protected directly or indirectly, loss of prestige, and increased danger to the installations or individuals that were protected. The greater the number of localities that are garrisoned permanently, the less is the mobility of the command; consequently, care should be taken to retain sufficient reserves properly located to take up the counter-offensive at every opportunity.

c. The necessity for bases of operation indicates that this method will be used to a greater or less extent in every operation, that is, irrespective of the plan adopted, this method will be used at least in part. The discussion in this paragraph is particularly applicable to those situations where this plan is the fundamental one for accomplishing the pacification.

d. Modification of this scheme wherein many detachments of regulars are encamped in infested localities and on or near hostile routes of movement, has been used successfully in combination with other courses of action.

5-20. **Patrols.**—*a.* These are detachments capable of operating for only a comparatively limited time without returning to a base. They vary anywhere from powerful combat patrols to small detachments performing police functions according to the situation and mission. They are usually controlled by the commander responsible for the area in which they operate, but in operations against well defined objectives they are often coordinated by higher commanders.

b. Patrolling is essentially offensive action. Accordingly its use in small-wars operations is universal even under conditions that require the strategical defensive.

c. When information of hostile forces is lacking or meager, recourse to patrolling for the purpose of denying the opposing forces terrain and freedom of movement may be the only effective form of offensive action open to the commander. In this case, patrols become moving garrisons and deny the opposing forces such terrain as they can cover by observation, movement, and fire. Extensive operations of this nature exhaust the command, but on the other hand are often more effective in the restoration of order than first appearances indicate.

5-21. **Roving patrols.**—*a.* A roving patrol is a self-sustaining detachment of a more or less independent nature. It usually operates within an assigned zone and as a rule has much freedom of action. As distinguished from other patrols, it is capable of operating away from its base for an indefinite period of time. Missions generally assigned include a relentless pursuit of guerilla groups continuing until their disorganization is practically complete.

b. This method is particularly applicable when large bands are known to exist and the locality of their depredations is approximately known. Such patrols are often employed in conjunction with other methods of operation.

5-22. **Zones of refuge.**—*a.* This system consists of establishing protected zones in the vicinity of garrisons. Their areas are so limited as to be susceptible of protection by the garrisons concerned. Peaceful inhabitants are drawn into this protected area together with their effects, livestock, and movable belongings. Unauthorized persons found outside of these areas are liable to arrest, and property that could be used by insurgent forces is liable to confiscation.

b. This procedure is applicable at times when, through sympathy with or intimidation by insurgents, the rural population is furnishing such extensive support to the resistance as to seriously hamper attempts at pacification. This is a rather drastic procedure warranted only by military necessity.

5-23. **The cordon system.**—*a.* This system involves placing a cordon of troops around an infested area and closing in while restoring order in the area.

b. The cordon may remain stationary while patrols operate within the line.

c. This system may be used when the trouble is localized or the regular force is of considerable size. Due to the limited forces usually available, the application of this system by a marine force will usually be confined to situations where the trouble is rather localized, or to the variation of the method where only a general or partially effective cordon is established.

5-24. **The Blockhouse system.**—The blockhouse system involves the establishment of a line of defended localities. In one way it is similar to the cordon system as both methods deny the opposing forces terrain beyond an established line. In principle it is defensive while the latter is offensive.

METHODS OF PACIFICATION

5–25. **Special methods.**—*a.* The peculiar nature of any situation may require the application of some special method in conjunction with and in accord with the general principles discussed in the preceding paragraphs. Two important phases of operation that may be used in any campaign of this nature are:

(1) River operations.

(2) Flying columns.

b. The tactics and technique of river operations are discussed in chapter X.

c. Flying columns are self-sustaining detachments with specific objectives. Their most common use is in the early phases of campaign such as the movement inland where large columns with important strategic objectives in view may temporarily sever their connection with the base, seize the objective, and thereafter establish lines of communication. (See par. 5–8.)

○

SMALL WARS MANUAL
UNITED STATES MARINE CORPS
1940

✦

CHAPTER VI

INFANTRY PATROLS

RESTRICTED

UNITED STATES
GOVERNMENT PRINTING OFFICE
WASHINGTON : 1940

TABLE OF CONTENTS

The Small Wars Manual, U. S. Marine Corps, 1940, is published in 15 chapters as follows:

SMALL WARS MANUAL

UNITED STATES MARINE CORPS

CHAPTER VI

INFANTRY PATROLS

(v)

Section I

SMALL WAR TACTICS

6–1. **Tactics during initial phases.**—During the initial phases of intervention, when the landing and movement inland may be opposed by comparatively large, well led, organized, and equipped hostile forces, the tactics employed are generally those of a force of similar strength and composition engaged in major warfare. If a crushing defeat can be inflicted upon those forces, the immediate cessation of armed opposition may result. This is seldom achieved. Usually the hostile forces will withdraw as a body into the more remote parts of the country, or will be dispersed into numerous small groups which continue to oppose the occupation. Even though the recognized leaders may capitulate, subordinate commanders often refuse to abide by the terms of capitulation. Escaping to the hinterland, they assemble heterogeneous armed groups of patriotic soldiers, malcontents, notorious outlaws, and impressed civilians, and, by means of guerrilla warfare, continue to harass and oppose the intervening force in its attempt to restore peace and good order throughout the country as a whole.

6–2. **Tactics during later phases.**—To combat such action, the intervening force must resort to typical small war operations, with numerous infantry patrols and outposts dispersed over a wide area, in order to afford the maximum protection to the peaceful inhabitants of the country and to seek out and destroy the hostile groups. The tactics of such infantry patrols are basically the military methods, principles, and doctrines of minor tactics, as prescribed in the manuals pertaining to the combat principles of the units concerned. The majority of contacts in small wars is in the nature of ambushes, or surprise-meeting engagements, in which the various subdivisions of a small patrol may be brought almost simultaneously under the opening hostile fire. This prevents the normal development and deployment of the command for combat. In larger patrols, however, most of the main body may escape the initial burst of fire and consequently may be developed and deployed for combat from the march column in an orthodox manner.

6-3. **Influence of terrain.**—The tactics employed by patrols in combat over open terrain are, in general, the same as those in open warfare operations in a major war. Since open terrain is more advantageous to regular troops than to irregulars, the latter usually try to avoid combat under these conditions. As a result, infantry patrols engaged in the later phases of small wars operations generally must cope with the military problems encountered in combat in mountainous, wooded terrain, with the attendant limited visibility and lack of centralized control. These tactics are analogous to those prescribed in training manuals for combat in wooded areas in major warfare.

6-4. **The principle of the offensive.**—So long as there is armed opposition to the occupation, the intervening force must maintain the principle of the offensive. If it adopts a defensive attitude by garrisoning only the more important cities and towns without accompanying combat patrols throughout the theater of operations, minor opposition to the force will soon increase to alarming proportions. A guerrilla leader, if unmolested in his activities, creates the impression among the native population that the intervening forces are inferior to him; recruits flock to his standard, and the rapid pacification of the country will be jeopardized. Such hostile groups will seldom openly attack the regular garrisons, but will pillage defenseless towns, molest the peaceful citizenry, and interfere with the systems of supply and communication of the force of occupation. The latter must, therefore, adopt an aggressive attitude in order to seek out, capture, destroy, or disperse the hostile groups and drive them from the country. (See also Section II, Chapter I, "Psychology.")

6-5. **The principles of mass, movement, surprise, and security.**— *a. Mass.*—In nearly every engagement, the hostile groups will outnumber the infantry patrols opposed to them. This superiority in numbers must be overcome by increased fire power through the proper employment of better armament, superior training and morale, and development of the spirit of the offensive.

b. Movement.—Infantry patrols of the intervening force must develop mobility equal to that of the opposing forces. The guerrilla groups must be continually harassed by patrols working throughout the theater of operation.

SMALL WAR TACTICS

c. Surprise.—Surprise is achieved by varying the route, dates, and hours of departure of combat patrols, by mobility, and by stratagems and ruses. The intelligence system of the guerrillas decreases in proportion to the mobility and number of patrols employed in the theater of operations.

d. Security.—The tendency of the force to relax its service of security during the later phases of small war operations must be carefully guarded against. Security on the march and at rest must be constantly enforced throughout the entire period of occupation.

Section II

ORDERS AND GENERAL INSTRUCTIONS

6–6. Written orders.—Whenever possible, orders to a patrol leader should be issued in writing. This is especially true when several patrols are operating simultaneously in the same general area. The patrol leader must assure himself that he understands the orders issued to him. Subordinate leaders and the other members of the patrol should be thoroughly informed of such parts of the order as will enable them to carry out the mission of the particular patrol, and of the force as a whole. For the purpose of secrecy it is sometimes necessary to limit the information imparted to individual members of the patrol. Written orders follow the general form of a regular operation order.

6–7. Verbal orders.—Because of the nature of small war operations, verbal orders will be issued to patrol leaders more frequently than written orders. Such verbal orders should be as complete as the situation permits, and will follow the general form of an operations order. Patrol leaders should reduce to writing any verbal orders or verbal modifications of written orders received.

6–8. General instructions.—The force commander should publish, in the form of general instructions, the policies which will govern the action of patrols in the theater of operations in regard to the following:

a. Firing upon suspicious individuals or groups before being fired upon.

b. Firing upon guerrillas accompanied by women and children.

c. The seizure of property and foodstuffs for the benefit of the patrol or to prevent its use by hostile forces.

d. The destruction of houses.

e. The destruction of crops which may be of value to the hostile forces.

f. Other pertinent instructions regarding general policy.

Section III

ORGANIZING THE INFANTRY PATROL

6-9. **Definition.**—An infantry patrol is a detachment of infantry troops dispatched from a garrison, camp, or column with the mission of visiting designated areas for combat or for other purposes. It is a military unit disposed in such a manner that its various subdivisions are in suitable formations to engage the enemy immediately after contact is made. In general, the infantry patrol in a small war differs from one in a major war in the following respects:

a. It is larger.

b. It is more capable of independent action.

c. It operates at greater distances, in miles and hours of marching, from its base or supporting troops; a distance of 50 miles or more is not uncommon.

d. It conducts its operations for a longer period of time; missions of 10 days or more duration are not unusual.

ORGANIZING THE INFANTRY PATROL

e. It is often encumbered by a proportionately large combat train.

6–10. **Factors which govern its organization.**—Some of the factors that govern the size and composition of an infantry patrol in a small war are:

a. Mission.

b. Information of the hostile forces.

c. The probability of combat.

d. Strength and armament of the enemy.

e. Nature of the terrain, with particular reference to its effect on the formation and length of the column, the number of men required on service of security, and the work to be done, such as cutting trails.

f. Proximity of friendly troops.

g. Aviation support, including reconnaissance, liaison, combat support, transportation of supplies and personnel, evacuation of wounded.

h. Personnel available for assignment to the patrol, their efficiency and armament.

i. Native troops available, their efficiency and armament.

j. Native nonmilitary personnel available, such as guides, interpreters, and transportation personnel.

k. Time and distance involved.

l. Problem of supply.

m. Methods of communication.

The above factors are considered in the estimate of the situation which precedes the organization of any patrol.

6–11. **Size of the patrol.**—*a. General.*—The patrol should be large enough to defeat any enemy force that it can reasonably expect to encounter in the field. It should be able to assume the defensive and successfully withstand hostile attacks while awaiting reenforcement if it encounters enemy forces of unexpected strength. It is desirable to keep the patrol as small as is consistent with the accomplishment of its mission. The larger the patrol the more difficult its control in combat, the more complicated its supply problems, and the more it sacrifices in the way of concealment and secrecy of movement.

b. Effect of mission.—The mission assigned an infantry patrol in a small war, such as reconnaissance, security, liaison, convoy, and combat, is analogous to the corresponding mission in major warfare, and will affect the strength of the patrol. In some situations it will be desirable to have the patrol sufficiently large to establish a temporary or permanent base in the theater of operations from which it can maintain one or more combat patrols in the field.

ORGANIZING THE INFANTRY PATROL

c. *Effect of terrain.*—The nature of the terrain in which the patrol will operate has a marked influence on its size and composition. In fairly open country, with roads available which permit the use of normal distances within the column, a reenforced rifle company or larger organizations can operate with reasonable control and battle efficiency. In mountainous, wooded terrain, where the column must march in single file over narrow, winding trails, the reenforced rifle platoon with its combat train has been found to be the largest unit that can be controlled effectively on the march and in combat. It is the basic combat unit in the later phases of small war operations. If, in such terrain, the situation requires a stronger patrol than a reenforced rifle platoon, it is advisable to divide the column into combat groups equivalent to a platoon, marching over the same route and within supporting distance (5 to 15 minutes) of each other. Liaison should be established between the rear leading patrols and head of the following patrol during halts and at prearranged time intervals during the day's march.

6–12. **Permanent roving patrols.**—It is sometimes desirable to organize a few permanent combat patrols with roving commissions throughout the theater of operations, irrespective of area boundaries or other limitations. These patrols should be as lightly equipped as possible commensurate with their tasks. Authority should be granted them to secure from the nearest outpost or garrison such replacements of personnel, animals, equipment, and rations as may be required. Aviation is normally their main source of supply while in the field.

6–13. **Selection of units.**—*a. Permanent organizations.*—Whenever possible, an infantry patrol should be composed of personnel permanently assigned to organized units, such as a squad, platoon, or company. This applies also to attached machine-gun units or other supporting weapons.

b. *Hastily organized patrols.*—In the rapidly changing situations encountered in wars, the operations may require the simultaneous movement of more patrols than can be furnished by a single organization. In some instances, two or more units from different posts will be combined into a single patrol for an emergency operation. Other situations will require that supply train escorts and special duty men be relieved and made available for patrol duty. The result of this pressing need for men is the intermingling of personnel from several different organizations, whose individual combat efficiency is unknown to the patrol leaders, or to one another. Although such hastily organized patrols should be avoided whenever possible, they are often necessary.

ORGANIZING THE INFANTRY PATROL

6–14. **Elimination of the physically unfit.**—Men who are physically unfit for duty in the field or whose presence would hinder the operations of a patrol should be eliminated from the organization. They include the following:

a. Those who have been recently ill, and especially those who have recently had malaria, dysentery, jaundice, or a venereal disease.

b. Those suffering from deformities or diseases of the feet, particularly flatfoot, hammertoes, bunions, corns, or severe trichophytosis (athlete's foot).

c. The old or fat, or those of obviously poor physique from any cause.

d. The neurotic or mentally unstable; and the alcohol addicts.

6–15. **Patrol and subordinate leader.**—*a.* Officers and noncommissioned officers assigned to the theater of active operations in small wars will generally command smaller elements than those assigned to them in major warfare, for the following reasons:

(1) A patrol on an independent mission is usually far removed from the direction and control of more experienced superiors.

(2) The suddenness with which action may break, and the necessity for rapid and practical employment of all the small elements in the patrol. An officer or experienced noncommissioned officer should be with each small group to facilitate its control during combat. This is especially true in wooded terrain where the limited visibility and short battle ranges usually restrict the patrol leader's control over the situation to his immediate vicinity.

(3) The possible dispersion of the troops in column at the moment of contact, and in the subsequent attack and assault.

(4) The possibility that the troops are not thoroughly trained.

b. Two commissioned officers should accompany every rifle platoon assigned to an independent combat mission. If this cannot be done, the second in command must be an experienced, capable, senior noncommissioned officer who is in addition to the regular complement. This requirement is necessary to insure a continuity of effort in the event the patrol leader becomes a casualty. The normal complement of officers is usually sufficient if the combat patrol consists of two or more rifle platoons combined under one commander.

6–16. **The rifle squad.**—Wherever possible, the rifle squad is employed in small wars in the same manner as in major warfare. In many situations in small war operations, however, it will be desirable to divide the squad into two combat teams of four or five men each, one of which is commanded by the corporal, the other by the second in command. Such combat teams can be profitably employed as the point for a combat patrol in close country, as flank patrols, and for reconnaissance or other security missions. In thickly wooded terrain, it is often impossible for the corporal to maintain control over the entire squad in combat. Under such conditions, the two combat teams must fight as independent units until the situation or better visibility permits the corporal to regain direction and control of the squad as a whole. Automatic and special weapons within the squad should be equally divided between the combat teams.

6–17. **The headquarters section.**—*a.* The headquarters section of a combat patrol, consisting of a rifle platoon or reenforced rifle platoon, must be augmented by certain personnel who are not organically assigned to it. Such personnel includes one or more competent cooks, a medical officer or one or more qualified hospital corpsmen, and a radioman when the patrol is equipped with a portable radio.

b. If the hostile forces are not complying with the "Rules of Land Warfare," the medical personnel should be armed for self-defense.

6–18. **Attached units.**—In the future, most combat patrols of the strength of a rifle platoon or more, operating in hostile areas, probably will be reenforced by attached supporting weapons. With the adoption of the semi-automatic rifle as the standard infantry arm or as a replacement for the Browning automatic rifle, a light machine gun squad or section and a 60-mm. mortar squad or section would appear to be appropriate units to accompany a rifle platoon assigned a combat mission. These organized units should be attached to the platoon from the headquarters platoon of the rifle company. A combat patrol consisting of a rifle company may require the support of a heavy machine gun section or platoon and an 81-mm. mortar squad or section. These should be attached to the company as intact units from the appropriate organizations of the battalion or regiment. (For further details, see Section III, Chapter II, "Organization.")

6–19. **Guides and interpreters.**—*a.* Native officials and foreign residents are usually helpful in securing reliable guides and interpreters whenever their employment is necessary. Local inhabitants who have suffered injury from the hostile forces and those having members of their families who have so suffered, often volunteer their services for such duty. The integrity of these men must be tested in the field before they can be considered entirely reliable and trustworthy. In many cases, their employment in any capacity makes them subject to hostile reprisal measures and the intervening force must assume responsibility for their protection.

ORGANIZING THE INFANTRY PATROL

b. **Troops** assigned to combat operations should learn the terrain and trails within their sectors, and gain a working knowledge of the local language as quickly as possible so that they may dispense with the employment of native guides and interpreters insofar as the situation permits.

6-20. **Native transport personnel.**—In most situations, the employment of native porters (carriers), muleteers, or other transport personnel will be required with each combat patrol. For further details, see Chapter III, "Logistics."

6-21. **Native troops.**—*a.* When native troops are available, they may be included in the patrol In addition to their combat duties, they will, if properly indoctrinated, do much to establish friendly relations between the peaceful inhabitants and the intervening force.

b. Native troops are especially valuable for reconnaissance and security missions. They will notice and correctly interpret those signs which indicate the presence of the enemy much more quickly and surely than will the average member of a foreign force unaccustomed to the country.

c. Work and guard duty must be divided and distributed proportionally between the regular forces and native troops, and friction between the two organizations must be avoided.

6-22. **Prominent native civilians.**—*a.* It is sometimes advisable to include prominent native civilians or government officials in the patrol. They can do much to explain the mission of the intervening forces in the community, spread the gospel of peace, friendly relations, and cooperation, and counter the propaganda of the enemy. The natives of the community are all potential enemies and many will become actively hostile if they are not convinced of the true objective of the occupation.

b. If political alignments and hatreds are virile in the area, the patrol leader must be very circumspect in the choice of civilians and government officials who accompany the patrol. If the patrol is suspected of political partisanship, the problems of pacification may be intensified.

c. Frequently prominent and well-informed civilians will furnish valuable information, provided their identity is not disclosed and they are not required to act as guides or otherwise openly associate themselves with the intervening force. Their wishes should be respected in order to gain their confidence and obtain the information which they possess.

ORGANIZING THE INFANTRY PATROL

6-23. **Transportation.**—*a.* The means and amount of transportation included in an infantry patrol will influence its composition, its mobility, the length of time that it can stay away from its base, and its combat efficiency. In general, infantry patrols should carry only the minimum equipment and supplies necessary to accomplish their mission. The more nearly they can approach the hostile forces in this respect, the more efficient they will become in the field. It is a common failing for troops engaged in small war operations to decrease their mobility by transporting too much equipment and too many varied, desirable but nonessential supplies.

b. The principal means of transportation employed by infantry patrols include:

(1) All or part of the equipment and supplies carried on the person.

(2) Native porters.

(3) Riding and pack animals.

(4) Airplanes for evacuation of the wounded and supply by plane drops.

(5) Motor transport.

c. In hot, tropical climates, the personnel should not be required to carry packs if it can be avoided. The weight of the rations which troops can transport in addition to their equipment will limit the range of a combat patrol unless it can subsist almost entirely off the country. On the other hand, a reconnaissance patrol whose members are inured to the local fare can often accomplish its mission more successfully if it is not encumbered with a train.

d. For further details concerning transportation, see Chapter III, "Logistics," and Chapter IX, "Aviation."

6-24. **Weapons.**—*a.* The weapons carried by an infantry patrol will normally be those organically assigned to the squad, platoon, or company, plus attached units of supporting weapons if the situation indicates the necessity therefor.

b. If the rifle units are completely equipped with the semi-automatic rifle, the inclusion of any full shoulder weapon in each squad is not warranted. If the basic arm in the patrol is the bolt-action rifle, the armament of each squad should include two semi-automatic, or two Browning automatic rifles, or one of each. This proportion of automatic shoulder weapons to bolt-action rifles should rarely, if ever, be exceeded. Ammunition supply in small wars operations is a difficult problem. Volume of fire can seldom replace accuracy of fire in a small war. The morale of guerrilla forces is little affected by the loss of a particular position, but it is seriously

ORGANIZING THE INFANTRY PATROL

affected by the number of casualties sustained in combat. The majority of the personnel in an infantry patrol should be armed, therefore, with weapons that are capable of delivering deliberate, aimed, acurate fire rather than with weapons whose chief characteristic is the delivery of a great volume of fire. The automatic weapons should be utilized to protect the exposed flanks, or to silence hostile automatic weapons.

c. Whether or not the bayonet is included in the armament of the patrol depends upon the terrain, the nature of the particular operation, the training of the men, and the opinion of the patrol leader. In jungle terrain, the bayonet impedes the movement of the individuals both on the march and when deployed for combat by snagging on vines and the dense underbrush; it is doubtful if it can be used effectively, even in the assault, in such terrain. In fairly open country, the bayonet should be carried and employed as in regular warfare. It is an essential weapon in night attacks. The bayonet is practically useless in the hands of untrained troops who have no confidence in it; it is a very effective weapon in small war operations when employed by troops who have been thoroughly trained in its use.

d. For further details regarding infantry weapons, see Section III, Chapter II, "Organization."

6-25. **Ammunition.**—*a.* In past small war operations, the average expenditure of small arms ammunition for a single engagement has seldom exceeded 50 rounds for each person in the patrol. There have been a very few instances where the expenditure has slightly exceeded 100 rounds per person. It is believed that the following is a reasonable basis for the quantities of ammunition to be carried for each type of weapon with infantry patrols assigned a combat mission in small war operations:

(1) *On the person*—the full capacity of the belt or other carrier issued to the individual.

(2) *In the combat train*—½ unit of fire.

These quantities should be modified as dictated by experience or as indicated by the situation confronting a particular patrol.

b. Emergency replacements of some types of ammunition can be dropped by plane.

c. If the regular ammunition containers are too heavy for the means of transportation in the combat train, the ammunition is repacked and the individual loads made lighter.

d. Cartridge belts and other carriers with the patrol must be in perfect condition to prevent the loss of ammunition.

6–26. **Signal equipment.**—*a.* The following signal equipment must be taken with every patrol:

(1) Airplane panels, Codes, and pick-up equipment.

(2) Pyrotechnics.

b. The following signal equipment should be carried with the patrol when it is available:

(1) Portable radio.

(2) Other special equipment demanded by the situation or the use of which can be foreseen. (See Section III, Chapter II, "Organization.")

6–27. **Medical supplies.**—*a.* The patrol leader, in conjunction with the medical personnel, must assure himself of the sufficiency of his medical equipment and supplies. If charity medical work among the native inhabitants is anticipated, additional supplies must be provided for that purpose.

b. Besides the regular medical kit carried by the hospital corpsman, reserve supplies should be made up into several assorted kits distributed throughout the column.

c. Sufficient ampoules should be carried for chlorination of water for the duration of the patrol. The Lyster bag, if carried, should be carefully inspected for leaks, particularly at the taps, and should be cleaned and dried. Four to six yards of muslin for straining trash from the water should be provided. The bag should be rolled and stowed so that it will not chafe in carrying.

d. A few "sanitubes" should be carried for prophylaxis and for the treatment of certain skin diseases.

e. Several additional first aid packets, tubes of iodine, and a small roll of adhesive should be carried with patrols to which medical personnel is not attached.

f. Preparations to carry the wounded must be made before the patrol leaves the garrison. In addition to the methods described in Chapter 14, "Landing Force Manual," USN, the canvas field cot cover is easily carried and can be quickly converted into a stretcher in the field.

6–28. **Miscellaneous equipment.**—Such of the following articles as may be necessary should be carried with the patrol:

a. Native machetes, for cutting trails, forage, firewood, fields of fire, and material for bivouac shelters.

b. Matches in waterproof containers, flashlights, candles, and a lantern for the mess force.

c. A quantity of hemp rope to assist the patrol in crossing dangerous streams. It can be stretched across the stream and used as a hand hold while crossing, or it can be used in building an improvised raft.

d. Entrenching tools or larger engineering tools as demanded by the situation.

e. A horse-shoer's kit, if the number of animals with the patrol makes it advisable.

6-29. **Personal clothing and accessories.**—*a. General.*—The personal clothing and accessories worn or carried by the patrol must be reduced to the minimum consistent with the length of time the patrol will be absent from its base, and the climate and season of the year. Clothing should be in good condition when the patrol leaves it base. It is better to rely on airplane supply for necessary replacements in the field than to overburden the patrol with too much impedimenta. Superfluous articles will increase the transportation problems, and decrease the quantities of essential ammunition and rations which can be carried. Personal comfort and appearance must always be of secondary importance as compared with the efficient accomplishment of the assigned mission. Officers should fare no better in these respects than the enlisted men of the organization. The inclusion of officers' bedding rolls, field cots, and similar equipment is unwarranted in patrols operating from a base in the theater of operations.

b. *Clothing worn by troops.*—Shoes should fit properly, be broken in, and in good condition. New shoes, though of the correct size, will usually give trouble on the march. Socks should be clean, free from holes and darns. Flannel shirts, which absorb perspiration, rain, and water freely, and still afford warmth and protection at night, are preferable to cotton khaki shirts even in the tropics. The scarf should not be worn. The value of canvas leggings in the field is questionable. The woolen sock pulled over the bottom of the trouser leg is a satisfactory substitute.

c. *Clothing and accessories carried in the pack.*—The following articles are considered reasonable quantities to be carried in the pack or roll of each individual with a patrol operating in a warm climate:

(1) A shelter-half, poncho, or light native hammock, depending upon the nature of the terrain, the season of the year, and the personal decision of the patrol leader. The shelter-half can be dispensed with if materials are available in the field for the construction of lean-to shelters. In this case, the poncho is utilized as a cover for the pack or roll. The poncho is primarily useful as protection

ORGANIZING THE INFANTRY PATROL

from the damp ground while sleeping at night. It interferes with movement of an individual if worn on the march, and is a distinct impediment if worn in combat. The hammock has many advantages, but it is bulky and adds considerable weight to the pack or roll. During the rainy season, two of these articles may be desirable. If the shelter-half is carried, the tent pole and pins are included when necessary.

(2) One blanket.

(3) A mosquito net is desirable in malarious countries. It is bulky and quite heavy. Some combat patrols in past operations in tropical countries did not carry the net in the field and did not incur any apparent harmful consequences.

(4) One change of underwear.

(5) At least two pairs of woolen socks; four pairs are recommended, if the patrol is to operate for 2 weeks or longer.

(6) One change of outer clothing.

(7) Toilet articles: soap, small bath towel, tooth brush and powder or paste, comb, and mirror. A razor, shaving brush, and shaving soap may be carried, although they are not considered essential items.

(8) Tobacco, as desired.

(9) Toilet paper, a small quantity to be carried by each individual, the remainder with the mess equipment.

d. Personal cleanliness.—A bath should be taken and soiled clothing should be washed as frequently as opportunity affords. Simply soaking clothes in water, wringing them out, and permitting them to dry in the sun, is better than not washing them at all.

6–30. **General preparations.**—Prior to clearing its base, the patrol leader of an infantry patrol personally verifies or arranges for such of the following as may be pertinent to the particular situation:

a. Aviation; support.—

(1) Liaison, reconnaissance, and combat support.

(2) Regular and emergency supply by plane.

b. Information.—

(1) A personal airplane reconnaissance over the area, if practicable.

(2) A map or sketch of the area, including the route or alternate routes to be followed. A rude sketch, however, inaccurate, is better than none.

(3) Airplane photographs of villages and important terrain features, such as stream crossings, possible or former ambush positions, etc., if practicable.

ORGANIZING THE INFANTRY PATROL

(4) The condition of the roads and trails, the attitude of the local inhabitants, and the possible food supply.

c. Inspection of.—

(1) Men; individual, combat, communications, and medical supplies and equipment; and animals, pack, and riding equipment.

(2) Cleaning materials for the weapons, especially oil for automatic arms.

d. Liaison with.—

(1) Native officials, when desirable.

(2) Native troops, or other persons not of the command, who are to accompany the patrol.

(3) Other friendly patrols operating in the area.

*e. Employment of.—*Native transportation personnel, intelligence agents, guides, and interpreters.

f. Money, in small denominations, for the purchase of supplies, emergency transportation, and information. In some countries, articles such as soap, salt, tobacco, etc., which are expensive and difficult to obtain locally, are more acceptable to the natives than money.

Section IV

FEEDING THE PERSONNEL

6–31. **Responsibility of patrol leader.**—The patrol leader should confer with the mess officer at the garrison or base from which the patrol will operate, and, in conjunction with the patrol's mess sergeant or cooks, determine what suitable foodstuffs are available for the patrol. Also, he must decide what kitchen equipment is required and procure it. Written menus for breakfast and supper for each day of the proposed operations are prepared. It is not desirable to make midday halts for the purpose of cooking a meal, although it may be desirable in some situations to prepare cold lunches which may be issued to the men prior to breaking camp in the morning. Based on these menus, a check-off list of the necessary rations is prepared, the rations drawn and carefully verified before loading. Thereafter, the rations are issued as required and notations made on the check-off list. The rations remaining in the train should be inventoried periodically while the patrol is in the field. Canned goods should be inspected for swelling of the top due to deterioration of the contents, for leaks, and for bad dents. Such cans should be rejected, or destroyed.

6–32. **Mess equipment.**—*a.* The amount of mess equipment carried by the patrol should be reduced to a minimum.

b. The cavalry pack kitchen is satisfactory for a large patrol which includes pack animals. The complete unit less hangers for the Phillips pack saddle, weighs 118 pounds and constitutes one pack load. It is adequate for feeding 200 men in the field. Patrols of between fifty and one hundred men can eliminate unnecessary pieces. It is questionable whether a patrol of less than 50 men should carry it.

c. If a regular pack kitchen unit is not used, issue or improvised cooking equipment will have to be provided. The following points are pertinent:

FEEDING THE PERSONNEL

(1) Although G. I. buckets are seldom used for cooking in garrison, they are useful for that purpose in the field. They can be set on a fire or suspended over it. They nest well and do not rattle if leaves or similar materials are packed between them.

(2) In comparison with tin boilers, buckets, and roasting pans, large iron kettles are not so fragile, do not burn food so quickly, hold heat better, can be used for frying, and pack better. Packed, one on each side of the animal, they can carry the cooked or uncooked foodstuffs necessary for the evening meal, thus expediting its preparation. Suitable iron kettles can generally be purchased locally in the theater of operations.

(3) A small metal grill about 2 feet square and fitted with four collapsible legs facilitates cooking in the field.

d. During rainy weather or in areas where many streams have to be forded, some provision must be made to protect such foodstuffs as sugar, salt, flour, coffee, etc. Bags made of canvas, leather, or of canvas material coated with rubber, and tarpaulins or pieces of canvas, have been used successfully in the past.

e. When a patrol is to be made into unfamiliar country where the existence of an adequate water supply is doubtful, drinking water may have to be transported in the train. 5-gallon cans may be used for this purpose in the absence of specially designed equipment.

f. A limited amount of soap should be carried as an aid in cleansing the cooking equipment.

6-33. Weight of rations.—*a.* The aggregate weight of the rations carried by a patrol is influenced by:

(1) Number of men in the patrol.

(2) Native foodstuffs available in the field.

(3) Issue foodstuffs available.

(4) Rations to be supplied by plane drop.

(5) Replenishments expected from outposts and other garrisons in the area.

(6) The ability of the cooks.

(7) The ability of the personnel to adjust themselves to diminished rations.

(8) The method of transport and the predetermined size of the combat train.

b. The normal field ration weighs approximately 3 pounds. The normal garrison ration weights about 4 pounds. The average pack animal found in most small-wars countries can carry 30 man-day garrison rations, computed on the assumption that no foodstuffs can be procured in the field, 40 man-day complete field rations, or 50 reduced field rations.

6-34. **The field ration.**—*a.* Every effort should be made to build up the supply of rations at the advanced patrol bases and outposts until they approach or equal the normal garrison ration in quantity and variety. A patrol operating from those bases, should never carry more, and may often carry less, than the components of the field ration, modified in accordance with the probable foodstuffs which can be obtained in the area. Emphasis should be placed on those articles which give the greatest return in food value for the bulk and weight carried, and the ease with which they can be transported. This may not result in a "balanced" ration, but the deficiencies encountered in the field can be compensated for upon the return of the patrol to its base. The general tendency of troops is to carry too great a variety and too large a quantity of foodstuffs with patrols in the field. Man should become accustomed to the native fare as quickly as possible. If properly led, they will soon learn that they can subsist quite well and operate efficiently on much less than the regular garrison ration. This is a matter of training and is influenced in a large measure by the attitude of the patrol leaders and other commissioned and noncommissioned officers.

b. The prescribed field ration is approximately as follows:

Component articles	*Substitute articles*
1 pound hard bread	1¼ fresh bread, or
	1⅛ pounds flour.
1 pound tinned	1¼ pounds salt meat, or
	1¼ pounds smoked meat, or
	1¾ pounds fresh meat, or
	1¾ pounds fresh fish, or
	1¾ pounds poultry.
¾ pound tinned vegetables	1¾ pounds fresh vegetables, or
	3 gills beans or peas, or
	½ pound rice or other cereal.
2 ounces coffee	2 ounces cocoa, or
	½ ounce tea.
1 ounce evaporated milk	1/16 quart fresh milk.
Salt and pepper.	

c. Suitable foodstuffs from the regular issue include: rice, rolled oats, hominy grits, dry beans, canned pork and beans, corned beef hash, salmon, corned beef, chipped beef, bacon, Vienna sausage, hard bread, dried fruits, cheese, sugar, coffee, tea, evaporated or dried milk, salt, black pepper, and limited amounts of canned potatoes and vegetables. In general, canned and fresh fruits should not be carried.

FEEDING THE PERSONNEL

Small sized cans are usually preferable to the larger sizes for issue to patrols. Generally a combat patrol should carry such foodstuffs that not more than one component, other than tea or coffee, requires cooking for each meal in order to reduce the number of cooking utensils to be carried and the time of preparation in the field.

d. Native foodstuffs sometimes found in inhabited areas include: beef on the hoof, fish, chickens, eggs, beans, rice, corn, coffee, and fruits and vegetables in season. To these may be added such wild game as may be killed by the patrol. If hostile groups are active in the area, the available supply of native food will be limited.

6–35. **Butchering on the march.**—*a.* Each patrol operating in the field should include a man familiar with the killing and dressing of livestock and game. If the patrol is dependent upon the country for its meat supply, suitable stock should be procured during the day's march unless it is definitely known that the desired animals will be available at or near the bivouac.

b. The animal should be butchered in such a manner that it will bleed profusely. It should be dressed, cut-up, and cooked while it is still warm. Meat cooked after rigor mortis has set in will be tough unless it is cooked in a solution of vinegar or acetic acid, or allowed to season for at least 24 hours. Excess beef may be barbecued and utilized the following day.

6–36. **Feeding native personnel.**—Native personnel attached to patrols may provide their own food and cooking arrangements. In certain situations they may be given a cash allowance which will permit them to eat with the local inhabitants. When circumstances require them to subsist with the patrol, they should receive their proportionate share of the available food. If the patrol is living off the country, equitable treatment given to the natives attached to the patrol will usually be more than repaid by their foraging ability and by assistance in preparing palatable dishes out of the foodstuffs which are indigenous to the locality.

6–37. **Emergency rations.**—Either a specially prepared, commercial emergency ration, or one composed of available materials, should be issued to each individual and carried on the person at all times while operating in the field. This ration should be eaten only on the orders of a responsible commander, or as a last resort if an individual becomes separated from his patrol. Frequent inspections should be made to insure troops are complying with these instructions.

Section V

THE MARCH

6–38. **General.**—The conduct of marches will vary considerably with the condition of the men, their state of training, the condition of the roads or trails, the climate, the weather, the tactical situation, and various other factors. Whenever it can be avoided, the men should not arrive at their destination in a state of exhaustion.

6–39. **Hour of starting.**—In small wars, breakfast usually should be served at dawn, animals fed and watered, camp broken, packs assembled and loaded, and the march begun as soon after daylight as possible. The march should begin slowly in order to warm up the men and animals, and to permit packs and equipment to settle and adjust themselves to both personnel and animals.

6–40. **Rate of march.**—*a.* The first halt should be made not later than three-quarters of an hour after the start, and should be of about 15 minutes duration, so that the men can adjust their equipment, check and tighten the pack loads in the train, and attend to the calls of nature.

b. Under normal conditions, troops usually halt 10 minutes every hour after the first halt. This cannot be accepted as doctrine in

THE MARCH

small wars operations, in which the rate of march is dependent upon the state of training and efficiency within the combat train. The column must be kept closed up at all times. Liaison should be constantly maintained throughout the column by word of mouth. Whenever a pack needs readjustment, or an animal becomes bogged in some mudhole, or any other delay occurs within the column, a halt should be called until the defect is remedied and the patrol ready to move forward as a body. If the regulation 50 minute march, 10 minute halt schedule is maintained, even a small patrol may become so elongated that several miles will separate the head and tail of the column at the end of the day's march. As the men become trained in such operations, forced halts will become more infrequent and of shorter duration, and the normal march schedule may be achieved. To avoid disaster, however, it is essential that liaison be maintained throughout the entire length of the column at all times, regardless of the frequency of the halts.

c. Under normal conditions, intervals of marching should be modified to take advantage of good halting places, especially those which afford proper security to the column.

6–41. **Factors influencing march formations.**—The march formation of a patrol in small wars is influenced by the following factors:

(1) The nature of the terrain.

(2) The strength, composition, and armament of the patrol.

(3) The size of the combat train.

(4) The necessity for security, observing the principle that security elements should increase proportionately in strength from the point to the main body.

(5) Ability to shift rapidly and automatically from a column to a line formation that will face the enemy, cognizance being taken of the possibility of the enemy being in several different directions.

(6) The necessity for dividing the patrol into small, mutually supporting, maneuver units, each one capable of developing its offensive power independently and immediately at short, battle ranges.

(7) Sufficient distance between elements to enable one or more units in the main body to escape the initial burst of hostile fire, thus assuring some freedom of maneuver.

(8) The distribution of supporting weapons throughout the column to facilitate their entry into action in any direction.

(9) The rapid development of maximum fire power.

(10) The necessity of withholding an initial reserve.

(11) The degree of darkness during night marches.

THE MARCH

6–42. **Influence of terrain on march formation.**—*a. Open terrain.*—In open country, the distribution of the troops in the column, and the distances between the various elements, will be similar to that employed by a force of comparable strength in major warfare.

b. Close terrain.—(1) In the mountainous, heavily wooded terrain in which the majority of small war operations occur, patrols are usually forced to march in a column of files. Underbrush encroaches upon the trails, which are narrow and tortuous, and visibility is often limited in every direction to only a few yards. As a result, the column is greatly elongated, the distances between security elements and the main body are reduced, and the patrol leader can personally see and control only a small portion of his command.

(2) There should be sufficient distance between subdivisions in the column to avoid the intermingling of units, to fix in the minds of each individual the maneuver unit to which he is attached, and to subject as few men as possible to the initial bursts of hostile fire delivered at short ranges. The distance between units should be sufficient to enable one or more of them to get free to maneuver, thus creating an opening for the employment of fire and movement. On the other hand, the various elements in the column must be mutually supporting as too much distance between them may enable an aggressive enemy to defeat the patrol in detail.

6–43. **Road spaces.**—*a.* Depending on the prevailing conditions, the distances between men within subdivisions of a patrol operating in thickly wooded terrain generally will be about as follows:

Subdivision	Distance between men
Point	10 to 40 yards.
Advance party	5 to 20 yards.
Support	3 to 10 yards.
Main body	2 to 5 yards.
Rear guard	2 to 20 yards.

b. The distances between the various subdivisions in the column will vary from 10 to 50 yards or more, depending upon the strength of the patrol and the nature of the terrain through which it is marching.

c. The distances given above should never be considered as fixed and immutable. They usually will be changed several times during a day's march on the orders of the patrol and subordinate leaders as required by the nature of the country.

d. The road space for a riding or pack animal is considered to be 5 yards. This includes the man assigned to ride, lead, drive, or guard the animal.

THE MARCH

6-44. Location of patrol and subordinate leaders in march formation.—*a. Patrol leader.*—The patrol leader will usually march with or at the head of the main body. This is particularly desirable in the case of large patrols. In small patrols, the leader may have to alternate with a subordinate as commander of the advance guard. The leader of the patrol should not make a practice of marching in the point unless necessity requires it. If he is at the head of the main body, he can always move forward to the point to indicate the route to be followed or to make some other important decision which cannot be assumed by the advance guard commander.

b. Subordinate leaders.—Subordinate leaders of all elements in the patrol, except the point, normally march at or near the head of their respective units or subdivisions. The point commander should march near the center of that group so that he may effectively control all of the men in the point. Leaders of supporting units, such as a machine-gun section or platoon, normally march close to the patrol commander.

6-45. Location of the combat train.—The location of the combat train in the column depends upon several factors. These include the strength of the patrol, the probability of combat, the normal tactics of the enemy, and the size of the train itself. Normally, the combat trains should follow the main body, preceding the rear guard of the column. If, as is often the case in small wars, attack may be expected from any direction, it may be advisable to place the combat train near the center of the column, or to split it into two or more sections interspersed with elements of the main body of troops. If the train is exceptionally large, it may be detached from the combat elements of the patrol and marched as a separate convoy (see ch. VIII, "Convoys and Convoy Escorts"). Whatever its location in the column, the reserve supply of ammunition should be distributed throughout the train so that some of it may reasonably be expected to escape the initial burst of hostile fire in the event of ambush.

6-46. Descriptive march formations.—*a. General.*—The march formations described in the next three succeeding paragraphs illustrate several of the principles previously described in this chapter. They should not be considered as the only formations which organizations of comparable size and composition may adopt. It is believed that they will be effective under the conditions assumed. Every experienced patrol leader will have his own opinion of how his patrol should be organized. He should not hesitate to modify the formation or redistribute the personnel of his command to meet the particular situation which confronts him.

b. **Assumptions as to terrain.**—In each instance, the terrain in which the following patrols are operating is assumed to be mountainous, heavily wooded country, with only narrow, winding trails available.

6–47. **March formations for a reenforced rifle company.**—*a. Situation.*—A reenforced rifle company consisting of: a headquarters platoon which includes a light machine-gun section (4 Browning automatic rifles, modified), a 60-mm. mortar section (2 60-mm. mortars), attached signal and medical enlisted personnel, a native guide, and an interpreter; 3 rifle platoons of 3 squads each, armed with semi-automatic rifles; an attached machine platoon(less 1 section) with 4 machine guns (2 of which are for defense only); an attached 81-mm. mortar section; an attached squad of native troops; and a combat train of 75 pack animals and 20 native muleteers; has been ordered to proceed to an outlying village to establish an advanced base and conduct further patrol activities therefrom. The village is 3 days march from the point of departure. The total strength of the patrol is 220 officers and men and 30 native soldiers and civilians. The road space for the patrol in column of files is estimated at 1,140 yards, of which the combat train (less 6 miles carrying organic machine-gun and 81-mm. mortar equipment), will occupy 350 yards. A hostile guerrilla force estimated at 600 men has been active in the area which must be traversed. That force is well led, and armed with bolt action rifles, automatic shoulder weapons, and some machine guns. In previous engagements, the enemy has attempted to ambush the leading elements of the main body, but there has been one instance in which he created a diversion at the head of the column and directed his main attack at the rear elements.

b. Formation "A."—

Element	Composition
Point	1 rifle sqd. plus ¼ sqd. native troops commanded by a Sgt.
Distance	
Advanced party	1 rifle plat. (less 1 sqd.) Lt. MG sect. (less 1 sqd.) 60-mm. sect. (less 1 sqd.) Commander by Lieut. "Rifle Plat."

THE MARCH

Element	Composition
Distance	
Main body	Patrol commander.
	Native guide.
	Native interpreter.
	Fwd. esch., Co. Hdqtrs.
	1 rifle plat.
	1 MG plat. (less 1 sect.)
	1 81-mm. sect.
Distance	
Combat train and train guard	Rear esch., Co. Hdqtrs.
	Supply personnel and ammunition sqd. from attached units.
	½ sqd. native troops.
	Commanded by Lt. "2d in command."
Distance	
Rear party	1 rifle plat. (less 1 sqd.)
	Lt. MG sqd.
	60-mm. sqd.
	Commanded by Lieut. "Rifle Plat."
Distance	
Rear point	1 rifle sqd.
	Commanded by a Sgt.

c. Formation "B."—

Element	Composition
Point	½ sqd.
	½ sqd. native troops
	Commanded by plat. Lt.
Distance	
1st section of main body	Patrol commander
	Native guide
	Native interpreter
	Fwd. esch., Co. Hdqtrs.
	1 rifle plat. (less 1 sqd.)
	Lt. MG Sect. (less 1 sqd.)
	60-mm. sect. (less 1 sqd.)
Distance	
Combat train and train guard	Approximately ⅓ combat train
	Rear esch., Co. Hdqtrs.
	½ sqd. native troops
	Commanded by Sgt.
Distance	
Rear point	½ rifle sqd.
	Commanded by Sgt.

THE MARCH

5 minutes marching distance

Element	Composition
Point	½ rifle sqd.
	Commanded by Sgt.
Distance	
2d section of main body	1 rifle plat. (less 1 sqd.)
	1 MG plat. (less 1 sect.)
	1 81-mm. sect. (less 1 sqd.)
	Commanded by Lieut. "Rifle Plat."
Distance	
Combat train and train guard	Ap. ⅓ combat train.
	MG ammunition supply personnel
	Commanded by Sgt.
Distance	
Rear point	½ rifle sqd.
	Commanded by Sgt.

5 minutes marching distance

Point	½ rifle sqd.
	Commanded by Sgt.
Distance	
3d section of main body	1 rifle plat. (less 1 sqd.)
	Lt. MG sqd.
	60-mm. sqd.
	81-mm. sqd.
	Commanded by Lieut. "Rifle Plat."
Distance	
Combat train and train guard	App. ⅓ combat train
	81-mm. ammunition sqd.
	Commanded by Lt. "2d in command"
Distance	
Rear point	½ rifle sqd.
	Commanded by Sgt.

NOTE.—Contact between subdivisions of the patrol is established once each hour as follows: 1st section halts. 2d section makes contact and halts. As 3d section gains contact with 2d section, word is passed forward to 1st section, which resumes march, followed at 5-minute intervals by 2d and 3d sections.

6–48. **March formation for a reenforced rifle platoon.**—*a. Situation.*—It is assumed that a reenforced rifle platoon, consisting of: One rifle platoon of three squads armed with semi-automatic rifles; one light machine section (4 Browning automatic rifles, modified); one 60-mm. mortar squad; an officer, a cook, and a hospital corpsman from company headquarters; a native guide; a native interpreter; and a combat train of 15 pack mules, 1 riding mule for wounded, and 4 native muleteers; has been ordered to proceed from its base for a 10-day combat patrol missions. The total strength of the patrol

THE MARCH

is 57 officers and enlisted men, and 6 natives.   Hostile guerrillas have been active in the vicinity.

b. Patrol formation.—

Element	Composition
Point	½ rifle sqd.
	Commanded by a Sgt.
Distance	
Main body	Patrol commander.
	Native guide.
	Native interpreter.
	Fwd. esch. Plat. Hdqtrs.
	1 rifle plat. (less 1 sqd.).
	Lt. MG sect. (less 1 sqd.).
	60-mm. sqd.
Distance	
Combat train and train guard	Rear esch. Plat. Hdqtrs.
	Lt. MG sqd.
	Commanded by Lt. "2d in command."
Distance	
Rear point	½ rifle sqd.
	Commanded by a Sgt.

6–49. **March formation for a rifle platoon.—***a. Situation.—*It is assumed that a rifle platoon consisting of three squads, each armed with bolt action rifles and two Browning automatic rifles or semiautomatic rifles; an officer, a cook, and a hospital corpsman from company headquarters; a native interpreter; and a combat train of 10 pack mules and 1 riding mule (for wounded) and 3 native muleteers; has been ordered to proceed from its base on a 10-day patrol into a section in which hostile guerrillas are known to be operating.   The total strength of the patrol is 33 officers and enlisted men, and 4 natives.

b. Patrol formation.—

Element	Composition
Point	½ rifle sqd.
	Commanded by a Sgt.
Distance	
Main body	Patrol commander.
	Native interpreter.
	Fwd. esch. Plat. Hdqtrs.
	1 rifle plat. (less 1 sqd.).
Distance	
Combat train and train guard	Rear esch. Plat. Hdqtrs.
	Commanded by Sgt.
Distance	
Rear point	½ rifle sqd.
	Commanded by Lt. "2d in command."

6–50. **March formation for a rifle squad.**—A rifle squad should rarely, if ever, be sent as a patrol on a combat mission. Its normal employment in small wars, as in a major war, is that of reconnaissance, security, or liaison. The duration of the patrol will seldom exceed 1 day's march. If it extends over 1 day, it will usually subsist off the country and should not be encumbered with a train. It may often be mounted, in which case the riding animals will carry the necessary impedimenta. The formations of a squad acting as an independent patrol are basically those prescribed in FM 21–45. The important points are: it must provide for all-around security by means of a point, rear point, and flank observation or flankers; the patrol leader should be near the head of the main body, rather than in the point, so that he can control the action of the entire patrol; the automatic weapons within the patrol should be located near the leader in order that he may control their initial action before they become committed or pinned to the ground in the first burst of hostile fire; a get-away man should be designated. The distances between the individuals in the patrol will depend entirely upon the nature of the terrain through which it is passing, bearing in mind that mutual support must be assured.

6–51. **March discipline.**—*a. Silence essential.*—A combat patrol operating in a hostile area must march in silence. The noises, including voices, made by the patrol at a halt should not be loud enough to be heard by the outguards.

b. Maintaining distances.—(1) The distances to be maintained between subdivisions of the patrol, and between individuals, are designated by the patrol leader. If these distances are temporarily decreased or increased due to the terrain or for other unavoidable reasons, they should be resumed as soon as warranted by the situation.

(2) Distances should be maintained with respect to the elements both in front and in rear. If an individual loses contact with the man next in rear of him, word should be passed forward and the rate of march decreased or the patrol temporarily halted until the gap is closed.

(3) It requires particular effort to prevent men from bunching at stream crossings, fallen trees, large mudholes, and similar obstacles.

(4) The arm signals "halt" and "forward" should be used freely to indicate to the men in rear what is happening to their front.

c. General rules.—All members of the patrol should comply with such of the following rules as pertain to the situation:

(1) No noise or "skylarking" to be permitted.

THE MARCH

(2) Weapons and ammunition carried by individuals will be retained on their persons. They will not be secured to riding or pack animals.

(3) A man leading an animal will not secure the lead line to his person or equipment.

(4) The riding animal for the sick will march at the rear of the train.

(5) Be alert at all times. Do not depend entirely on the leading elements for reconnaissance.

(6) No smoking except when authorized.

(7) Do not leave articles foreign or strange to the locality on the trail or in camp sites.

(8) Only the patrol leader will question natives encountered on the trial for information about distances and directions. When he does so, he should ask for data about several places so as to disguise the route to be taken.

(9) No conversations will be entered into with natives except by the patrol leader, designated subordinates, or interpreter.

(10) The native guide will not talk to other natives except in the presence of the interpreter.

(11) When passing or halting in the vicinity of dwellings occupied by peaceful natives, do not take fruit, eggs, or other things without fair payment; do not gamble or drink with natives; do not enter native houses without clearly understood invitation; do not assume a hostile attitude.

(12) All distances will be maintained at temporary halts as when marching.

6–52. **March outposts.**—March outposts should be established at every temporary halt. The advance party, or, in small patrols, the main body should halt on ground which can be easily defended. The point should proceed at least a hundred yards along the trail and take up a position in observation. Other routine security measures are followed, such as reconnoitering and observing lateral trails, reconnaissance of commanding ground to the flanks, and security to the rear. These requirements are fully as important in small war operations as in major warfare.

6–53. **Camp sites.**—*a.* If the patrol is to bivouac on the trail, the day's march should cease at least 2 hours before sundown.

b. When the location of the camp site is not definitely known, the patrol leader should begin looking for a favorable site at least 3 hours before sundown. In peaceful territory, inquiries may be made of friendly natives but this is inadvisable in a hostile region. Too much reliance cannot be placed in the information received. Usually the natives accompanying the patrol as guides, interpreters, or muleteers will be able to give fairly definite information regarding good camp sites.

c. The camp area should be a level or slightly rolling, cleared, dry, well-drained field with firm turf free from stones, stubble, and brush, and ample in size to accommodate the command without crowding. Water is essential. Fuel and **forage** should be available. The vicinity of swamps, marshes, and native houses should be particularly avoided because of the danger of insects and disease. Camp sites recently used by other troops are undesirable unless they have been left in good police.

d. Dry stream beds and ravines are undesirable because of warmth, poor ventilation, and the danger of floods.

e. Part or all of the desirable features for a camp site may have to be disregarded in hostile territory when proper defense of the bivouac will be paramount.

6-54. **Making camp.**—When the patrol is halted for the night, march outpost security is immediately enforced until the regular outguards can be formed, instructed, and posted. Reconnaissance patrols should be sent over all trails radiating from the camp site for a distance of at least a half mile, including the route which has just been traversed. Outguards will usually be detailed from the unit which has furnished the advance guard for the day. In small patrols it is often necessary to detail some personnel from the main body for this duty. Plans for the defense of the bivouac should be formulated, and every element of the patrol instructed accordingly. Squads and other units should be bivouacked as organizations and in relation to their respective sectors in the defense. Working details are assigned to procure water and fuel, to dig latrines, and to perform other necessary tasks.

6-55. **Shelter.**—*a. The shelter tent.*—In good weather it is often better to sleep in the open rather than to construct temporary shelter. If some shelter is desirable, the shelter tent is generally the best type for troops in bivouac.

b. The lean-to.—When necessary materials are available lean-tos can be constructed almost as quickly as shelter tents can be erected. They are roomier than the shelter tent and afford better protection during heavy rains. The lean-to consists of two forked uprights, a cross pole, and a rough framework which is thatched with large leaves, such as manaca, banana, palm, etc., or with grass or reed tied in bunches. The uprights may be two convenient forked trees or saplings. After the cross-pole is secured in place, the framework is leaned against it, and the covering secured in place. The various parts of the lean-to are lashed together with vines which are usually found in the vicinity. A well made lean-to will last for 3 or 4 weeks before it has to be recovered. (See Plate I.)

c. The canvas lean-to.—This combines certain desirable features of both the shelter tent and the thatched lean-to. It consists of a frame of light poles with a tarpaulin, tent fly, or several shelter halves or ponchos thrown over it and staked down on one edge. The two ends are enclosed. The front is left open like a lean-to. This shelter is strong, quickly built, and makes use of various sizes of canvas.

d. Native buildings.—Native buildings generally should not be used by patrols for shelter. Most of them are unclean and infested with insects. They are usually more difficult to defend than bivouac which can be selected with its defense in view. Sometimes, vacant buildings may contain mines or bombs laid by the hostile forces to explode on contact.

6–56. Bivouac beds.—*a.* Men should not sleep on damp ground. In temporary camps and in bivouac, they should raise their beds with leaves, boughs, or makeshift bunks in addition to placing the poncho between them and the ground. Satisfactory bunks can be made from small poles placed together on crosspieces raised about 6 inches from the ground. The poles should be covered with leaves or similar material. Bamboo can be split lengthwise, the joints cracked, and the piece flattened out to make an excellent, spring-like bunk when laid on crosspieces at the head and foot.

b. Native hammocks made of light material are of practical use in some operations. The sleeper can protect himself from the rain by stretching a line between the hammock lashings and hanging a shelter-half or poncho over it.

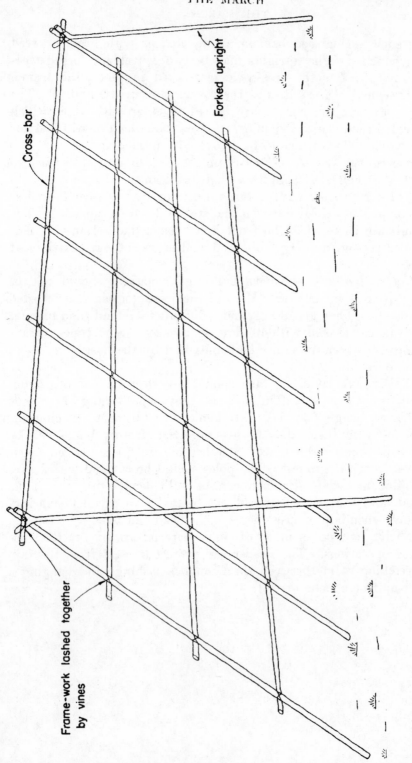

Forked upright

Cross-bar

Frame-work lashed together
by vines

PLATE I.

RD 2826

Section VI

RECONNAISSANCE AND SECURITY

6–57. **Methods of reconnaissance.**—The various methods of reconnaissance and security employed by patrols in small wars do not vary in principle from those used in major warfare. Because of the nature of the terrain in which most small war operations occur, the difficulties of reconnaissance and security are increased. Several months of active operation in the field are required to train average individuals as efficient scouts, and only a small proportion will acquire the ability of a native to interpret correctly the things observed along the trail.

6–58. **Reconnoitering by scouts.**—*a.* The most certain method to uncover an enemy is to send scouts to visit suspected positions. The disadvantages of this method include:

(1) It slows up the progress of the patrol.

(2) Dense underbrush, and mountains, broken terrain are difficult to negotiate and will rapidly exhaust the personnel.

(3) Scouts are likely to be in the line of fire when the battle commences.

In spite of its disadvantages, the results of this method of reconnaissance are so reliable that it should normally be employed.

b. Scouts sent to reconnoiter positions which may be occupied by the enemy at any time during the passage of the patrol, such as commanding positions, and roads and trails intersecting the route being traversed, should remain in position until the patrol completes its passage. This is the principle of "crowning the heights."

6–59. **Careful visual reconnaissance.**—The careful visual reconnaissance of suspected positions while approaching and passing them enables a patrol to march more rapidly, but it is not as certain to disclose the presence of an enemy as the method of reconnaissance by scouts. Excellent field glasses are essential for efficient observation.

6–60. **Hasty visual reconnaissance.**—*a.* In some situations reconnaissance will consist only of a hasty visual inspection of dangerous and suspicious places. Hasty visual reconnaissance may be employed when:

(1) Patrols are operating in supposedly peaceful areas, or in areas which have been recently vacated by the enemy.

(2) Airplane reconnaissance has indicated that the area is free of the enemy.

(3) Military necessity requires the patrol to expedite its march.

b. It must be understood that to carefully reconnoiter every commanding position and suspected ambush site will, in some terrain, almost immobilize the patrol.

6–61. **Reconnaissance by fire.**—*a.* Reconnaissance by fire attempts to inveigle the enemy to disclose his position by returning fire directed against a suspected hostile position. This method should never be used by patrols assigned to aggressive or offensive missions.

b. Some of the disadvantages of reconnaissance by fire are:

(1) It discloses the presence and location of the patrol to the enemy within range of the sound of the gunfire.

(2) It prevents the capture of guerrillas who may be traveling alone or in small groups.

(3) It expends valuable ammunition, the supply of which may be limited and all of which may be needed in a crisis.

(4) It has a tendency to make the men on service of security less observant.

(5) There is always the chance that a well-controlled enemy force in ambush will not return the fire.

(6) The members of the patrol have difficulty in distinguishing between the reconnaissance fire and the initial shots fired from an ambush.

c. Reconnaissance by fire may be reasonably employed by:

(1) Liaison patrols which are too weak to engage in combat with hostile forces.

(2) Patrols whose mission requires them to reach their destination as quickly as possible.

6-62. **Reconnaissance by aviation personnel.**—*a.* Reconnaissance by plane is invaluable in small wars operations. It has the following disadvantages, however:

(1) Difficulty of detecting the enemy in wooded country.

(2) It may divulge the location of friendly ground patrols.

(3) The difficulties of maintaining continuous reconnaissance.

(4) It does not relieve the ground patrols of their responsibility for continuous close reconnaissance, although it often gives them a false sense of security.

b. For further details, see Chapter IX, "Aviation."

6-63. **Airplane reconnaissance by patrol leaders.**—Patrol leaders should make an airplane reconnaissance of the area of operation at every opportunity in order to study terrain features. This is especially important if accurate maps of the area are not available.

6-64. **Intelligence agents.**—Reliable intelligence agents can sometimes be employed to reconnoiter an area prior to the arrival of a patrol, and to continue their reconnaissance in conjunction with the patrol's activities.

6-65. **Questioning inhabitants for information.**—Patrol leaders must evaluate cautiously information obtained by questioning inhabitants encountered on the trail. A person who resides in a community overrun by guerrillas generally is sympathetic towards them or fearful of their reprisals.

6-66. **Dogs on reconnaissance.**—Dogs may sometimes be profitably employed with outguards and security detachments on the march to detect the presence of hostile forces. Unless they are carefully and specially trained, their usefulness for this purpose is doubtful.

6-67. **Security on the march.**—*a. General.*—Whenever practicable, the methods of security employed in normal warfare are used by patrols in small wars.

b. Breaking camp.—Security measures must not be relaxed when breaking camp. The exit from the camp should be reconnoitered and the patrol should be vigilant when getting into its march formation.

c. Duties of the point.—The primary function of the point is reconnaissance, to disclose the presence of hostile forces on or near the route of march before the next succeeding unit in the column comes under fire. It is a security detachment rather than a combat unit. There is a tendency in small war operations to overlook this important principle. If a patrol leader assigns too large a proportion of his force to the point, he sacrifices his freedom to maneuver in combat. The leading man of the point should never be armed with an automatic rifle. He, more than any other man in the patrol, is likely to become a casualty in the initial burst of hostile fire from ambush. Point duty is dangerous and fatiguing. Men assigned to the point should be relieved every 2 or 3 hours during the day's march, and more frequently in dangerous localities.

d. Flank security.—The most difficult feature of security for a patrol marching through wooded terrain is adequate protection against ambush and attack from the flanks. It is usually impossible or undesirable to maintain flank patrols continuously in such country. An experienced patrol leader will often detect the presence of a hostile force in the vicinity by signs along the trail. At that time, he should establish flank patrols abreast or slightly in rear of the point, even though the rate of march will be adversely affected. Except under these conditions, flank security is generally maintained by observation, and reconnaissance of intersecting trails.

6–68. **Security at rest.**—*a.* See paragraphs 6–52 and 6–54.

b. Camp fires should be screened at night to prevent the personnel from being silhouetted against them in the event of a hostile attack.

c. Not more than 50 percent of the patrol, including the mess detail, men washing or bathing, and working parties, should be separated from their weapons during daylight hours. During the night, all men should keep their weapons near their persons.

Section VII

LAYING AMBUSHES

6–69. Definition.—An ambush is the legitimate disposition of troops in concealment for the purpose of attacking an enemy by surprise. The laying of a successful ambush in hostile territory in a small war is a difficult operation.

6–70. Selection of position.—*a. Offensive ambush.*—An offensive ambush should be so located as to facilitate the assault after the initial burst of fire.

b. Defensive ambush.—A defensive ambush presupposes an inability to assault and the probable necessity of a rapid withdrawal. It should be so located as to facilitate defense, with natural obstacles between the position and the enemy, and routes of withdrawal should be carefully planned, reconnoitered, and prepared, if necessary. These requirements usually limit the location of a defensive ambush to the military or geographical crest, where the withdrawal will be protected by the reverse slope.

c. Direction of wind.—The ambush site should be selected so that the odor and noises of the men will be carried away from the enemy's route of approach.

d. Obstacles.—Stream crossings, large mudholes, or fallen trees across the trail, are all useful obstacles. They generally cause the ambushed troops to bunch up before the firing starts, and hinder their movements afterwards. Intersecting stream beds and trails at the position should be enfiladed by fire.

6–71. Usual characteristics of an ambush.—Every ambush must provide suitable firing positions and concealment in close proximity to the hostile route of march. The usual position is located on the forward slope and at a bend in the trail. Automatic weapons or machine guns are placed in prolongation of the probable direction of march at the bend in order to take the enemy in enfilade. The main body of the ambuscade is placed parallel to the hostile route of march to facilitate the assault after the initial burst of fire. The security elements of the enemy should be permitted to pass by the

LAYING AMBUSHES

position in order to secure the maximum effect against the hostile main force. A position that permits engaging the enemy column from both flanks simultaneously is possible only if the trail lies in a deep ravine. Even then there is considerable danger that ricochets and wild shots from one flank will cause casualties to the other flank.

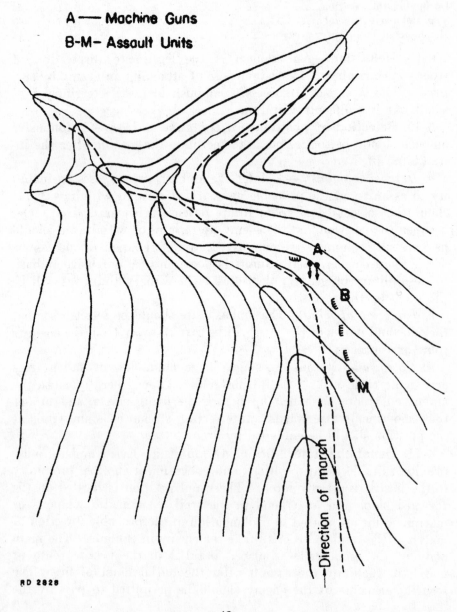

A —— Machine Guns

B-M- Assault Units

Direction of march

RD 2829

6–72. Occupying the position.—*a. To ambush a pursuing force.*— In the event a combat patrol wishes to ambush a hostile force known to be following over the same trail, it must proceed well beyond the ambush position selected. At a suitable point, such as a stream, it is led off the trail and counter-marched, parallel to, but clear of the trail, until it reaches the reverse slope immediately in rear of the selected ambush. The men then move individually, as carefully as possible, into their firing positions and remain motionless.

b. To ambush a meeting force.—It is more difficult to lay a successful ambush against a meeting force than one which is pursuing, unless the patrol leader is thoroughly familiar with the terrain and has definite information of the approach of the hostile party. The ambuscade must leave the trail some distance in advance of the selected position. It then moves into firing position as before and awaits the approach of the enemy. Any movement along the trail in advance of the ambush will disclose its location by footprints, or other tell-tale signs.

c. Night ambush.—In some situations it may be desirable to occupy an ambush position at night. This maneuver requires a definite knowledge of the terrain, and good guides.

d. Ambush outposts.—An outpost must be established at the point the patrol leaves the main trail to intercept and capture any person traveling the trail who might inform the hostile force of the location of the ambush.

e. Observation post.—An observation post should be established in a position that will enable the observer to give timely warning of the enemy's appearance. The most desirable position is some distance from the ambush and in the direction of the enemy's approach.

f. Firing positions.—Each man should select a good firing position as close to the trail as is consistent with complete concealment.

g. Tell-tale signs.—Every effort must be made to avoid moving foliage or earth for purposes of cover, shelter, or camouflage. The keen eyes of the enemy may detect the turning of a leaf, the breaking of a twig, or the appearance of a handful of new dirt.

6–73. The ambush engagement.—*a. The enemy approaches.*—As the enemy approaches the ambush, the men lie face downward and remain motionless until the signal to "commence firing" is given. If they raise their heads, the position will usually be disclosed by the outline of the heads or headgear, by movement, or by the reflection of light from their eyes and faces. Too often, some man will become so excited that he cannot resist firing prematurely at the first enemy he sees.

LAYING AMBUSHES

b. The signal to commence firing.—The patrol leader should give the signal to commence firing. An excellent method of doing this is by opening fire with an automatic weapon. He should be on the flank toward the enemy, or in a commanding position that will enable him to observe the entire enemy force.

c. Action after opening fire.—Depending upon the situation, the heavy initial fire will be followed by an assault, a defense, or a withdrawal.

6-74. **Employment of infantry weapons.**—Machine guns should be sited to enfilade a portion of the trail. Trench mortars, hand grenades, and rifle grenades are difficult to employ in an offensive ambush because of the danger of such projectiles falling among friendly assaulting troops. They are of value in defensive ambushes.

Section VIII

ATTACKING AMBUSHES

6-75. **Mental preparation.**—The principal objective of an offensive ambush is to take advantage of surprise. The closeness and suddenness of the attack is supposed to disorganize and demoralize the enemy. A necessary protection against complete disorganization, and possible demoralization, is to prepare the troops mentally for the shock of ambush. They must be steeled to withstand a sudden blast of fire at close quarters and to react to it in a manner that will unnerve the enemy. To accomplish this, the troops must have a thorough understanding of what is likely to happen if they are ambushed.

6-76. **Prearranged schemes of maneuver.**—*a. General.*—Since the great majority of ambushes have certain similar characteristics, the nature of an ambush attack can be anticipated. Usually there will be a burst of automatic fire from the front that will enfilade the column, combined with an attack from one flank. Both of these attacks will be delivered at short ranges and from positions located in thick cover on commanding ground. With this situation in mind, the patrol can be indoctrinated with simple prearranged schemes of maneuver to combat such attacks.

b. Actions of the train and train guard.—(1) In the event the train is not under fire when the engagement commences, it should be closed up on the forward elements in the column. As soon as closed up, or when endangered by hostile fire, the animals are driven into positions affording cover or shelter. When possible, the animals are tied to trees to prevent them from running away. This enables the train guard to use its weapons to protect the train, and assists the native muleteers to control the train. In some situations, particularly where the area is heavily wooded, the animals tired, and the enemy aggressive, the ration and baggage animals are abandoned until after the battle.

ATTACKING AMBUSHES

(2) Men leading animals carrying weapons and ammunition retain possession of them. In desperate situations, it may be necessary to shoot these animals to prevent them from bolting into the enemy positions.

(3) If heavy machine guns or 81 mm. mortar units are attached to the patrol the crews take their loads from the pack animals and, moving "by hand," prepare to go into action.

(4) The train guard keeps the muleteers and animals under control. It is assembled under the train commander and is available to augment the patrol reserve.

c. Patrol reserve.—A patrol reserve should be withheld from the initial action. An alert enemy may fire upon the leading elements from one flank only and, once the patrol has been committed, launch an unexpected attack from the rear or some other direction. As soon as the hostile position has been fully developed, however, the reserve may be employed to envelop his flank or as otherwise required by the situation. In many situations the rear guard will constitute the patrol reserve. Certain automatic weapons should be definitely assigned to the patrol reserve.

d. The rear guard.—If the patrol is ambushed from either flank, the rear of the column becomes an exposed flank. The primary function of the rear guard is to protect this flank, and it should not be committed to action until the situation makes it mandatory. The rear guard commander may, if necessary, send part of his unit to assist the train guard in controlling the animals and native muleteers. If the rear guard constitutes the patrol reserve, it may be employed after the train is secured and the train guard has been assembled to assume the functions of the reserve.

e. Anticipated action against an attack from the front and a forward part of the right (left) flank.—In the majority of an ambush the attack will be delivered against the front and forward part of the right (left) flank of the column. The point (advance guard) and the leading elements of the main body usually are immobilized by the initial burst of fire. They return the fire and act as a holding force, developing the hostile position. The rear elements of the main body immediately maneuver to envelop and overrun the exposed hostile flank and capture the enemy's automatic weapons. As the attack progresses, the hostile force will begin to withdraw, and the point (advance guard) and leading units of the main body will be enabled to participate in the final assault of the position. The

ATTACKING AMBUSHES

patrol reserve may be employed to extend the envelopment in order to intercept the enemy's line of retreat. The action of the rear guard and train guard is as outlined above.

f. Anticipated action against an attack from the front and entire length of the right (left) flank.—(1) The point (advance guard) builds up a firing line facing the enemy and makes a holding attack, developing the enemy's position, until able to participate in the assault.

(2) The main body advances the attack as rapidly and aggressively as possible in order to penetrate the hostile position. When the ambush is penetrated, a flank attack is delivered in one or both directions, overrunning and capturing the enemy's automatic weapons.

(3) The train guard builds up a firing line facing the enemy and makes a holding attack, developing the hostile position, until able to participate in the assault.

(4) Attached weapons, if present with the patrol, are unpacked and put into action.

(5) Enlisted mule leaders secure their animals under cover and then assemble under command of the train commander as the patrol reserve.

(6) The rear point builds up a firing line facing the enemy and makes a holding attack, developing the hostile position, until able to participate in the assault.

g. Direction of fire.—When operating along winding trails in hilly country the selection of targets and direction of fire must be well controlled to avoid killing or wounding friendly personnel.

h. The bolo attack.—In certain theaters of small wars operations there is the possibility that a patrol may be ambushed and rushed from both sides of the trail by an enemy armed only with bladed weapons. Such attacks are launched from positions located a few feet from the sides of the trail. The use of rifle fire in the general melee which results is fully as dangerous to friendly personnel as to the enemy. The experience of regular forces which have encountered such tactics in the past has indicated that the bayonet is the most satisfactory weapon to combat an attack of this nature.

6–77. **Spirit of the offensive.**—*a.* Troops engaged in small war operations must be thoroughly indoctrinated with a determination to close with the enemy at the earliest possible moment. A rapid, aggressive attack is necessary to overrun the hostile positions and

ATTACKING AMBUSHES

seize his automatic weapons. It often happens that some slight movement, or the reflection of light from hostile weapons, will disclose the location of an enemy ambush before the first shot has been fired. If this occurs, immediate action of some sort is imperative. To stand still, even momentarily, or simply to attract the attention of the person next in column, is usually fatal. If the individual or unit who observes the ambush rushes forward immediately, not in a straight line but in a zigzag course depending upon the nature of the terrain, the enemy may break from his position. In any event his opening burst of fire will be erratic and comparatively ineffective instead of deliberate and well aimed. The rush should be accompanied by a yell to warn the remainder of the patrol, which will also disconcert the enemy. This action is effective even though the bayonet is not carried or fixed to the rifle, nor is it any more dangerous than taking up a firing position near the trail which is almost certain to be within the beaten zone of some hostile weapon. It is analogous to the final assault which is the objective in every combat.

6-78. **Fire and movement.**—*a.* If an immediate assault is not initiated against the hostile position, the ambushed patrol must seek cover and engage in a fire fight. Even though the patrol is armed with superior weapons and is better trained in combat firing than the enemy, it is at a disadvantage in a purely passive fire fight. The hostile foces have the advantage of commanding ground and concealed positions. So long as they are not forced to disclose their individual positions by actual or threatened personal contact, they are free to break off the engagement at any time. The eventual loss of ground means nothing to the guerilla. If he can withdraw with no casualties or only minor ones after delaying and harassing the patrol, the engagement has been a success. The objective of the patrol must be, therefore, to inflict as many casualties upon the enemy as possible. This can be accomplished in a fire fight only if the spirit of the offensive, with movement, is employed. During the fire fight members of the patrol must move forward at every opportunity in order to close with the enemy as quickly as possible, or make him disclose his position so that he may become a definite target. The culmination of the action is the assault, to overrun the enemy, capture his weapons, and pursue him by fire to the limit of visibility.

6-79. **Authority of subordinates to act on own initiative.**—*a.* Considerable authority must be granted all leaders to act independently and on their own initiative. In the absence of orders, action on

ATTACKING AMBUSHES

the part of the patrol's subdivision is preferable to inaction. Subordinate leaders must remember that the action which they initiate should furnish mutual support in the action to the hostile force.

b. Leaders must make every effort to gain direction and control of the elements of their own units. They must not hesitate to influence the action of subordinate leaders of nearby elements which have become separated from the control of their normal superiors.

Section IX

ATTACKING HOUSES AND SMALL BIVOUACS

6–80. **Attacking houses.**—*a.* In small war operations, it is frequently necessary to seize individuals or attack hostile groups known to be at a certain house. Although the task may appear to be simple, it is often difficult to accomplish successfully.

b. The following instructions are generally applicable in planning an attack against a house:

(1) Secrecy is essential. Relatives, sympathizers, or intimidated natives may warn the enemy of the patrol's approach. In some instances, they have been warned before the patrol cleared its home station.

(2) The location of the house and the nature of the terrain surrounding it must be definitely known, either by personal reconnaissance, a sketch, or through the medium of a guide.

(3) The patrol should usually approach and occupy its position under cover of darkness.

(4) Do not use a larger patrol than necessary to carry out the mission. A large patrol is hard to control, difficult to conceal, and makes too much noise.

(5) The approach must be made quietly and cautiously. Barking dogs often warn the inhabitants of the approach of the patrol.

(6) Utilize all available cover.

(7) Cover all avenues of escape, either physically or by fire.

(8) Bayonets should be fixed. The patrol is sometimes unable to open fire due to the presence of women, children, or unidentified persons, or because of instructions received from higher authority.

(9) If the mission is the capture of the occupants and armed resistance is not expected, surround the house and approach it from all sides.

(10) If the mission is to attack the house, and armed resistance may be expected, the patrol must be located so that every side of the building will be covered by fire. Particular care must be taken to make certain that no member of the patrol will be in the line of fire of any other individual in the patrol.

ATTACKING HOUSES AND SMALL BIVOUACS

6-81. **Attacking small bivouacs.**—*a.* A successful attack on a hostile bivouac often has a more demoralizing effect than a defeat in ordinary engagements.

b. Many of the instructions for attacking houses are applicable to attacking bivouacs. In addition, the leader of a patrol making a surprise attack on a small enemy force in bivouac should be guided by such of the following instructions as may be applicable in the particular operation:

(1) Secure a guide who knows the exact location of the bivouac. If he is a reliable, friendly native, an effort should be made to have him reconnoiter a good approach to the bivouac.

(2) Require the guide to make a sketch of the bivouac and its approaches. This can be traced on the ground. The leader should study it carefully, but should be prepared to find the actual situation quite different from that expected.

(3) Attack with few men. A leader and two teams of four men each is a suitable group for most situations.

(4) Arm the majority of the patrol with automatic or semiautomatic weapons.

(5) Leave the trail as soon as convenient and approach the bivouac from unexpected direction. When in the vicinity of the bivouac, approach slowly and cautiously.

(6) After sighting the bivouac, the leader should make a careful reconnaissance of it. Determining the exact location of the principal groups of the enemy force is generally difficult. When confident of the location of the major portion of the enemy, the leader builds up a final firing line.

(7) When the firing line is in position and prepared to open fire, the leader orders the enemy to surrender. In the event they refuse, the leader signals, "Commence Firing." All men direct their fire into the bivouac, firing rapidly, but semi-automatically.

(8) The possibility of an assault and a pursuit should be considered, but the lack of bayonets, the nature of the terrain over which the enemy will flee, and the agility of the enemy, will often make such efforts futile.

6-82. **Destroying captured bivouacs.**—The value of a bivouac as a known enemy camp site should be considered before destroying it. Guerrillas have a weakness for occupying camp sites they have previously found satisfactory, particularly if shelters have been constructed. The burning of bivouac shelters rarely serves any useful purpose unless they contain military stores of some value.

Section X

STRATAGEMS AND RUSES

6–83. **Rules of land warfare.**—Patrol operations always furnish opportunities for the employment of stratagems and ruses; however, such as are used must be in accordance with the accepted Rules of Land Warfare.

6–84. **Clearing the station.**—A patrol should be able to clear its stations without that fact being transmitted to the hostile forces by their intelligence agents. The following methods may be used to deceive the enemy as to the plans of the patrol:

(1) Having decided to attack a certain hostile bivouac, a rumor is started among the natives that the patrol is to march to some place in the opposite direction. The patrol clears the town in that direction and eventually circles at some distance from the station and marches towards the objective.

(2) Clearing the station late at night.

(3) The members of the patrol filtrate out of the camp during the day or night and assemble at a rendezvous some distance from the station.

6–85. **Apprehending informers.**—Guerrilla spies who live near the garrison are a constant menace. While they are often suspected, it is very difficult to apprehend them in a guilty act. One way of doing this is to establish sentinels on the trails leading from the station to the hostile areas. After the sentinels have reached their posts, organize a patrol and allow word to be passed among the natives that the patrol is clearing for the hostile area. This information may cause the informers to start for the enemy camp in order to warn them, thus permitting the sentinels to intercept and detain them for questioning.

6–86. **Spies following a patrol.**—Hostile intelligence agents may follow a patrol, but at a safe distance. One way to capture them

STRATAGEMS AND RUSES

is to leave an outpost several hundred yards in rear of a selected camp site. The men in the outpost take cover and capture any suspicious persons following the patrol.

6-87. Guerrilla ruses and stratagems.—For purposes of protection, a careful study must be made of the ruses and stratagems practiced by the enemy, and the facts learned should be published for the information of the regular forces concerned. Ruses and stratagems practiced in warfare between forces of irregulars, or by irregulars against regulars, include:

(1) Inveigling the enemy into an attack and pursuit and then, when he is disorganized and scattered, make a violent counter-attack A modification of this method includes abandoning animals and supplies and then, when the attacking force is more interested in booty than pursuit, to counter-attack.

(2) A group of the enemy may retreat before the attacking force and lure it into a carefully prepared ambush.

(3) Disguising themselves to resemble their foes, sometimes wearing a similar uniform.

(4) Having their men on service of security disguised like their foes.

(5) On one pretext or another, to lure a small enemy force into an exposed position and destroy it. Examples:

(*a*) Cutting a telegraph wire and then destroying the repair party.

(*b*) Raiding a community with a small group and then striking the patrol sent to its relief with a stronger force.

(6) A guerrilla group surprised in an area may hide its firearms and assume the appearance of a peaceful group of citizens busy in their fields or clearing trails.

Section XI

RIVER CROSSINGS

6–88. **Introduction.**—*a.* The passage of a patrol across a stream in small wars operations is similar to the passage of a defile. It should be assumed that every crossing will be opposed by the enemy, and necessary precautions should be taken to effect the passage with a reasonable degree of safety. The security measures taken and the tactics employed to force a crossing against opposition in small wars do not differ from those of major warfare.

b. All streams act as obstacles to a greater or lesser extent. Some means must be devised to get the troops and material across without disorganization and in condition effectively to resist an enemy attack before, during, and after crossing. Crossing may be opposed or unopposed. The probability of opposition is frequently the determining factor in the choice of a crossing site. Poorly adapted sites may have to be used by the reconnoitering parties. Time is a factor in every crossing and the means employed must be those that will permit a crossing in the minimum of time to avoid continued separation of the parts of the patrol by the river.

6–89. **Availability of means.**—Means of crossing may be divided into fords, boats (including rubber boats of the collapsible variety), rafts, ferries, permanent and temporary bridges, and swimming. It may be necessary to make use of several or all of these expedients for the crossing of a body of troops with its supplies. It may be said that fording and swimming will be the normal means of crossing in small wars operations.

6–90. **Swimming.**—Most unfordable streams, especially in small wars operations, will have to be crossed, initially at least, by swimming, until protection for the main crossing has been established on the opposite bank. Most men will be incapable of swimming even a short distance with their rifles, belts, and clothing and would be helpless and naked when they landed on the opposite bank if they discarded this equipment. Swimming is therefore usually combined with some such method as the use of individual floats or rafts, or the use of the few boats available for the transport of arms and supplies. Such floats may assist the swimmers in keeping afloat while crossing relatively wide rivers. When the men land, they establish themselves in a position that will protect the crossing, assist on the far side in construction of bridges, ferries and similar means of getting the remaining troops and supplies across, or proceed on their assigned mission, which may be to drive an enemy detachment from a more suitable crossing place. Life lines may be stretched across the river, one above and one below the crossing, as safety measures to prevent men from being swept away downstream by the current.

6–91. **Bridges.**—*a.* The construction of bridges for the passage of all arms requires considerable time and material, and a certain amount of technical engineering training. Bridges are most useful at crossings on the line of communications and have the advantage of providing a permanent means of crossing a river. Only those forms of bridges easily constructed with materials or tools available at the bridge site will be considered here.

b. Felled tree.—Very narrow streams may be bridged by felling a large tree across the stream, staking down or otherwise securing the ends, and then cutting off the branches above the water so that troops can walk across. A line can be stretched as a handrail if necessary from bank to bank over the fallen tree trunk.

c. Foot bridge of two or more trees.—Where one tree will not reach across the stream, two trees may be felled on opposite sides of the stream, and their branches and tops secured together in midstream. Lines should be made fast near their tops before felling, and the line snubbed to other trees well upstream and eased downstream by the ropes until the two trees intertwine. Then the tops should be securely lashed together and the branches cleared to provide a footpath over the tree trunks. Stakes on either side of the felled tree trunks may be used to strengthen this bridge. A third tree, felled so that its top falls on the point where the other two meet in midstream, but at an angle to them, will strengthen the bridge, as it provides a tension member against the force of the current.

RIVER CROSSINGS

d. Floating bridges.—Floating bridges may be constructed of rafts or boats, with planks laid across the gunwales. These boats should be securely anchored or moored with their bows upstream. Sections of the bridge may be constructed and floated downstream into place. It is desirable to have all the floats of about the same capacity, to avoid extra strain on the planks or other flooring. The larger boats or rafts may be placed at greater intervals than the smaller, to accomplish the same purpose. Rafts or floats may be made of timber, casks, barrels or anything at hand that will float.

6–92. **Boats.**—*a.* In the countries in which small wars operations usually occur, some native boats will be found in the vicinity of river crossings which are too deep to be fordable. These will normally be of the dug-out type which are quite unstable, but unsinkable. Their capacity will range from small, two-man boats, to bateaux of 30- or 40-man capacity. If the crossing is opposed, it is desirable to have a sufficient number of boats available to execute the crossing of the patrol promptly. If the passage is unopposed, even one small boat will be found of inestimable value.

b. If no boats can be found in the locality, it is sometimes necessary to construct makeshift boats from available materials. In recent small wars operations, boards found in a local dwelling were used to construct a 6- by 22-foot boat, caulked with gauze from the hospital corpsman's kit. It was employed to ferry a patrol of 135 officers and men and 70 animals across a stream swollen with torrential rains which made the customary fording impossible. The current was so swift that the use of a raft was impracticable.

c. Collapsible rubber boats, which can be carried by patrols in the field or dropped by aircraft in case of necessity, will probably be used extensively in future small wars operations.

6–93. **Ferries.**—*a.* Ferries may be either "flying" ferries or "trail" ferries, and in most streams the current can be used to propel them. If the current is too sluggish, a "rope" ferry may be used; i. e., the boat or raft is drawn across the stream by pulling on the rope or by poling. "Trail" and "flying" ferries are propelled by the current by holding the boat or float at an angle to the current, either by means of ropes or by rudders, or even by paddles held in the water. The angle and the speed of the current controls the speed of the ferry, and since the angle may be varied, the speed can also be controlled.

b. **Trail ferry.**—A good trail ferry may be constructed in most streams that are not too wide by stretching a rope or cable (several telegraph wires twisted together will do for a small boat) across the stream and rigging a pulley so that it will travel on this line ("sheer line"). A line is then fastened to the pulley from the bow of the boat. Some ferries, especially in slow streams, must be controlled by a bow and a stern line ("maneuvering ropes") attached to the pulley. By hauling in on the bow line and slackening the stern line, or vice versa, the bow may be set at such an angle to the current that the force of the current acting on the hull will cause the boat to move across the stream. In a swifter current, where the boat points more nearly upstream, a paddle may be held over the downstream side at the stern, or a rudder may be used.

c. **Flying ferry.**—A flying ferry uses the same means of propulsion, but in place of the "sheer line," there is an anchorage upstream to which a long line is fastened from the bow of the boat, enabling the boat to swing like a pendulum across the stream. The line is supported by floats at intervals, so that it will not trail in the water and slow up or even stop the ferry. In narrow streams it may be possible to find a curve in the stream that will permit the anchorage to be on land, but still above the center of the stream at the point where the ferry crosses. An island is also a convenient location for an anchorage. The cable must be long enough so that the ferry will not have difficulty in reaching both banks.

d. Rafts as ferries.—Rafts may be used in all types of ferries, but they should be so constructed that the current will act against them efficiently, and so that they may be easily maneuvered but hard to swamp.

6–94. Fords.—*a.* The requisites of a good ford are low banks, no abrupt changes in depth of water, the bottom offering a firm footing for men and animals and the current moderate. One of the first duties of the reconnaissance parties of a patrol or larger detachment on arriving at a stream is to reconnoiter for fords and bridges in the vicinity, and when found, to test them and define their limits. Dangerous fords should be marked before use by the main body. Fords that show signs of use are likely to be passable, but *care must be taken to allow for the height of the water* above normal. This may be ascertained by an examination of the banks, especially of the vegetation on the banks and of small trails parallel to the water. When fording swift shallow streams with native pack animals, each animal should be led and not herded across. When the water is deep enough to reach the pack, the cinch may be loosened and two men accompany each animal, one on either side to raise the sugar, coffee, and similar loads to their shoulders in the deep water. Rain in the uplands, which are drained by rivers, may cause sudden floods. These floods rapidly descend the rivers, and make it dangerous to use fords until after the swollen streams subside.

b. Fords may be improved in very swift streams or during freshets, by felling trees across the stream and lashing their ends together (narrow streams only) or by fastening a line of floating obstacles such as logs or barrels above the ford, to cut down the current, at least on the surface. A life line should be stretched across the stream for the crossing of large numbers of troops, and men should hold on to this line. Crossings should be guarded by good swimmers and by boats downstream if available, to take care of those who may be swept off their feet or who may stray from the ford. Infantry may ford a stream in a column of squads, by men in each rank holding on to each other abreast. The distance between ranks is greatly lengthened to avoid the increased resistance which might be caused by partially damming the stream. Mounted men pass over the ford in column of twos or files.

6–95. **Rafts.**—Rafts may be constructed of any materials that will furnish sufficient buoyance, and which are available at or very near the point of crossing. Many woods do not have enough buoyance even when dry, and this is especially true of green hard woods. "GI" cans with burlap or other cloth under the covers, to make them watertight, casks and barrels, gasoline drums and other containers may be used to give buoyancy to a raft. In general the construction of rafts to effect a crossing will be inadvisable except for the heavy articles that cannot be conveniently crossed by other means. Rafts drift more than boats, and must therefore be started across farther upstream than boats. Rafts may be used for carrying equipment of men who cross by swimming or for the assistance of those who do not swim well enough. These individual rafts may be made of two logs with a board or two across them, reeds bound together (or bound banana stacks), or inflated rubber bags. Large rubber bags have been carried by patrols operating on or near the rivers. These bags were made from coffee sacks by coating the sacks with crude rubber. They were larger than seabags, and would hold one man's personal belongings and equipment. When partially filled with air and the mouth tied securely, they floated indefinitely, with sufficient buoyancy to support a man in the water.

6–96. **Crossing unfordable streams with usual infantry equipment.**—*a.* Recent experiments by the Philippine Scouts have resulted

RIVER CROSSINGS

in a method of stream crossing making use of little else besides the equipment usually carried. This method is believed to be superior to most methods of crossing by swimming. It is suitable for crossing in the face of opposition, its rapidity gives it great tactical value, and should cut down casualties considerably.

b. The two-man rifle float.—This float can be prepared by two men in 7 minutes. The two shelter halves (one on top of the other) are placed on the ground, and the remainder of the two packs and the clothing of the two soldiers are placed in the center of the canvas. Now the rifles are placed (crossed to give rigidity) on top of the packs and clothing. The float is completed by binding the 4 corners of the outside shelter half to the four extremities of the rifles by means of the shelter tent ropes.

TWO-MAN FLOAT, USING PONCHO

FIGURE 1.—First stage.

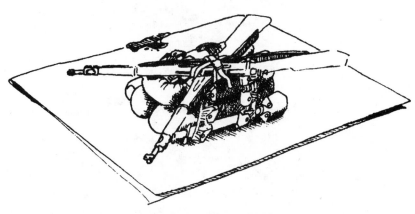

FIGURE 2.—Second stage.

RIVER CROSSINGS

"In a similar manner, using two 3-foot sticks or shelter tent poles instead of rifles, a machine gun complete can be floated in a shelter tent." ("Infantry Journal" for March and April, 1933.)

c. In the construction of the above float, bayonets are attached to rifles, with the scabbards on, to give greater length. A very little untying will make the rifle available to the man after landing on the opposite side since slip knots are used. As no additional materials are required this method is suitable for almost any stream crossing, even for deep fords, and is available to detachments who must go up or down the stream to make a land attack on forces opposing the use of a ford or ferry. Canvas is more nearly waterproof if wetted before making up the float.

d. Use of poncho in two-man float with horseshoe rolls.— (1) A method of constructing a two-man float using ponchos and the equipment usually carried in small wars operations was developed at the Marine Corps Schools along the lines stated in subparagraph *b* above, and has met tests satisfactorily. The float is made up of the rifles, ponchos, and shelter tent guy ropes. It contains: the horseshoe rolls consisting of one blanket, spare suit of underwear, toilet articles, food, the cartridge belts, canteens, haversacks, and

FIGURE 3.—Finished float.

FIGURE 4.—Cross section of float.

RIVER CROSSINGS

ayonets of two men, as well as their clothing and shoes which they emove. The horseshoe roll has the poncho on the outside of the lanket with the shelter tent guy rope used for lashing.

(2) *To make the float.*—(*a*) Remove the ponchos and shelter tent uy ropes from the horseshoe roll. Fold the poncho the long way rith the head hole at the side, and lay it flat on the ground.

(*b*) Lay the two cartridge belts on top of the center of the oncho, ammunition pockets down, so that one bayonet and one anteen will be on each side.

(*c*) Lay the horseshoe rolls on top of one another on the cartridge elts with their long dimensions parallel to the longest side of the olded poncho. Buckle the two cartridge belts around both rolls and lip the end of each bayonet under the other cartridge belt to hold he sides of the bundle rigid. See Diagram No. 1.

(*d*) Take the haversacks and place them on edge in the space n the center of the horseshoe rolls. Men remove their shoes and lace one pair on each side of the bundle so as to build out the sides.

(*e*) Lay the two rifles, crossed to form an X, on top of the bundle rith their ends pointed at the four corners of the folded poncho. Lash the rifles where they cross, with a waist belt to hold them in lace. See Diagram No. 2.

(*f*) Then the two men, starting at opposite ends of the same rifle, old the corner of the poncho up over the rifle end, and wrap the ides of the poncho corner up around it, lashing the poncho around he rifle end with one end of the shelter tent guy rope.

(*g*) The two men, next pass to the other rifle and wrap its ends in he two remaining corners of the poncho, lashing them in place with he bight of the shelter tent guy rope. The remaining end of the ope is then lashed to the next rifle end so that the float will be held n place by lashings between all four rifle ends. Care should be aken to wrap the corners of the poncho around the rifle ends and ash them so that water will not enter the float readily if one corner lips under. Diagram No. 3.

(*h*) Place the other poncho, folded twice, over the top of the quipment in the float and tuck the sides down around the edges of he equipment.

(3) When making up this float, stow the equipment and lash the rifles, in such a way that the float will be regular in shape, flat on the bottom, and not too high, so that it will float on an even keel and not tend to upset easily. The two men swimming

with the float should be on opposite sides of it and use a side stroke to swim, leaving one hand free to guide the float. If desired, the men can take turns pushing the float ahead of them with both hands, the float keeping their heads out of water while they paddle with their legs.

(4) Trained men need about 5 minutes to make these floats, exclusive of the time they spend removing their clothing. Care should be taken in lashing the float to use slip hitches that can be removed quickly when the line is wet, so that the rifles can be used instantly once the river is crossed.

(5) Similar larger floats may be improvised, by using wall, etc., tent flies with the upright poles or ridge poles.

e. Other canvas floats.—All canvas floats should be wide enough to prevent their capsizing easily, and the loads should have sufficient bulk in proportion to their weight so as to give buoyancy to the whole float. A light frame of boards or sticks will help in the case of heavy articles. Patrols and larger detachments may easily carry extra line for use in establishing lifelines for crossing streams, for starting animals into the water, etc., by wrapping about 30 feet of line, one-fourth inch in diameter, about a man's body just below the belt. (Many natives habitually carry such a line on their bodies when contemplating crossing streams.) Stronger lines may be carried by patrols using pack animals carrying the lines as top loads. These lines serve as picket lines during halts, and may be used in the construction of ferries. Extra canvas may be carried in the form of tent flies, which are used to construct floats, and as shelters for galleys or other purposes in camp. The inclusion of these few pounds of extra equipment may actually increase the mobility of a patrol. If strong vines grow near the river, they may be used as ropes in lashing trees together to obstruct the current, in lashing rafts together, as lines for small flying ferries, or as life lines. It is presumed that all patrols will include some machetes in their equipment when operating in the small wars situations.

6–97. **Crossing horses and mules.**—*a.* A stream that is too deep to be crossed by fording presents a very serious obstacle to a unit which includes riding or pack animals, and particularly so if the unit be operating in hostile country. Horses and mules can ford with relative ease streams that are difficult if not impossible to ford by men on foot. To cross animals, their cargo loads and equipment over a stream too deep or too swift to be forded is an opera-

RIVER CROSSINGS

tion to be undertaken only when the situation permits of no other course of action. The difficulties of such a crossing increase with:

(1) Width of the stream.

(2) Swiftness of the current.

(3) Size of the command (No. of animals).

(4) Slope of the banks, particularly on the far side.

(5) Hostile opposition encountered.

The width and the current are difficulties which are correlated. For any given width of stream, the animal will be carried farther downstream as the current is increased and likewise for any given current, the animal will be carried farther downstream as the width is increased. Since some animals will naturally swim faster than others, they will arrive on the far bank dispersed over a wide front. This front will sometimes be several hundred yards wide. When the animals arrive on the far bank, they must arrive at a point where the bank is not too steep or the footing too poor for them to get out of the water. Many horses will be drowned if a good landing place of suitable slope, width, and footing is not available on the far bank. Unfordable streams have been and can be crossed by single riders and by very small patrols but even the most daring and boldest leaders have hesitated to cross such streams. Many of the smaller streams rise and fall very rapidly. The unit leader should bear this in mind. Frequently it will be advisable to increase the rate of march so as to arrive at a crossing before an expected rise, or to decrease the rate so as to take advantage of an expected fall before crossing.

b. When the patrol includes riding or pack animals, these are usually taken across the stream by swimming. Mounted patrols in small wars usually cross by swimming with full equipment. Sometime the saddles and all equipment except the halter and snaffle bridle are removed and ferried across by boat or raft, and the rider removes most of his clothing. Pack animals are unpacked and unsaddled. The rider then mounts and rides the horse into the water at a point well upstream from a good landing place on the other side. The reins should be knotted on the neck, and only used when necessary to turn the horse, and before the horse gets beyond his depth in the water. Horses that are unwilling to take the water may be ridden behind other horses or tied behind boats or rafts. When the horse enters deep water, the rider slides out of the saddle,

RIVER CROSSINGS

or off the animal's back, and hangs onto the mane or halter on the downstream side, keeping low in the water and stretching out. Care must be taken to avoid being struck by the horse's front feet, and not to throw the animal off balance by putting too much weight on his head and neck. A swimming animal may be easily guided across a stream and prevented from turning back by the rider resting slightly forward of the withers and pushing the horse's head to the front with either hand if he attempts to turn. If the horse gets too excited, it may be necessary to release the halter or mane, and then catch the horse's tail and hold on. The horse should be kept headed upstream, especially in a swift current.

c. A much quicker and easier method of swimming animals is to herd them across, but animals not trained or practiced in swimming are often difficult to herd into water, and often turn around and come back when halfway across. (Herding should not be attempted in a swift current.) Sometimes the appearance of a few horses or mules on the opposite bank will help to start the herd across, and those which are seen to turn back in midstream may be kept out of the herd and crossed individually later. Some animals will have to be forced into the water. This may be done by passing a strong line across the rump of the animal and manning each end, a good rider then mounting the animal, and having a long lead line pulled by a man in a boat or on the other side of the stream. In dangerous streams it may be necessary to run a strong line across, and have a lead of at least 10 feet long tied to a pulley or sliding ring on the line and tied to the horse's halter with two leads of smaller line on the ring from each bank. The horse is then put into the water, head upstream, and drawn over to the other side by men pulling on the line and by the horse swimming. The ring is then drawn back for further operations. If the ropes do not fail, no horse can be washed downstream, and if done promptly, none should be in danger of drowning, as the tendency of this pulley method is to force the horse's head up. Sheer lines for trail ferries may be used in this manner, but it is usually more economical to lead the horse behind the ferry. A few difficult horses may be crossed by simply carrying a line from the horse to the other side of a narrow swift stream, holding it at a point upstream from the landing, and letting the horse swing across the stream and land on the other side (as is described in par. 27–7 e, for Flying Ferries). As this requires more

RIVER CROSSINGS

trouble than any of the other methods, it should be used only when there are few such difficult animals, and when other methods fail.

d. Mounted Marines should be thoroughly trained in swimming their mounts and in fording. Practice of this kind makes the animals willing to enter the water, and saves much time in emergencies. With light equipment, mounts may be crossed by swimming with the saddles on. The cantle roll aids buoyancy rather than detracts from it.

Section XII

SPECIAL OPERATIONS

6–98. **Trail cutting.**—*a.* Some situations will require the cutting of trails through wooded terrain. One cutter at the head of a small column will suffice. When it is desirable to open a trail for pack animals, there should be three or four cutters.

b. The leading cutter is charged with direction. In general the trail will follow a compass azimuth, with necessary variations to the right and left as determined by the terrain features, the ease of cutting, etc. As a prerequisite to trail cutting, the patrol leader must have a general knowledge of the area, the direction of flow of the more important stream lines and the intervening ridge lines, the distance to his objective and its general direction from the point of origin. A fairly accurate map, an airplane mosaic or a previous air reconnaissance over the route would be of inestimable value, but probably none of them will be available. Native cutters should be employed, if possible, and they can usually be relied upon to select the best and shortest route. This does not relieve the patrol leader of his responsibility of checking the general direction of the trail from compass bearings.

c. The second cutter widens to the right, the third to the left, and so on depending upon the number of cutters and width desired. If the trail is to be used by mounted men, a second group of cutters should follow the first, increasing the height of cut. All cutters should be equipped with suitable machetes. The cutters have a fatiguing task and should be relieved after from 10 to 30 minutes, depending upon the speed of movement desired, the thickness of the underbrush, and whether native or enlisted cutters are employed. The speed of cutting will vary from an eighth of a mile per hour for a large trail through the worst sort of jungle, to a mile or more an hour for a hasty trail through lighter growth. The heaviest brush will be found in the river bottoms, while the ridge lines will usually be comparatively open.

SPECIAL OPERATIONS

6–99. **Night operations.**—*a. General.*—Night operations present, in general, the same problems that are associated with such operations in major warfare.

b. Night marches.—Night marches are extremely exhausting for both men and animals. The rate of march is approximately half that of a day march. The distances between men and subdivisions in the column will have to be less than during day operations. In heavily wooded terrain, a night march is impossible on a dark night unless artificial lights are used. When marching over bad trails on clear or moonlight nights in such terrain, it will be necessary to march slowly and with only comfortable walking distance between men and subdivisions. Night marches are practicable, and desirable under some circumstances, when conducted over known trails. They are of very doubtful value if made over routes which are being traversed for the first time.

c. Night attacks.—A night attack, to be successful, implies an accurate knowledge of the location and dispositions of the enemy, of the routes of approach, and of the terrain in the vicinity of the hostile position. Each man participating in the operation should wear a distinctive white arm band, or other identification marker. A night attack should be an assault only, with the bayonet. Indiscriminate firing by the attacking force is fully as dangerous to friendly personnel as it is to the enemy. The problems of control are greatly increased. The favorable outcome of an attack is so doubtful that this operation should be attempted only after the most careful consideration. These remarks do not apply to attacks that are launched at dawn, but only to those that are made during the hours of darkness.

O

SMALL WARS MANUAL
UNITED STATES MARINE CORPS
1940

✦

CHAPTER VII

MOUNTED DETACHMENTS

RESTRICTED

UNITED STATES
GOVERNMENT PRINTING OFFICE
WASHINGTON : 1940

TABLE OF CONTENTS

The Small Wars Manual, U. S. Marine Corps, 1940, is published in 15 chapters as follows:

SMALL WARS MANUAL

UNITED STATES MARINE CORPS

CHAPTER VII

MOUNTED DETACHMENTS

v

Section I

INTRODUCTION

7-1. **Purpose.**—This chapter is designed to present in convenient form the minimum information required by Marines to procure and handle pack and riding animals in the theater of operations, to organize temporary and permanent mounted detachments, and to set forth the uses, limitations, and characteristics of such detachments in small wars. Exhaustive treatment of the subject of animal management, training and conditioning of animals and men, and mounted tactics, is beyond the scope of this chapter. However, sufficient detail is included to provide the basic essentials necessary for the successful handling of animals in the field. If the principles enunciated in this chapter are applied in small wars, many of the difficulties associated with the employment of animals will be obviated.

7-2. **Use of animals an expedient.**—The use of animals in small wars by the Marine Corps is a move necessitated by expediency. The tables of organization of the Marine Corps do not include animals for any purpose whatever, but the probable theaters of small-wars operations present transportation and tactical problems which usually require the use of animals for their successful solution. In those theaters where animal transport forms a basic part of the native transportation system, it generally will be found necessary for Marine Corps forces to employ animals, at least for transportation of supplies, and, generally, to some extent, for mounted work.

7-3. **Need for training in animal care and employment.**—The value of animals for military purposes is directly proportional to the skill and training of the personnel charged with their handling. It is therefore essential in our small-wars operations that considerable attention be devoted to the manner in which animals are handled and employed and to the proper training of personnel.

1

INTRODUCTION

7–4. **Some difficulties in employing animals.**—*a.* The employ-ment of animals by Marine Corps units should be attended by careful planning, intensive training, and close attention to detail. The diffi-culties and responsibilities of an officer commanding a unit employing animals are multiplied and such commander must realize these re-sponsibilities and make proper preparations to overcome these difficulties.

b. Some of the handicaps which must be faced by units employing animals are—

(1) Lack of personnel experienced in handling animals and animal equipment.

(2) Lack of the necessary specialists, i. e., horseshoers, veteri-narians, stable sergeants, pack masters, etc.

(3) Absence of any animal-procurement or remount service.

(4) Necessity for subsisting animals off the country due to prac-tical difficulties of providing an adequate supply of grain and forage to units.

(5) Lack of personnel experienced in the tactical handling of mounted units.

(6) Possible necessity for transporting animals to theatre of operations.

SECTION II

CARE OF ANIMALS

7–5. **Knowledge of animal management required.**—Successful handling of animals in the field demands at least a limited knowledge of the basic principles of animal management. This subject includes a general study of the framework and structure of the horse and mule, the colors and markings, feeding and watering, grooming, conditioning, first-aid treatment of diseases and injuries and care of the animal's feet.

7–6. **Nomenclature.**—*a.* The regions of the horse and mule are shown in plate No. 1 (p. 4).

b. The principal bones and joints of a horse and mule are shown in plate No. 2 (p. 5).

c. All officers and noncommissioned officers charged with the handling of animals should be thoroughly familiar with the matter contained in these 2 plates as they constitute what might be called the "nomenclature" of the military animal. All members of regularly organized mounted detachments and men in charge of pack animals should be instructed in this "nomenclature."

7–7. **Identification.**—*a.* There are certain prominent and permanent characteristics by which an animal may be identified. These characteristics are, color, markings, and height. A proper execution of the animal descriptive card (Form NMC 790) required by article 21–2, MCM., necessitates a familiarity with these characteristics.

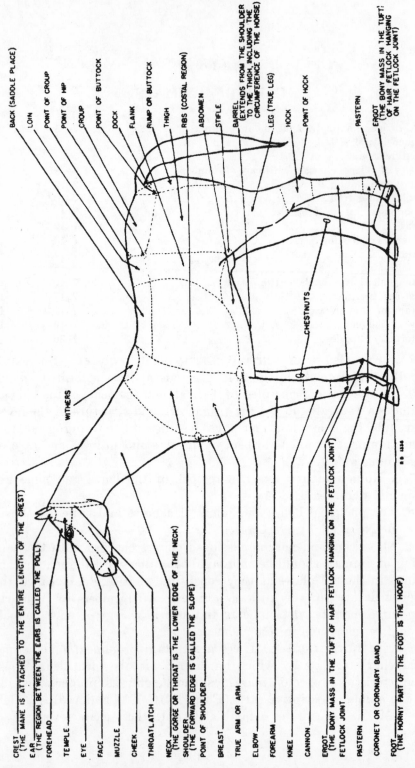

PLATE 1.—Regions of the horse and mule.

CARE OF ANIMALS

b. The colors of horses and mules are shown below:

(1) Black is applied to the coat of uniform black hairs.

(2) Chestnut is a medium golden color.

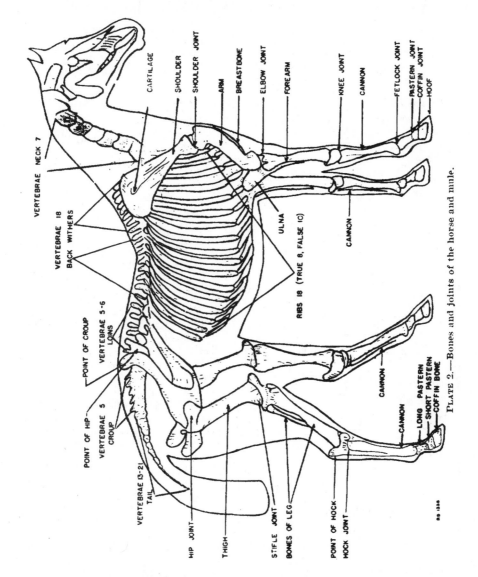

PLATE 2.—Bones and joints of the horse and mule.

(3) Bay is a reddish color of medium shade.   Black points.

(4) Brown is the color of the coat almost rusty black and distinguished therefrom by the reddish coloration around the nostrils, elbows, flanks.

CARE OF ANIMALS

(5) Gray is applied to a coat of mixed white and dark colored hairs, about equal in numbers.

(6) Mouse is an ash gray shade resembling the color of the mouse.

(7) White is an absence of pigment. Skin is white.

N. M. C. 700 QM

DESCRIPTIVE CARD OF PUBLIC ANIMALS

Place or organization ....

....

Date of purchase ....

Date of receipt ....

Horse: Riding, Driving, or Draft ....

Mule: Draft, Pack, or Riding ....

Hoof No. .... Name..............................

    Sex .... Age .... years

    Weight .... Color............................

    Height .... hands.................. inches

Remarks: ....

....

SPECIAL DESCRIPTION

4-6771

SERVICE

(1) Transferred from ....
                (Organization)

..............................., ...
    (Place)   (Date)

to ....
           (Organization)

..............................., ...
    (Place)   (Date)

.... Quartermaster

(2) Transferred from ....
                (Organization)

..............................., ...
    (Place)   (Date)

to ....
           (Organization)

..............................., ...
    (Place)   (Date)

.... Quartermaster

(3) Transferred from ....
                (Organization)

..............................., ...
    (Place)   (Date)

to ....
           (Organization)

..............................., ...
    (Place)   (Date)

.... Quartermaster

(4) Transferred from ....
                (Organization)

..............................., ...
    (Place)   (Date)

to ....
           (Organization)

..............................., ...
    (Place)   (Date)

.... Quartermaster

FINAL DISPOSITION

Died ...., 19......, at ....

Cause ....

Surveyed, condemned, and sold ...., 19......

at .... Price, $..............

Cause ....

Remarks: ....

....

....

....

....

U. S. GOVERNMENT PRINTING OFFICE: 1937   4-6771

(8) Roan is applied to a coat composed of red, white, and black hairs, usually red and white on body with black mane and tail.

(9) Buckskin is applied to a coat of uniform yellowish colored hairs.

(10) Piebald is applied to the coat divided into patches of white and black only.

CARE OF ANIMALS

(11) Pied Black, Pied Bay, and Pied Roan are terms used to designate the patched coats of white and black, white and bay, or white and roan. If the color other than white predominates, the term pied should follow the predominating color, as black pied, bay pied, or roan pied.

(12) Dapple is prefixed to the designation of any color when spots lighter or darker about the size of a silver dollar overlay the basic color.

c. The following are the principal white or other contrasting hair markings found on horses and mules:

(1) White Hairs is a term used to designate a few white hairs on the forehead, at the junction of the neck and withers, on the shoulders, the coronet, over the eyes, etc.

(2) Star designates a small, clearly defined area of white hairs on the forehead.

(3) Race designates a narrow stripe down the face, usually in the center and further described as "short" when it does not reach the nose.

(4) Snip designates a white mark between the nostrils.

(5) Blaze designates a broad splash of white down the face. It is intermediate between a Race and White Face.

(6) White Face means that the face is white from forehead to muzzle.

(7) Silver Mane and Tail designates the reflection of white in these appendages.

(8) White Pastern means that the white extends from coronet to and including the pastern.

(9) Quarterstocking means that the white hairs extend from coronet to and including the fetlock.

(10) Halfstocking designates that the leg is white from the coronet to an inch or two above the fetlock.

(11) Three-Quarterstocking means that the white hairs extend to midway between fetlock and knee or hock.

(12) Full Stocking designates the leg white to or including the knee or hock.

(13) Cowlick is a term applied to a tuft of hair presenting an inverse circular growth. They are permanent distinguishing characteristics, which should be recorded.

(14) Black Points means black mane, tail, and extremities.

(15) Ray designates the dark line found along the back of some horses, and many mules.

CARE OF ANIMALS

(16) Cross designates the dark line over the withers from side to side.

(17) Zebra Marks designates the dark, horizontal stripes seen upon the forearm, the knee, and the back of the cannons.

d. The height of horses and mules is expressed in "hands." A hand is 4 inches. The animal is measured by first placing him on level footing and causing him to stand squarely on all 4 feet. The perpendicular distance from the highest point of the withers to the ground is then measured with a stick that is graduated in hands and inches.

7–8. Duties of officers charged with care of animals.—*a.* Officers having animals attached to their units should keep them in such training and health as will enable them to do their work to the best advantage. This requires careful instruction of the men in the treatment, watering, feeding, grooming and handling of animals, and such continuous supervision and inspection by officers as will insure that these instructions are understood and carried out.

b. Officers in charge of animals should know the symptoms and treatment of common diseases, first aid treatment of injuries and should be familiar with the principles of horseshoeing. This information can be obtained from TM 2100–40, "The Horseshoer."

7–9. Rules for handling animals.—*a.* All men connected with the care and handling of animals must be taught, and must thoroughly understand, the following rules for the care of animals:

(1) Animals require gentle treatment. Cruel or abusive treatment reduces the military value of animals by making them difficult to handle.

(2) When going up to an animal speak to him gently, then approach quietly.

(3) Never punish an animal except at the time he commits an offense, and then only in a proper manner—never in anger.

(4) Never kick an animal, strike him about the head, or otherwise abuse him.

(5) Never take a rapid gait until the animal has been warmed by gentle exercise.

(6) Animals that have become heated by work should not be allowed to stand still but should be cooled down gradually by walking.

(7) Never feed grain or fresh grass to an animal when heated. Hay will not hurt a heated animal.

(8) Never water an animal when heated unless the march or exercise is to be immediately resumed.

CARE OF ANIMALS

(9) Animals must be thoroughly groomed after work.

7-10. Stables and corrals.—Stables need not be provided in tropical climates, if some type of shed is available for protection from rain. If stables are used, they should be well ventilated, but without draughts. Stalls should be so constructed that the animals can lie down in comfort. Stables must be well drained, and stable yards and corrals must be so situated that even heavy rains will drain off. Sand is a good standing for animals, either in stalls or corrals. A corral should have a strong fence and gate, and have some shade, either natural or artificial, at all hours of the day. A manger should be provided if grain is to be fed. It is highly desirable to have water available at all times in corrals.

7-11. Grooming.—a. Thorough and efficient daily grooming has a very close relation to the good condition which is so essential in animals being used in military operations. Proper grooming aids greatly in maintaining the skin of the animal in a healthy condition, prevents parasitic skin diseases and infections, and reduces to a great extent the incidence of saddle sores.

b. The following points are important in the proper grooming of animals:

(1) The currycomb should not be used on the legs from the knees and hocks downward, nor above the head.

(2) First use the currycomb on one side of the animals beginning at the neck, then chest, shoulders, foreleg down to the knees, then back, flank, belly, loins and rump, the hind leg down to hock. Proceed in similar manner on the other side.

(3) Next brush the animal in the same order as when currycomb was used except that in brushing legs go down to the hoof.

(4) In using the brush, stand well away from the animals, keep the arm stiff, and throw the weight of the body against the brush.

(5) The value of grooming is dependent upon the force with which the brush is used and the thoroughness of the work.

(6) Wet animals should be dried before grooming.

(7) The feet should be cleaned and the shoes examined.

(8) Sponge out the eyes, nose, and dock.

(9) Officers and noncommissioned officers should, by continual and personal supervision, see that the grooming is properly done.

7-12. Forage.—Forage can be conveniently divided into two classes; roughage, including such types as hay, grass, sugarcane tops, leaves of trees, etc., and grain, including such as oats, corn, and Kaffir corn. Grain is not necessary to the animal's existence if he is doing no work, but for military animals is a concentrated energy-producing food, which enables the animal to do the sustained work required of him. It requires much more time for a horse or mule to eat enough grass or hay to support life and keep in condition than if the roughage is supplemented with a reasonable amount of grain. An economical method of feeding is to make the fullest possible use of pasture. Bulk is an essential for the diet of horses and mules. Concentrated foods, no matter how nourishing, cannot alone maintain an animal in condition. An unlimited supply of grain cannot take the place of roughage.

7–13. **Principles of feeding.**—a. The following principles of feeding are the results of long experience and should be adhered to as closely as the circumstances will permit:

(1) Water before feeding.

(2) Feed in small quantities and often. The stomach of the horse and mule is small in comparison with the rest of the digestive tract and therefore cannot digest large feedings. Three or more feedings a day are desirable.

(3) Do not work hard after a full feed.

(4) Do not feed a tired horse a full feed. Failure to observe this principle frequently results in the most severe colic, in laminitis (feed founder), or both.

(5) Feed hay before grain. This is not necessary for the first feeding in the morning because hay has been available all night and has therefore taken the edge off the animal's hunger.

7–14. **Watering.**—a. The following rules for watering should be adhered to:

(1) In corrals it is desirable that animals should have free access to water at all times. If this is impossible animals should be watered morning, noon, and evening.

(2) Water before feeding, or not until 2 hours after feeding.

(3) Animals may be watered while at work but, if hot, they should be kept moving until cooled off.

(4) On the march the oftener the animals are watered the better, especially as it is not usually known when another watering place will be reached.

(5) In camp, where water is obtained from a river or stream, animals must be watered above the place designated for bathing and for washing clothes.

(6) Animals should be watered quietly and without confusion.

CARE OF ANIMALS

7–15. **Conditioning.**—*a.* Condition as applied to animals used for military purposes means health, strength, and endurance sufficient to perform without injury the work required of them.

b. Good hard condition is the best preventive against loss of animals from any causes except accidental injury. The importance of proper conditioning cannot be overestimated. More animals are incapacitated or die in military operations from lack of proper conditioning than from any other cause. This is especially true in those countries where native animals are procured locally. As a general rule, such horses or mules are of the grass-fed variety accustomed to working 1 day and grazing the next 3 or 4 days.

c. There is only one way to condition animals, whether they are required for riding, pack, or for draft. The only method is a judicious combination of sufficient good feed, and healthful work, continued over an extended period. The transformation of fat, flabby flesh into hard, tough muscle cannot be forced. A regular program of graduated work is the only way to accomplish it. Some animals require longer periods of conditioning than others depending upon their age and the amount of previous work they have performed. Individual attention to each animal is required in conditioning. All work should be light at first and gradually increased.

7–16. **Management of animals on the march.**—*a.* Without condition, it is impossible for animals to undergo the fatigue and exertion incident to any prolonged effort. When military necessity requires the marching of uncondition animals, unless the situation is such as to override all thought of loss, the space and time must be comparatively short, otherwise exhaustion, sore backs, and sore shoulders will shortly incapacitate the majority of the animals and the mobility and efficiency of the unit will be very greatly reduced. From the viewpoint of the animal's welfare, the length of the march is to be estimated not only in miles but also with regard to the number of hours that the load has to be carried. This latter consideration is frequently the more important of the two. Particularly does this latter consideration apply to pack animals. The advance of a column in small wars is sometimes as slow as 1 mile an hour and even less. Under such circumstances a short march may, in reality, demand extreme endurance of the animals in the column.

b. Prior to starting, a special inspection of the saddles, packs, harness, and shoeing should be made to insure that all is in order. After the halt for the night all animals and equipment should be in-

11

spected, necessary treatment and repairs made, and all gear placed in order so as not to delay the hour of starting in the morning.

c. With any considerable number of animals in the column, it is seldom advisable to start before daylight except for purely military reasons. In the dark, feeding and watering cannot be satisfactorily handled; saddles, packs, and harness may not be properly adjusted and it is practically impossible to properly inspect the adjustment of equipment. Many sore backs will result from saddle blankets and pads being improperly folded in the darkness. Night marches, with any considerable number of animals in the column, will prove most difficult and unsatisfactory unless the personnel is thoroughly experienced in handling, saddling, bridling, and in packing animals.

d. A first halt should be made after being under way for ten (10) or fifteen (15) minutes to allow men and animals to relieve themselves. At this time, equipment is adjusted, the girths tightened and an inspection made to insure that the saddles, packs, and harness are all correct. This halt is most important, especially for the purpose of rechecking the saddling and packing. Subsequently a short halt of about 5 minutes should be made hourly. At each halt each man should look over his animal, examining the feet and the adjustment of animal equipment and loads. Noncommissioned officers should be indoctrinated to enforce this inspection.

e. Animals should be watered within reason whenever an opportunity occurs, especially on hot days. The principle of watering before feeding is of course adhered to on the march but, if a stream is crossed an hour after feeding, they may again be allowed to drink if circumstances permit the delay. Bits, especially the curb, should be removed when it is intended to give a full watering. While watering, overcrowding must be prevented, and plenty of time given every animal to drink his fill. Groups of animals should come up to the watering place together and leave together. If they are moved away individually, others cease drinking and try to follow them. At the end of the day's march it is best to relieve the animals of the weight of their equipment without delay and before watering.

f. Feeding of animals on the march in small wars has always been a difficult problem. On an extended march it is very difficult to keep animals in condition. The problem of providing sufficient forage requires constant attention and is one that will tax the ingenuity of the leader of any column containing a considerable number of animals. The first step in the solution of this problem is to

CARE OF ANIMALS

indoctrinate thoroughly the marine in the principle that the animal under his care is going to suffer if he does not think of and for his animal on all occasions. It must become a matter of simple routine for him to care for his animal at every possible opportunity. At every halt, if it is at all possible, he must permit his animal to graze. Many standing crops provide excellent filler and may be used for feeding the animals, or if circumstances permit, a sufficient quantity may be cut and carried to provide for the next halt.

g. Feed bags should be provided for animals for the purpose of feeding grain on the march. The feeding of grain from the ground is not only highly wasteful but is the frequent cause of severe colic. These bags can be used for carrying grain on the march by filling them and securing the open end. They are generally carried attached to the pommel of the saddle on riding animals and as a top load on pack animals. A supply of corn or other grain along the trail should never be passed without refilling the empty feed bags. With the feed bag, the principle of feeding little and often can be adhered to with greater facility. The best type for military use is that at present issued by the Army. If no feed bags are available they should be improvised from canvas or other durable material.

h. It frequently will be possible to halt for the night at places where small enclosed pastures are available. If such be the case the animals should be permitted to graze throughout the night. It will, of course, be necessary to provide an adequate pasture guard to prevent the seizure or destruction of the animals in case of a night attack. When possible a site within or adjacent to a suitable enclosed pasture should be selected. Otherwise ground flat enough to provide level standing for a suitable length of picket line should be selected. Marshy ground should be avoided if possible. A nearby water supply suitable for men and animals is essential. It should be as near the camp as possible but some sacrifice may have to be made in this respect in order to occupy a position suitable for night defense. In some areas it will not be possible to graze animals at night. In such cases some form of restraint must be used.

i. The most satisfactory method of securing animals at night is to secure them to a picket line raised 3 or 4 feet above the ground. This line should be stretched taut between trees suitably spaced or between other suitable supports. After the animals have been secured to the picket line it is essential that a man be kept on watch to prevent the animals from becoming entangled and injuring themselves. Hay, tall grasses, sugarcane, etc., should be procured and

13

CARE OF ANIMALS

placed within reach of the animals along the line as it will make them stand more quietly and provide them with nourishment.

7-17. First-aid treatment.—*a.* The evidence indicates that in small wars casualties among animals have occurred in the following order of frequency:

(1) Wounds and injuries. (Pack and saddle injuries account for most of these.)

(2) Loss or want of condition and exhaustion.

(3) Intestinal diseases. (Colic.)

(4) Contagious diseases.

b. Loss or want of condition, and pack and saddle injuries, account for the bulk of the losses in small wars. These are to a large extent preventable. While the animal is fit and in condition, hardship and exertion can be borne without injury, but the unconditioned animal soon becomes unfit and a handicap because of injury or disease. The prevention of injuries and disease is far more important than their treatment. And particularly so as there are no veterinarians available. Injuries and emergency cases must be dealt with in a common-sense manner. If veterinary service is obtainable it should be utilized.

c. The healthy animal stands with the forefeet square on the ground; one hind foot is often rested naturally. The pointing, or resting of one forefoot or the constant shifting of the weight on the forefeet indicates a foot or leg ailment. The pulse of the healthy animal is thirty-six (36) to forty (40) beats per minutes; the respiration at rest nine (9) to twelve (12) per minute; the temperature ninety-nine (99) to one hundred (100) degrees Fahrenheit. The temperature is taken by placing a clinical thermometer in the rectum for about three (3) or five (5) minutes. The droppings of the healthy animal should be well formed but soft enough to flatten when dropped.

d. Loss of appetite, elevation of temperature, accelerated breathing, listlessness, dejected countenance, stiffness, profuse sweating, nasal discharge, cough, diarrhea, pawing, excessive rolling, lameness, reluctance to move, and loss of hair or intense itching are some of the most common indications of disease.

e. Pressure on the back will often cause swellings, which by further rubbing and pressure become open sores. While unbroken, swellings may be cured by removing the pressure and soaking with cold water packs. If the animal is to work, fold or cut away the saddle pad so that the place is left free. If open, saddle sores, like other wounds, must be kept clean and flies kept out of them.

f. Wounds will heal naturally if they are kept clean and well drained. Almost any grease that is unappetizing to flies will help to heal the wound. Screwworms and maggots, the larval form of certain flies, are frequently found in wounds. Lard with a little sulphur mixed in it is usually available. If the ingredients can be obtained, the following mixture will keep the flies out of wounds:

½ ounce creolin.

1 ounce linseed oil (or oil of tar).

10 ounces olive oil (or salad oil).

If close inspection shows the presence of worms or maggots, or if there is a thin reddish discharge from the wound or sore, the following treatment is indicated: swab out thoroughly with a soft cotton swab dipped in creolin. The edges and especially the lower edge where the wound drains should be greased, to prevent burning by the creolin or the spreading of the sore by the discharges coming in contact with the skin. This treatment will kill the worms or maggots and they will slough off with the dead flesh. All wounds should have some opening through which they may drain, at the lowest point in the wound, and grease should be used to prevent the drain causing running sores. Pus in the feet is drained off through the sole, and treated like any other wound.

g. Colic is the term given to the symptoms shown by animals with abdominal pains. This pain may be caused by any of numerous conditions. The predisposing causes are: small size of stomach compared to the size of the animal and capacity of digestive tract, and inability of the animal to vomit. The chief exciting causes of colic are:

(1) Over feeding.

(2) Feeding or watering exhausted animals.

(3) Feeding wilted grass.

(4) Sudden changes of food.

(5) Working hard after a full feed.

(6) Lack of sufficient water.

(7) Eating hay or grain on sandy soil.

(8) Eating mouldy hay or grain.

(9) Eating green grain.

(10) Intestinal tumors, abscesses, etc.

h. The symptoms of colic are uneasiness, increased perspiration depending upon the degree of pain, pulse and respiration accelerated,

CARE OF ANIMALS

pawing, turning head towards flanks, lying down, sometimes rolling and rising frequently, and excessive distension of abdomen.

i. A compliance with the principles of feeding and watering as set forth in this section will reduce the incidence of colic to a minimum. Prevention is far more important than treatment, and therefore it is most important that the principles of feeding and watering be adhered to closely. Military necessity may sometimes prevent a strict adherence to these principles.

j. Bed down a space with hay or dry grass and tie the animal with just enough shank to allow him to lie down comfortably. In ordinary cases give one aloes ball or the following drench: raw linseed oil one (1) pint, turpentine one (1) ounce; if not relieved repeat the drench in 1 hour. Induce the animal to drink but withhold food until the acute symptoms subside.

k. Any ordinary long-necked bottle properly wrapped to protect it from breaking may be used in giving a drench. The animal's head should be raised until the mouth is just slightly higher than the throat to provide a gravity flow to the throat. The neck of the bottle is inserted in the side of the mouth and a small amount of the drench administered. This must be swallowed before more is administered. Repeated small amounts are administered in this way until the required amount has been given. If the animal coughs or chokes his head should be immediately lowered to prevent strangling.

7-18. **Communicable diseases.**—*a.* Prevention is again the prime aim. Proper conditioning and seasoning, plenty of wholesome food, good grooming, and protection from undue exposure to the elements and mud, keep the animals strong and in such a state of health that they can resist considerable exposure to infection. When a disease appears among a group of animals, there are certain rules of procedure that have been found absolutely necessary in checking the spread to healthy animals and in stamping out the disease. These measures are:

(1) Daily inspection of all animals in order to detect new cases. This insures the prompt removal of the sick as a source of infection and the initiation of the proper treatment or destruction.

(2) Quarantine of exposed animals.

(3) Isolation of sick animals.

(4) Disinfection of infected premises, equipment, and utensils.

b. The treatment of the various communicable diseases of the horse and mule are beyond the scope of this chapter.

CARE OF ANIMALS

7-19. Care of the feet.—Animals that have to travel over hard and stony roads should be shod, at least in front, to prevent excessive wear of the horn of the hoof. On soft going and in pasture, shoeing is neither necessary nor desirable. Shoes should be flat, without calks, and should fit the outline of the hoof. When the hoof wears too much, the animal goes lame, but a rest in pasture allows the hoof to grow again. Shoes left on too long do not permit the natural growth and wear of the hoof, and they should be removed from animals which are expected to remain out of service for a considerable time. A properly fitted shoe allows the animal to stand flat on a level floor without strain. Neither commercial shoers nor members of the detachment should be allowed to cut away any of the sole or frog or to rasp or trim the outer surface of the hoof. The thick callous of the frog and sole protects the animal's foot from bruises, and the compression of the frog circulates the blood through the foot. The natural varnish of the outside of the hoof retains the moisture in the hoof, and keeps it from rotting. Hoofs should be cleaned out at grooming time and before starting out, to make sure that there are no stones or small sticks caught in the hoof between the frog and the sole. Feet should be inspected by riders and drivers at each halt.

7-20. Veterinary supplies.—Standard surgical and medical supplies for the treatment of common ailments are obtained by requisition to the quartermaster or in emergencies from the medical department. (See art. 21-5, MCM.) In small wars where any considerable number of animals is employed, it is highly desirable that Stable Sergeant's Veterinary Chests, supplied by the Medical Department, U. S. Army, be requisitioned in sufficient number. These chests are quite complete and contain sufficient medicines and instruments for all ordinary veterinary cases.

Section III

PROCUREMENT OF ANIMALS

7–21. **Necessity for local purchase.**—*a.* Animals have seldom been transported with expeditionary forces to the theater of operations. Mounted detachments and pack transportation have been employed only in those areas where horses and mules were used locally and in which the supply of animals was reasonably adequate. A careful study of the theater of operations should indicate whether the type and number of animals available will be adequate for the military needs in that particular area. For special operations and for special cargo loads the type of animal procurable in the theater of operations may not be suitable. In some operations it has been necessary to bring into the occupied country United States bred mules for use in transporting pack artillery. Present pack artillery weapons cannot be broken down into loads sufficiently light to be transported on mules weighing much under 1,000 pounds.

b. The great majority of animals used in small-wars operations will probably be procured within or near the theater of operations. This section will deal, therefore, primarily with the problem of procurement in the theater of operations.

7–22. **Procurement agents.**—*a.* While procurement is primarily a supply or quartermaster function, the general practice has been to authorize units in the field to procure their own animals. Under this system the quartermaster has simply set a maximum price and advanced the necessary funds to the unit commander in the field and he, or an officer appointed by him, has acted as the purchasing agent. While this system is frequently necessary to provide animals quickly

PROCUREMENT OF ANIMALS

for immediate use in the early stages of the operation, it has the effect of rapidly running up the price of animals due to the fact that there will be a number of purchasing officers in the market competing for the more serviceable animals.

b. As soon as practicable, it is advisable to establish a single procurement agency with the function of selecting, purchasing, distributing and accounting for all animals procured by the Force. This should be done with a view to providing a better selection of animals, to facilitate accountability, and to effect a saving in animals and in money.

7-23. **Native dealers.**—It is well to utilize native dealers in procuring animals. The average cost per head thus will possibly be higher than it would be otherwise but this is offset by a saving in time and energy. Moreover, the native dealer will know where the desired types are to be found. It is most essential however, to convince the dealer at the outset that he must deliver for inspection only such animals as conform to the minimum standards.

7-24. **Purchasing from native dealers.**—The purchaser of native animals should determine by thorough inquiry what prices have been usual among the natives for the best types of animals. Although economy in expenditure of government funds may be temporarily subordinate to military necessity, it is never wise to pay excessive prices for animals.

7-25. **Minimum specifications for animals.**—*a.* All animals delivered for inspection as saddle animals or pack animals should have all of the following qualifications:

(1) Be reasonably sound.

(2) Have been worked under pack or been ridden enough to require little or no further training.

(3) Be mature; immature animals are useless for military purposes, no matter how sound they may be.

(4) Be of the size required for the purpose for which intended; this should be fixed only after careful consideration of the types and capabilities of the mounts available, but it will be found unprofitable to go below the minimum, once fixed.

(5) Be as nearly as possible in condition for use; it is not necessary to require that an animal be in perfect condition, but he should be able to carry his load the day of purchase.

7-26. **Height qualifications.**—In most small war theaters the native animals are undersized according to United States standards. Moreover, the average size varies somewhat in the various countries.

PROCUREMENT OF ANIMALS

An average as low as 13.1 hands has been used in past expeditions. The purchaser can soon determine, however, a fair average in height from observation of the animals he finds in use. He should then set a height standard and use this in his selection as the first step in eliminating undesirable animals from the herd the dealer is showing.

7–27. **Age qualifications.**—*a.* The dealer should be given definite age limits, as a guide toward satisfactory animals. Six to twelve years is the most satisfactory period. The animal under 6 years of age cannot endure the exertions and privations incident to taking the field. Since few matured native animals can be found that will not show the white hairs and scars of sores and injuries, the purchaser will be strongly tempted to choose immature animals in order to obtain animals free from blemishes.

b. It is better to choose older horses because they will generally have become thoroughly broken. This factor is most important at the outset because the majority of men detailed to handle them will have had little recent practice with animals.

c. The twelve (12) front (incisor) teeth afford the easiest and most reliable means of determining the age of a horse or mule. These teeth consist of six uppers and six lowers and, from side to center, are known as the "corners," "laterals," and "centrals." The horse (mule) has two sets of teeth:

(1) The temporary or colt teeth which are cast off when the permanent teeth erupt. This shedding of the temporary teeth begins with the "centrals" at about the age of three and is completed when the "corners" are shed at about five years.

(2) The permanent or second set of teeth. The permanent teeth have all erupted and are in wear at the age of 5½ years. The temporary teeth can be distinguished from the permanents by their milk-white color. The permanent teeth stain very quickly and generally have the dull appearance of old ivory.

d. The incisor tooth in cross section is shown below:

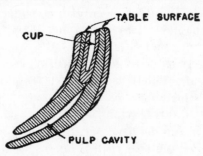

INCISOR TOOTH OF HORSE

PROCUREMENT OF ANIMALS

e. Wear of the incisor teeth. The principles of age determination from 6 years upward are based primarily upon the wear of the incisor teeth. The tooth attains its greatest length the second year after eruption. During the succeeding years, teeth do not grow but undergo a regular process of destruction from wear and from the receding of the bony socket margin. As the tooth wears down, the cup finally disappears. The table surface of the tooth changes from oval to triangular and finally becomes rounded.

f. Some of the more important means used in age determination are:

(1) Loss of temporary teeth and eruption of the permanent teeth. This has been completed at the age of 5.

(2) Disappearance of the cups. All cups have disappeared at 8 years.

(3) Shape of table surface of teeth. At 10 to 12 years the centrals and laterals are triangular. At 16 to 20 the table surfaces are round or flat from side to side.

(4) Angle of incidence. The angle of incidence between the upper and lower incisors becomes more and more acute as the age increases.

(5) Space between the teeth at the gums increases as the animal grows older.

g. Procedure in determining the age. The angle of incidence and the presence or absence of temporary teeth should be noted. The mouth should then be opened and the teeth examined for cups and the shape of the table surfaces noted. The lower incisors are more reliable as a guide than the uppers.

h. The following classifications are listed:

(1) Animals under 4½ years; temporary teeth present; permanent teeth erupting.

(2) Animals between 5 and 8 years; cups present, table surfaces oval; angle of incidence about 180°; all teeth are permanent.

(3) Animals over 8 years; absence of cups; table surfaces triangular or round; angle of incidence more acute; space between teeth at the gums begins to show as the animal gets older. Old horses begin to show gray hairs around the eyes and nose, and the depressions over the eyes become more sunken.

7-28. Examination for soundness.—*a.* The examination for soundness should be as thorough as the circumstances, the availability of trained men, and the knowledge of the buyer permit. However, the average officer can buy horses and mules without having had any special training, if he uses good judgment and buys what he

PROCUREMENT OF ANIMALS

considers to be sound animals. The following points are suggested as guides, and will assist the inexperienced horse buyer until he acquires experience:

(1) Observe the horse at a halt, noting whether he is very lean or obviously crippled.

(2) Examine the head and neck. Check for blindness by looking at the eyes and by passing the hand suddenly over each eye in turn. If the horse does not blink, he is blind in that eye. Pass the hand over the head and face, and see if there is evidence of any sores or injury. See if the mouth and nostrils look healthy, and if the animal breathes freely.

(3) Examine back, noting scars or sores caused by packs and saddles. Many animals will be found with scars, but these need not be rejected, if the scar tissue appears to be healthy. In many cases, animals without them will turn out to be young and untrained, and actually less desirable. All animals with actual puffs and sores should be rejected.

(4) Examine the legs. If any variation in symmetry between the legs of a pair be found, it is safer to reject the animal unless you know enough to differentiate between temporary and permanent disabilities. Legs should be reasonably near to the same vertical plane, fore and aft, and the animal should put his weight on all of them. Joints should not be swollen.

(5) Examine feet. The feet and pasterns should not be sore to the touch. There should be plenty of horn on the hoofs, and they should not show any split or crack. The frog should rest on the ground, but since it is quite usual to pare the frog and sole, do not reject on this account, but trust the frog to grow later. Examine the coronet and press it with the fingers. A prick in the sole of the foot will sometimes result in pus breaking through at this point (just above the horny part of the foot).

(6) Examine the hindquarters, sheath, tail, anus, etc.

b. Animals which have passed these tests should be segregated from those already rejected and those awaiting examination. If no serious and obvious defects have been noted, and the animal has a general healthy appearance, an alert bearing and a reasonable amount of flesh, then have him led on a loose rope directly away from you and then directly toward you, at the walk and trot. If the animal is lame or has badly formed legs, this will usually be apparent. A lame animal "favors" the lame foot, adjusting his weight so as to put very little weight on it. Since his head is his principal means of doing this,

PROCUREMENT OF ANIMALS

you will see his head drop as he puts the sore foot down. If all feet are lame, the animal will trot very short and reluctantly. Most lameness is in the feet.

c. Have each animal saddled and mounted, and ridden for a short distance, to demonstrate that he can be handled. Do not require that a horse have any particular gait, except the walk, as many of the best will have been gaited in a way not suitable for military purposes, but which may be changed by training. The bargaining may well be begun now. Have the animal worked at a trot or canter for a few minutes, to see if he appears to have good wind. In this connection, the animal with a broad muscular breast is to be preferred to one with a very, narrow breast, as its wind is usually better.

7–29. **Marking of purchased animals.**—As soon as an animal has been purchased, the animal descriptive card (Form NMC 790) required by article 21–2 MCM should be completely and accurately filled in showing all markings and pertinent data and he should be immediately branded. The customary method is to brand the animal on the left shoulder with the letters "U. S." (See par. 7–7 IDENTIFICATION.)

7–30. **Use of United States animals in small wars.**—When it is necessary to transport United States animals to the theater of operations, a period of recuperation and acclimatization after the sea travel will be necessary. The unnatural environment and the lack of adequate exercise incident to sea travel debilitate animals to an extent dependent upon the length and character of the voyage. After a period of recuperation and acclimatization, and after the animals have become gradually adjusted to any changes of food necessitated by the local forage supply, they should thrive practically as well as the native animals. The larger United States bred animal, being required to carry a greater load, requires a greater quantity of grain and roughage to keep him in condition. Such losses in United States animals as have been suffered in the past operations have been due principally to lack of feed and to unskilled handling.

Section IV

MOUNTED DETACHMENTS

7–31. Value of mounted detachments.—*a.* It is reasonable to expect that small-war operations of the future, like those of the past, will require the use of mounted detachments. The value of mounted detachments will depend upon the nature of the terrain, the character of the resistance, the extent of the operations, and, finally, the missions assigned to them.

b. The nature of the terrain has a direct bearing upon the value and use that is to be made of mounted detachments. The more open flat and rolling terrain is more favorable for the successful employment of mounted detachments. As the country becomes more mountainous, more given to jungle growth or marsh lands the general use of mounted detachments becomes less practicable. Furthermore, the more unpopulated and uncultivated areas are less favorable for the general use of mounted detachments. As the country is more given over to waste lands it becomes increasingly difficult to maintain mounted detachments in the field.

c. (1) The character of resistance encountered will have considerable effect both on the general effectiveness of mounted detachments and on the manner in which they should be employed. If the enemy tends to retain a well organized unity, the mounted detachment can be employed by attaching it to the force or column sent in the field to destroy this resistance. If, on the other hand, the enemy retains no definite organization but uses guerrilla tactics the mounted detachment can be best employed as an area or district reserve to be available for independent action on special missions. Again, in a situation where the enemy operates to any extent mounted, it may be necessary to use mounted detachments in place of foot patrols for regular combat patrolling.

(2) The effectiveness of mounted detachments also varies with the type of armament in use by the enemy. Prior to the advent of automatic weapons, mounts were of some value after contact was gained. They could be used for shock action or for maneuver even in the immediate presence of the enemy. The modern high powered automatic weapon in the hands of small war opponents has made the horse not only of little use, but an actual handicap once contact is gained. The mounted detachment is extremely vulnerable to ambush by guerillas armed with automatic weapons.

d. The more extensive the operations, the greater will be the value of mounted detachments. If the operations include the occupation of the seaports and a few of the important inland towns only, the need for mounted detachments will be limited. As the operations extend farther and farther inland and over wider areas, the need for these detachments will become greater.

e. Regardless of the individual efficiency of mounted detachments, their value will depend upon their employment by the higher commanders who assign them their missions. A thorough understanding of the capabilities and limitations of mounted units and due consideration of the factors which affect their combat value, is required for the proper assignment of missions to these units. The mounted and foot patrols should be assigned missions that enable them to work together and not in competition.

7–32. **Basis for organization.**—A definite basis for transition from the normal dismounted organization to a mounted organization status should be adopted. To this end the dismounted organization given in Tables of Organization should be adopted as a basis for transition with such obvious modifications as may be necessary. The conversion of an infantry unit to a mounted status requires more than the simple addition of horses and equipment. The converted organization, even with the minimum of necessary modifications, presents difficulties of training, administration, and tactical use. The officer assigned to organize a mounted unit will find himself so beset with unfamiliar details that the adoption of some system is practically mandatory. The deficiencies which become apparent may be remedied as the organization progresses, without disturbing the general scheme of organization.

7–33. A mounted rifle company.—Assume that a rifle company is to be organized as a mounted detachment. The Tables of Organization provide for a company headquarters and three platoons of three squads each. This organization is suitable for a mounted company. The platoons are small enough, even though mounted, to be handled in most situations by one officer. The addition of the necessary horses and equipment, together with the additional training and upkeep incidental to the transformation from dismounted to mounted status, will require some essential changes in the enlisted personnel provided in the organization tables. A stable sergeant, a horseshoer, and a saddler, all being necessary for a mounted organization, must be added to the company headquarters. It is also necessary to provide about five drivers in company headquarters for the necessary kitchen, cargo, and ration pack animals. The company should be able to operate independently; it must therefore be organized to carry such supplies as will enable it to remain away from its base for at least 3 days, which period can be taken as a minimum patrol period. For longer periods away from the base provision will have to be made for additional drivers, arrangements made for ration drops, visits to friendly outposts planned; or for the unit to subsist itself off the country, or some combination of these methods.

7–34. Machine-gun and howitzer units.—*a.* It is not contemplated that machine-gun companies or howitzer platoons will be mounted as units in small war operations. Unquestionably, however, subdivision of such units will have to be mounted and attached to the mounted rifle detachments. The attachment of two or more machine-gun squads to each mounted company will almost invariably have to be made, and in some situations, it may be necessary to attach 37 mm. and mortar sections. For this purpose, it is simply necessary to mount the attached units with their weapons placed in pack, the weapon crews acting as drivers for their own weapon and ammunition pack animals.

b. It is absolutely essential that the attachment of these units to the mounted companies be made as early as possible so that the personnel and animals can be properly trained and conditioned for their mounted duties. The attachment of these units will not require additional specialists except possibly one additional horseshoer per mounted rifle company.

7-35. Animals for mounted detachments.—The better animals of the occupied country will not be available upon landing. Great effort and ingenuity will be necessary to obtain suitable animals in sufficient numbers. The best animals obtainable will be necessary for mounted organizations. The purchase of animals should be undertaken as early as possible in order to condition the animals.

7-36. Spare mounts.—Mounted service in small war expeditions is especially trying upon the mounts. Experience indicates that the number of mounts should exceed the number of men authorized for the organization by from 20 to 30 percent. The excess should furnish replacements for the lame, sick, sore-backed, wounded, or debilitated mounts, certain to develop in hard field duty. This figure may decrease as men and animals become accustomed to each other, and as the condition of the animals improves.

7-37. Assignment of mounts.—*a.* Every officer and man in the mounted organization should be assigned a horse. Two horses for each officer will usually be required. The assignments of horses should be kept permanent. Changes should be made only upon the decision of the organization commander in each case. Sickness and injuries to animals will require changes from time to time. Such changes should be understood to be distinctly temporary. Men whose animals are sick or injured should be temporarily mounted from the spare animals of the organization.

b. The maintenance of animals in constant fitness for duty is one of the most difficult tasks of the commander of the mounted organization. He cannot do this effectively unless he holds every individual under his command responsible in turn for the animal he rides. This individual responsibility most certainly will be evaded by enlisted men if two or more riders are permitted to use the same mount.

c. In changing horses a definite loss in efficiency results because the man who knows a certain horse will, as a rule, secure the best performance from that particular horse. Also a man will normally become fond of his horse after he becomes acquainted with him. This in turn prompts greater interest in the welfare and training of the animal.

d. It sometimes happens that a certain man and a certain horse will not get along well with each other. The commander of a mounted organization should be constantly on the lookout for such a situation and, after assuring himself that a bona fide case of mutual unsuitability exists, correct it by reassignment of mount and man.

e. The officers and the senior noncommissioned officers must have the best horses available to the organization. Their duties require them to exert their horses to a greater degree than is required of men in the ranks.

7–38. **Horse equipment.**—*a.* The following is the minimum necessary equipment, one set of which, modified to suit the conditions of the operations and its availability, is issued to each man:

> 1 saddle, McClellan.
> 1 blanket, saddle.
> 1 bridle, with snaffle bit.
> 1 headstall, halter.
> 1 halter tie rope.
> 1 surcingle.
> 1 pair saddlebags.
> 1 feed bag.
> 1 grain bag.
> 1 currycomb (preferably one equipped with a hoof hook).
> 1 brush, horse.
> 1 pair spurs.
> 1 pair suspenders, cartridge belt, pistol.
> 1 machete.

b. Grain and feed bags are carried strapped to the pommel. The feed bag should cover the filled grain bag, to protect the grain from rain and from other animals chewing through the bag.

c. The snaffle bit is listed, but the curb bit may prove more satisfactory for some horses.

d. The machete should be carried in a sheath attached to the saddle on the off (right) side, in a horizontal position, hilt to the front. If issue saddlebags are carried, it may be necessary to attach the machete to the off (right) pommel and let it hang. The machetes are not intended for use as weapons, but are provided for cutting trails, clearing camp sites, building shelters, and even more important, for cutting forage, such as grass and cane tops.

7–39. **Individual equipment.**—*a.* There are three general ways of carrying emergency rations, mess gear, grooming kit, toilet articles, etc., each having certain advantages, all being practical, as follows:

(1) Saddlebags (standard equipment) are two large leather pockets, fastened together, in a size approximately to a full-sized cavalry mount rather than to a small horse (mule) which fits on the cantle of the McClellan saddle. As they are large, they must not be over-

loaded, thus preventing pressure on the flanks and consequent chafing of the stifles and hips.

(2) Use of infantry equipment as issued, but attach the blanket roll to the saddle. The canteen should always be carried on the belt.

(3) Use of one or two NCO haversacks per man, fastened to the cantle by their hooks through cantle rings. These may be placed one on each side. This method carries much less than the saddle bags, but is much easier on the horse, especially if he be short coupled.

b. A "cantle roll" will ordinarily be carried on patrol or on the march. It should include those articles not needed until camp is made for the night, which are not easily carried in the saddlebags. Care must be taken that rolls remain small and light, and that the weight is divided equally between the sides of the mount. The roll should be smaller in the center, so that it may bend easily. It is carried strapped up tight on the cantle of the saddle, the ends extending down about as far as the cantle quarterstrap D-rings, no part of it touching the horse, but all its weight held up by the saddle. The following list is not exhaustive, nor need all these things be carried on every patrol:

1 blanket, wool.
1 pair socks.
1 suit underwear.
1 poncho.
1 mosquito net.

Canned rations may be placed in the ends of the roll, and will be more easily carried there than in the saddlebags, especially if in round cans. If the roll is carefully made up, and the opening formed by the edge of the poncho turned so that it will not catch water, the roll is rainproof and nearly waterproof.

7-40. **Arms and ammunition.**—*a.* The arms and ammunition carried by each man are regulated in the same manner and by the same considerations as for dismounted troops. It will be noted that pistols and rifle scabbards have not been included in the minimum requirements for issue to mounted detachments. If the rifle scabbard is issued, care must be taken that troops are so trained that there will be no danger of their being separated from their rifles when they dismount. The rifle scabbard has the disadvantage of interfering to some extent with the seat of the rider and the normal action of very small horses on rough going and at increased gaits, and hastening rust under field conditions by retaining moisture. The rifle scabbard should not be used in territory where contact is at all probable.

MOUNTED DETACHMENTS

Mounted men should be armed with the bayonet and indoctrinated in its use. The rifle, automatic rifle, and submachine gun may be carried by the mounted men slung in the same manner as they are carried dismounted, or the butt may be rested on the thigh, or the rifle may be held by the right hand at the small of the stock, the balance resting on the pommel of the saddle.

b. All grenades and other ammunition should be carried on the persons, not only to save the horse ("live loads" are easier to carry than "dead loads"), but also to have them always available in an emergency. All such loads should be supported on the shoulders of the man, carried high enough so as not to interfere with his seat in the saddle; that is, nothing should extend lower than the bottom of his belt in front or rear. If the 50-round drums for the Thompson submachine guns are carried, some form of sling should be provided for them, or they should be attached to the left side of the belt, to keep them off the saddle. Carriers for grenades should be as high on the body as possible.

7-41. **Pack equipment.**—The Phillips packsaddle, which is coming more and more into general use, should always be used by mounted detachments if it is obtainable. This saddle can be used at the walk, trot and, when necessary, at the gallop without injury to the animal or derangement of the load. The mobility of the detachment, therefore, is not reduced when accompanied by pack horses using the Phillips saddle. If this saddle is not obtainable, a special study of native equipment available will have to be made to determine the type most suitable for military use. If the Phillips saddle is used, the necessary hangars for weapons, ammunition, pack kitchens, and other special loads should be obtained. (See art. 3-30.)

7-42. **Training, general.**—For the general training of mounted detachments see U. S. Army Training Regulations 50-45. "The Soldier; Instruction Mounted without Arms."

7-43. **Training for specialists.**—*a*. For the training of specialists such as the stable sergeant, packmaster, horseshoer, packers, and saddlers, the following publications should be referred to:

(1) Animal Management, the Cavalry School.

(2) FM 25-5, "Animal Transport."

(3) TM 2100-25, "The Saddler."

(4) TM 2100-30, "The Packer."

(5) TM 2100-40, "The Horseshoer."

(6) BFM, Vol. I, Chapter 3, "Equipment & Clothing, Mounted and Dismounted Organizations."

MOUNTED DETACHMENTS

(7) Department of Agriculture Pamphlets. These give the names and kinds of feed found in foreign countries, with their nutritive ratio to oats, the form in which they are usually fed, and other useful information.

b. All of the above except ANIMAL MANAGEMENT may be secured by the Quartermaster from the Government Printing Office at from 5 to 15 cents per copy. The Department of Agriculture pamphlets may be secured from that Department direct. In this connection, application should be made for the pamphlet or pamphlets applicable to the country in question.

7-44. **Time required for training.**—Sufficient time for thorough training in all details will seldom if ever be available. The mounted unit commander is usually ordered to be ready to take the field within a short time after organization. Whatever the situation, the mounted unit commander must adapt his training schedule to the time available. He makes every effort to secure a reasonable time for training. Six weeks may be considered a minimum requirement after the order for mounting is received. Failing this, he conducts his initial operations in the field with due regard to the limited training of his men and the conditioning of his animals.

7-45. **Combat training.**—The combat training of the mounted detachment cannot be neglected. This training is all important and must be carried on concurrently with the mounted training. Since the mounted detachment will habitually fight on foot its small wars combat training will be practically identical to that of a foot patrol. When contact is made the mounted unit will habitually dismount, turn over its mounts to horseholders, and thereafter fight on foot. The combat training of mounted detachments should include numerous and varied combat exercises, which require the men to dismount rapidly and without confusion and to go instantly into dismounted action against a simulated or outlined enemy. Only by repeated exercises of this type will the mounted unit become indoctrinated in the schemes of action for combat.

7-46. **Tactical uses of mounted detachments.**—*a.* Some of the tactical uses of mounted detachments are:

(1) For normal patrolling in pacified areas. Smaller numbers of troops can patrol larger areas with greater facility when organized into mounted detachments. A show of force in these pacified areas can be made almost continuously over wide areas and with a small force by the judicious use of mounted detachments.

MOUNTED DETACHMENTS

(2) As a mobile unit attached to a large combat column. The commander of a large column in some small wars situations may require a mounted detachment for the execution of special missions such as distant reconnaissance to the front and flanks, escort for evacuation of wounded, foraging, investigation of towns or district along the route of march, etc.

(3) As an area or district reserve. Such a reserve can be used for the accomplishment of special urgent missions such as the relief of towns which have been attacked, the rescue of the personnel of planes making forced landings in hostile territory, as an escort for area and district commanders on inspection trips, and for other special missions of a similar nature.

(4) To augment aerial reconnaissance.

b. From a study of the above tactical uses of the mounted detachment, it can be seen that it is not contemplated that the mounted detachment will perform the normal patrol work in small wars. In bush and jungle warfare where the situation is always vague and the enemy never definitely located, the foot patrol is more effective for combat patrols and particularly so when the enemy habitually fights on foot. The foot patrol, whose primary combat training is that of infantrymen, will give a better account of itself in this type of warfare when contact is made and such a patrol is far less vulnerable to ambush. The superior mobility of the mounted detachment means very little if there is no definite objective on which to move. But there are special missions in most small wars operations in which strong and boldly led mounted detachments, well organized, trained and equipped, will be of great value. When such detachments are available to a commander in a hostile area they may be moved rapidly on a definite objective when aeroplane, radio, or other communication or intelligence agency indicated its use at a particular point.

7-47. **Conduct of mounted patrols.**—a. In country that is open enough to permit marching at the trot, patrols may move at better speed mounted than dismounted, and arrive at the destination or point of contact with the hostile force, in better condition to fight. A greater load can be carried without undue fatigue by the mounted man than by the dismounted, but loads should not be such as to cut down mobility. Over average dirt roads, with few steep grades and with small horses in fair condition, a platoon should march about 30 miles in 1 day, or 80 to 85 in 3 days. Longer daily marches may often be made, but losses in condition must be made up by rest after the march. These figures will not be found accurate under all circumstances, but falling much below them makes the mounting of the men unprofitable, as seasoned infantry can move in small bodies for limited periods at rates nearly approaching these.

b. Patrols required to remain out for long periods should take advantage of all facilities of friendly garrisons, so as to preserve their mobility. Sick should be left at the first post passed through, supplies replenished if they can be spared by the garrison, and information exchanged at every opportunity.

c. Timely preparations should be made for any march, to insure that men and animals are in the best condition possible, that the required equipment and supplies are present and loaded as required and that provision is made for the care and evacuation of the disabled.

d. The strict observance of march discipline is most essential in mounted units. It is maintained only by frequent and rigid inspections by officers and noncommissioned officers both on the march and at all halts. The object of these inspections are:

(1) To keep equipment, especially saddles and packsaddles correctly adjusted at all times.

(2) To require all riders to maintain the correct seat in the saddle. Slouching in the saddle has a tendency to injure the animal's back.

(3) To maintain the prescribed gaits within subdivisions of the column.

(4) To require all riders to dismount, when there is no need for remaining mounted. This is especially important. A horse standing still, and with his rider sitting on him is not able to relax and rest.

(5) To permit individuals to leave the column only in case of urgent necessity.

(6) To police halting places and camp or bivouac areas.

7-48. **Combat patrols.**—*a.* Most patrols sent out in small wars must be ready to accept combat, even if not sent out primarily with the mission of seeking it. Usually psychological considerations will require that no patrol give ground, and patrols are habitually made strong enough to repel expected attacks. Reconnaissance and other special patrols, therefore, are considered with combat patrols, and not as requiring special formations. The essentials required of mounted formations are the same as those of dismounted formations. There must be adequate control by the leaders of parts of the patrol, mutual support, power of maneuver must be preserved as long as possible by holding out supports, and the patrol must be protected from surprise from any direction. The principles of the dismounted patrol formations may therefore be followed, modifying distances and intervals to conform to the different capabilities of the mounted man. For short distances, a horse can easily travel twice as fast as a man, and thus support can be furnished from a greater distance.

b. When the point comes under fire, the men of the leading squad should dismount at once and take firing positions. There will seldom be time for these men to secure their horses. Other squads, not actually engaged, may have time and opportunity to secure their mounts by having one man hold four of them, and it will sometimes be possible, especially in patrols larger than two squads, for units not engaged to make a mounted dash to a position from which they can make a dismounted attack on the enemy's flank or rear. Units not actually engaged should maneuver, either mounted or dismounted, to take the opposition in flank or rear, but always attacking dismounted. Actions of this type may be prearranged and practiced, but must be kept so simple as to be flexible in application, and must not permit any part of the patrol to go beyond effective control of the leader without definite orders from him.

c. Distances are shortened in woods and lengthened in more open country. Details for flank reconnaissance are usually arranged before the march is begun, so that a signal by the leader will be sufficient to start the reconnaissance.

d. Mounted patrol formations are identical with those of foot patrols with the exception that allowance must be made for the greater road space required by mounted units. For patrol formations, distribution of weapons, tactics, and other details see Infantry Patrols, chapter VI.

7–49. Ambushes.—*a.* Many areas afford innumerable good ambush positions. If all such positions are carefully reconnoitered by mounted patrols operating in such areas the rate of march will be reduced to that of foot troops. The mounted detachment when employed on an urgent mission requiring rapid movement on a definite objective avoids being ambushed not so much by cautious movement and careful reconnaissance as by rapid and secret movement and by radical changes of direction to deceive the enemy. The mounted unit leader, for this reason, must have as thorough knowledge of the terrain as possible and must have the best guides obtainable.

b. The above paragraph is not to be construed as relieving the mounted detachment commander of the responsibility of providing

MOUNTED DETACHMENTS

reasonable security for his column when on the march and of carefully investigating any position which he has reason to believe is occupied. If the attack be from a flank on the center of the column, the leading and following elements do the maneuvering, and the attack is still normally driven home on the flank or rear. The horses of the elements caught in the initial burst of fire will generally have to be temporarily abandoned. Tired horses will not stray far and those not wounded or killed can be recovered as soon as the enemy position is taken. The horses of elements not caught under fire should be turned over to the appointed horse holders of the squad who will get them under such cover as is available.

SECTION V

HASTILY ORGANIZED MOUNTED PATROLS

7–50. **Definition and uses.**—*a.* Hastily organized mounted patrols are units mounted for immediate patrolling without prior training in riding or animal care. This expedient has been resorted to frequently in the past. It usually suggests itself in areas where suitable riding animals are available in considerable numbers. Whether for increasing the mobility of the patrol, or conserving the strength of the men, due consideration should be given to the advantages and disadvantages of such procedure as indicated in this section, prior to the adoption of this expedient.

b. The leaders of infantry units often err when they decide to execute combat patrolling on mounts instead of on foot. Past operations have definitely indicated that there are certain advantages and disadvantages of hastily organized mounted patrols. Some of the considerations which bear upon the advisability of organizing hastily mounted patrols are set forth below:

(1) The mounted patrol is more formidable in appearance and it affords an easy way to make a display of force in fairly peaceful territory.

(2) For a march of not over 1 or 2 days' duration and with suitable terrain, the hastily mounted patrol can travel more rapidly and cover a greater distance than on foot.

(3) Mounted men can, for short marches, carry heavier individual loads than men on foot.

(4) A mounted man can give all his attention to observing the terrain and looking for signs of the enemy. The mount will select its own footing on the trail. A man on foot on bad trails must spend much time looking at the trail to pick his route.

(5) A man on horseback presents a smaller target than a man standing. His body appears shorter and is partly protected by the horse.

HASTILY ORGANIZED MOUNTED PATROLS

(6) A mounted man is higher than a man on foot and he can see farther. On some terrain his eyes will clear brush over which the man on foot cannot see. This sometimes enables a mounted man to detect a waiting enemy at a distance denied to a man on foot; particularly if the enemy is not well schooled in lying in ambush.

(7) The mounted patrol is easier on the men. They will arrive at their destination in a less exhausted condition.

(8) The rapidity of movement and the distance covered will decrease materially as the size of the patrol is increased.

(9) For extended marches the foot patrol, untrained in riding, and in the care and handling of animals, will make better progress on foot day in and day out than if mounted on animals.

(10) For large patrols, even marches of only 1 or 2 days' duration can be made with greater facility on foot than if hastily mounted.

(11) If the march is to be extended, it is essential that the load on the ridden animals be kept as light as possible. The average load carried by the ridden animal in the field is about 250 pounds. It can readily be seen that this load places considerable burden on small native animals and every effort should be made to reduce this load by the use of accompanying pack animals.

(12) The animal casualties in hastily organized mounted patrols will be excessive because of poor handling and lack of condition in the animals.

(13) Lack of training in dismounting and securing animals, places the hastily organized mounted patrol at a distinct disadvantage once contact with the enemy is made.

(14) The mounted man is generally more conspicuous, and more clearly outlined, and he cannot hit the ground, take cover, and return the hostile fire as rapidly as a man on foot. At any given range, therefore, he is more exposed to the hostile initial bursts of fire.

(15) A foot patrol has the advantage over a mounted patrol when it runs into an enemy ambush. In a mounted patrol the animals will suffer severely and the men will be more exposed. If any men or horseholders have to concern themselves with the animals they are not much help against the enemy.

(16) A mounted patrol is more visible from a distance than a foot patrol, particularly if the colors of the animals do not blend with the background. Natives on one mountain can see mounted men marching on another mountain then, under the same conditions, they could not see foot troops.

(17) A foot patrol can make and break camp more rapidly than a mounted patrol and is not as much concerned about a site that will furnish forage and water. Hastily organized mounted patrols of any size are notoriously slow in breaking camp.

(18) A mounted patrol is more expensive than a foot patrol. The expenditure for animals and animal equipment, in past operations, has been extremely high for hastily organized mounted patrols.

(19) At the end of the day's march the work of the mounted man has just begun. The animals, pack and ridden, have to be groomed, watered, fed, the sick and injured treated, and, in hostile territory, guarded during the night.

(20) The general consideration that riding is less fatiguing than walking is apt to outweigh other more important considerations in tropical countries where there is ever present a strong tendency to avoid bodily exertion.

7–51. **Discussion.**—The above considerations indicate that the use of the hastily organized mounted patrol in hostile territory is rarely justified. Only for small patrols when equipment and conditioned animals are immediately available, when the march is not in excess of about 2 days, when the patrol is to be made in fairly peaceful territory, and when rapid movement is desirable and practicable are the conditions suitable to justify the organization of a hastily mounted patrol.

7–52. **Type of animal to employ.**—In some localities there may be a choice between horses and mules. The characteristics of the mule, as set forth in article 3–27, make him more suitable for riding and handling by untrained men who for the most part make up hastily organized mounted patrols. However, for small patrols on short urgent missions the horse can well be used to advantage.

7–53. **Other details.**—The organization, assignment of duties and animals; marches, equipment and other details of a hastily organized mounted patrol should be based upon a study of the preceding section, section IV, and should approximate the standards set for regular mounted detachments.

○

SMALL WARS MANUAL
UNITED STATES MARINE CORPS
1940

CHAPTER VIII

CONVOYS AND CONVOY ESCORTS

RESTRICTED

UNITED STATES
GOVERNMENT PRINTING OFFICE
WASHINGTON : 1940

TABLE OF CONTENTS

The Small Wars Manual, U. S. Marine Corps, 1940, is published in 15 chapters as follows:

SMALL WARS MANUAL
UNITED STATES MARINE CORPS

CHAPTER VIII

CONVOYS AND CONVOY ESCORTS

8–1. **Definitions.**—*a*. When a number of pack animals, carts, wagons, or trucks are to be sent from one place to another, they are formed into a train. The train is called a Convoy.

b. When carts, wagons, or trucks are used at least one man armed with rifle or automatic weapon generally goes with each vehicle, or men so armed are otherwise disposed throughout the length of the train. These men constitute what is known as the Train Guard. They stay with the convoy and fight only for the close protection of the convoy. They are under the direct command of the convoy commander.

c. When operating in dangerous territory troops are furnished for the protection of the convoy. These troops are called the Convoy Escort. The officer in charge of the escort is in command of, and is responsible for, the convoy. He defers to the recommendations of the convoy commander insofar as is consistent with the necessary safety precautions.

8–2. **Mission.**—The mission of the convoy escort is to insure the uninterrupted march and safe arrival of the convoy it is detailed to protect. All formations adopted and all plans of action must be in accordance with this mission. This does not and should not preclude offensive action, but merely requires that the attack must have as its purpose the safety of the convoy, rather than the destruction of the enemy. Pursuit that carries the escort, or any part of it, to such distance that it can no longer act in defense of the convoy violates the mission of the escort and such action may be just the opportunity for which the enemy seeks. The rate of march is limited by the rate of the slowest element of the convoy or the convoy escort.

8–3. **Organization.**—*a*. Escorts vary in strength and composition with the size and importance of the convoy, the length of the march, the nature of the terrain, and the type of resistance expected. Detachments from the rifle companies with the necessary machine guns

1

attached will usually constitute the escort. When the rate of march of the convoy exceeds that of infantry, the escort should be provided with a means of transportation which is no slower than the convoy. For other than convoys mounted detachments may be employed very effectively for this duty.

b. For large trains the vehicles or pack animals should be formed into sections, each under a responsible officer or noncommissioned officer, before the march begins. The train guards placed on the individual vehicles or within the pack-animal sections, should come under the control of the section commanders.

8-4. Convoy types.—*a.* Convoys of pack animals are difficult to protect because of the vulnerability of the animals themselves and because of the extended road space required on narrow trails. When possible, the pack animals should be herded in groups of about five animals but generally it will be necessary to have them travel in single file. Pack trains of untrained or poorly trained animals will take more road space and will require more men. In extreme cases, a man to lead each animal may be necessary, but usually the animals may be led in strings of two or three, fastened together, "head to tail." If train guards are placed in the pack-animal sections, they may be employed by the section leader to lead animals, particularly ordnance animals. Comparatively speaking, pack convoys are not economical and should be used only in very difficult terrain.

b. Trains of native packmen (porters or carriers) are sometimes necessary. The average useful load of a packman is about fifty pounds. These trains, like all other trains should be organized into sections, each section under its native leader, when any considerable number of packmen is employed. Trains of packmen have the disadvantage of extending to great length on the line of march and of requiring strong escorts. The packman, unlike the pack animal, fully appreciates danger, and, at its first appearance, is likely to abandon his load and seek safety in flight.

c. A train of 100 bull carts or wagons when well closed up occupies a road space of about 1 mile. The difficulties of control and protection of such trains will naturally increase with their length. In dangerous territory it will seldom be advisable to operate very large trains because of the difficulties of protecting them. Close protection should be provided by having train guards armed with rifle or automatic weapon ride on the individual carts or wagons. If marine drivers are employed they should be armed with the rifle or an automatic weapon (no pistol or revolver). Communication should be maintained between sections of the train by mounted messengers. The prompt discovery of breaks in the column is dependent upon continuous communication between sections of the train by mounted messenger. Each element of the convoy must guide on the element next in rear. If contact with a rear element is broken the leading element must immediately halt on ground that will afford the animals rest and wait for the rear element to close up. If this is habitually done, serious breaks in the column can be averted. Proper consideration for defense must be given in the choice of such halting places.

d. Motor convoys are made up of military motor transport units or of hired or unrequisitioned motor vehicles. In any case, the convoy should be properly organized before starting the march. The escort may be carried on the train or in separate vehicles capable of at least the same speed as the convoy. The slower elements of the convoy should be placed in the lead. With suitable roads, motor transportation is peculiarly advantageous for convoys because it is less vulnerable to small arms fire, and because of its speed and ability to travel long distances with few halts. Fast tanks, armored cars, or trucks converted to serve as armored cars, may be employed to great advantage in escorting motor convoys. If the enemy has made a practice of using land mines, it may be advisable to have a pilot cargo truck precede the point.

e. For river convoys and escorts see chapter X.

8-5. March of convoys.—*a.* Convoys should be assembled in sufficient time prior to the march to organize them into sections and to appoint and properly instruct the section leaders in their duties. Carts, wagons, and motor vehicles should be loaded the day prior to starting the march, and should be so located the night before that there will be the minimum of confusion in forming the column in the morning for the march. Newly organized pack and bull cart trains should be marched only a very short distance the first day, preferably only a few miles beyond the limits of the town from which the march originates. This first day of march can then be used to perfect the organization of the convoy, to correct any deficiencies in equipment, and to indoctrinate the escort in their duties. Spare vehicles or carriers should be provided to carry the burden of those that may be disabled enroute. When this is not practicable, the loads of disabled vehicles or pack animals must be distributed amongst the convoy, or destroyed.

CONVOYS AND CONVOY ESCORTS

b. It may happen, as it has in past operations, through misinformation, inability to foresee road conditions or other circumstances, that large convoys will begin a march greatly overloaded. When such a condition develops, the escort commander is faced with a serious problem which requires immediate action to save the convoy from serious difficulties, or, in extreme cases, from complete break-down. If the convoy has not proceeded far from its base, it may be advisable to return and reorganize the convoy with lighter loads or to procure additional animals or vehicles from the base or in the vicinity of the halt. If this is not practicable, it will be necessary to lighten the loads by the establishment of a dump or by destruction of the excess cargo. It is far better to make a radical reduction of the loads as soon as it becomes apparent that the loads are excessive than to make small reductions from day to day as the animals become worn and exhausted from carrying excessive loads.

c. The available route should be considered carefully. Long distances, poor roads, steep grades, many stream crossings, sharp turns, defiles, close country, and exposure to view from considerable distances along an open ridge are objectionable features. The loss of a few animals shot down by a handful of men in ambush, or the disabling of a few trucks or tractors might seriously delay the convoy. Generally, the best road, even though it may not be the most direct, is to be preferred. The route should be selected with a view to avoiding probable hostile forces, and a detour is justifiable if it be reasonably certain that an ambush may thus be avoided. Provision should be made for temporary repair of roads and bridges and for the crossing of fords.

8–6. Disposition of the escort on the march.—*a.* The escort commander, after assigning the necessary train guards to the sections of the convoy, should divide his force into an advance guard, a main body, a rear guard, and flank guards as necessary. The functions, armament, and tactical disposition of these elements are the same as those corresponding elements in combat patrol formations.

b. The advance guard precedes the train in the convoy in the usual patrol formation. Critical places on the route of march, such as fords, defiles and trail crossings should be reconnoitered and commanding positions occupied before the convoy is committed to them. Over some terrain, the convoy can be protected best by having the advance guard proceed by bounds from one position to another. The head of the convoy is never permitted to enter a defile until the advance guard is in possession of the farther end.

CONVOYS AND CONVOY ESCORTS

c. The main body is charged primarily with the defense of the convoy. It is therefore likely to fight on the defensive, this being forced on it by its mission. The main body may be marched ahead of the convoy; but, if this is done, adequate train guards must be provided within the sections of the train and the rear guard must be sufficiently strong to fight independently until support can reach it from the main body. For very long convoys, it may be necessary to split the main body and have these divisions of the main body march between sections of the convoy.

d. The rear guard marches a short distance in rear of the convoy with the usual rear-guard formation. Rear guards should not maneuver in support of advance elements of the escort if by doing so the rear of the convoy is left totally unprotected. Conditions may sometimes warrant the detachment of a part of the rear guard to maneuver in support of advance elements of the escort, but sufficient force should always be retained by the rear guard commander to protect the train from attack in that direction.

e. The flanks of a convoy are most vulnerable and are particularly difficult to protect in heavy brush, jungle, or mountainous country where parallel roads or trails do not exist. In such terrain it is often not practical to employ flank guards as they will slow the rate of march to too great an extent. Whenever practical, adequate flank guards should be provided. In terrain not practical for the employment of flank guards, the flanks will generally have to be protected by a careful reconnaissance by the advance guard and by the occupation of commanding positions and intersecting roads or trails by elements from the advance guard. These elements join the rear guard when the convoy has passed.

8–7. **Defense of a convoy.**—The escort fights only when necessary. The mission of the escort is to protect the train and to insure its uninterrupted progress. If the enemy is discovered holding a commanding position or a defile on the line of march, he should be dislodged and driven off before the convoy is permitted to proceed. In most small war situations, hostile forces attempt to ambush the convoy. Prearranged schemes of maneuver, as described in chapter VI should be prepared to meet such ambushes. The sections of the convoy under the immediate protection of the train guards should seek any available cover that will prevent the hostile troops from firing into the train. If it becomes evident that the capture of the train by the enemy cannot be prevented, the transport and its contents should be destroyed.

CONVOY AND CONVOY ESCORTS

8-8. Attack of a convoy.—The most favorable time for an attack against a convoy is when it is passing through heavy woods, jungle growth, defiles, or stream crossings; when it is ascending or descending steep slopes, or passing over bad sections of the road; when the convoy is making or breaking camp; and when animals are being watered. The objective of the attack is always the transport and not the escort. The attacking force should first bring the convoy to a halt and then throw it into confusion by attacking from an unexpected direction. The fire of automatic weapons and the attacks of airplanes flying at low altitude are very effective. When a convoy is captured, the parts that cannot be carried off should be destroyed.

8-9. Security measures at the halt.—*a.* During short halts, elements from the escort should be so disposed as to afford protection to the convoy for the period of the halt. Commanding positions and intersecting trails particularly should be guarded and the train guards should be kept on the alert.

b. For long halts and halts for the night, the train must be so parked that it will afford the maximum protection to the convoy escort. At the same time the train should be ready to move out without delay when march is resumed. In dangerous territory, when no suitable defensive positions are available, the cargoes and the pack saddles of pack convoys may be so placed at night as to serve as breastworks. In very close country, this use of the cargoes is particularly valuable as it provides a definite line on which the escort can form in case the convoy is rushed in a night attack. Carts and wagons can be arranged in a circle or hollow square with the escort located within the enclosed area. This disposition should not relieve the escort commander from the responsibility of establishing outposts. The animals should be afforded such protection as is possible. Motor vehicles should be so parked for night halts as to provide maximum protection for the radiators, motors, and other vital parts of the vehicles from small-arms fire in case of night attack.

○

SMALL WARS MANUAL
UNITED STATES MARINE CORPS
1940

✦

CHAPTER IX

AVIATION

RESTRICTED

UNITED STATES
GOVERNMENT PRINTING OFFICE
WASHINGTON : 1940

TABLE OF CONTENTS

The Small Wars Manual, U. S. Marine Corps, 1940, is published in 15 chapters, as follows:

SMALL WARS MANUAL

UNITED STATES MARINE CORPS

CHAPTER IX

AVIATION

Section I

INTRODUCTION

9–1. General.—*a.* The opposition usually encountered in small war operations consists of scattered bands of irregular troops, well armed and extremely mobile, but deficient in disciplined morale. Logistical arrangements for such forces are apt to be very primitive and sketchy, offering no substantial target for bombing aviation. Air opposition is usually nonexistent or negligible. The Marine air force is thus able to concentrate almost entirely on the close support of ground units.

b. In order to secure the full measure of cooperation between the air and ground forces, it is necessary that each understands the problems of the other. The aviator must know something of the tactics of the ground patrol, and he must be ready and willing to assume any justified risk to assist the ground commander. On the other hand, the ground commander should understand the hazards and limitations imposed on aviation operating over difficult terrain, and should not expect the impossible.

9–2. Special air tactics involved.—*a.* The employment of aviation in small wars is characterized by the operation of many small units, two or three plane patrols, over a wide area. Normal scouting missions will in most cases be modified to search attacks, performed by airplanes of the scouting or observation class armed with light bombs and machine guns.

b. If attack or light bombing units are included in the force, the tactics of their employment will not differ greatly from normal procedure. They should constitute an aerial reserve, to be dispatched only against definitely located targets, and in such force as may be necessary. Occasions will arise where one six-plane division may be ample force for the task at hand; in fact, the employment of small striking units will be frequent, and independent missions for the division the rule rather than the exception. The usual absence of air opposition in small wars gives to an air force a freedom of action, and the ability to employ small units independently, not enjoyed in major conflicts. If air opposition should exist, it must of course be countered by fighters in the normal way.

1

INTRODUCTION

c. In the past, Marine air forces have been equipped generally with dual-purpose airplanes of the two-seater type, suitable for observation or scouting, and equipped with the armament necessary for limited ground attack. The observation and light bomber types were so similar that they were used indiscriminately on whatever mission came first to hand. While it is true that such diversion and substitution is still possible for emergency situations, modern design of airplanes and engines is along specialized lines and does not permit the wide latitude of tactical employment practiced with the more simple machines of former years.

SECTION II

COMPOSITION AND ORGANIZATION

9–3. **Types.**—The composition of an air force organized for small wars operations cannot be definitely prescribed, nor can its comparative strength in relation to the ground force be determined prior to a careful estimate of the situation in each case. Much depends upon the character of the campaign, and upon the nature of the theater of operations. The final choice will be influenced by the type of air units immediately available. The discussion contained in this chapter assumes a typical situation wherein an independent brigade or force is supported by a composite group of aircraft.

9–4. **Reconnaissance aircraft.**—Primary consideration should always be given to reconnaissance types in the organization of a small wars air force. Due to the advisability of operating in small formations and to the frequent calls for air reconnaissance to be expected from the commanders of independent columns and patrols, at least twice the number of observation or scouting airplanes will be required for the support of a force engaged in a campaign of this nature as would suffice for normal operations.

9–5. **Combat aircraft.**—The inclusion of combat types of aircraft in addition to the dual-purpose scouts may be advisable or necessary in many small wars situations. In making a decision as to what types to include in the air force, consideration should be given to the existence of objectives which are beyond the capabilities of the dual-purpose scouting airplane.

9–6. **Transport aircraft.**—This type of aircraft has proven indispensable for small wars operations. The lack of railroads, improved motor roads, and navigable waterways in some of our probable theaters of operation makes the supply and transportation of troops by air more or less mandatory. Two types of transports are standard: the multiengined cabin land plane; and the multiengined cabin

COMPOSITION AND ORGANIZATION

amphibian. Both should be included in the air force; both are included in the organization of the present utility squadron. The ratio of land planes to amphibians will depend upon operating conditions to be encountered.

9-7. **Organization.**—The present squadron organization of the Marine Corps is satisfactory for small wars operations. The only problem of organization is the selection of the units which are to compose the group. One headquarters and service squadron, one utility squadron, and two scouting squadrons may be considered as the minimum basic force for the support of a brigade or similar unit. To these should be added such additional transports and combat units as the situation demands. The composite group is flexible and can take care of several operating squadrons without additional overhead.

9-8. **Movement to the theater of operations.**—Aircraft should always be flown to the theater of operations whenever distance and the situation will permit. Air units so transported arrive in the minimum of time with less hazard of damage en route and are ready for immediate action upon arrival. This method presupposes available landing fields within the theater of operations protected by Marine detachments from naval vessels, or by friendly native troops. In most cases intermediate refueling stops must also be available, either on foreign airdromes or on board own aircraft carriers.

SECTION III

SELECTION AND PREPARATION OF BASES

9–9. **Main airdrome.**—*a.* The main airdrome within the theater of operations should be located within a reasonable distance of Force Headquarters and must be accessible by motor transport or on a navigable waterway. The air commander must be able to maintain close personal contact with the Force staff, and, conversely, the various departments of Force Headquarters should have easy access to the airdrome facilities. The main airdrome should be of such size as to permit heavily loaded transports to operate during adverse weather and field conditions. Existent landing fields which meet all of the requirements will seldom be encountered, and provision must be made for labor and construction materials to clear and prepare landing surfaces.

b. The ground activities of a main airdrome can be conducted under canvas, but the use of permanent or temporary buildings will greatly facilitate shop work and improve the general efficiency of the organization. Provision must be made for the underground storage of bombs and fuzes. Protected areas for the storage of gasoline and oil must be selected, and preferably fenced off from other airdrome activities. Should there appear to be danger of sabotage, it may be advisable to fence off the more vulnerable areas of the airdromes with barbed wire entanglements. Airdrome guards, in addition to those furnished by the air units themselves, may be necessary. Should the opposing forces possess aircraft, antiaircraft protection must be provided for the airdrome. For defense against sporadic air raids which might be expected from a weak and poorly trained opposing air force, the air units would be able to organize their own antiaircraft machine gun crews for emergency protection, provided equipment were made available. In other cases, it would

SELECTION AND PREPARATION OF BASES

be necessary to arrange for a stronger defense by regular anti-aicraft units.

9-10. **Auxiliary airdromes.**—In small wars situations the use of Auxiliary airdromes is contemplated, not for the dispersion of air units for protection, but to facilitate the provision of air support for semi-independent commands. Territorial departments are organized and garrisoned by subordinate units of appropriate size. The headquarters of these departments may be situated in isolated regions with indifferent transport facilities, and so remote from the main airdrome as to seriously curtail air support during periods of unfavorable weather. Auxiliary airdromes established in the vicinity of department headquarters, lightly stocked with supplies of fuel, bombs, ammunition, and spare parts, and staffed with skeleton ground crews, enable the air commander to detach small units for the close support of departmental operations. Furthermore, the uninterrupted transportation of troops and supplies by air is dependent upon the existence and maintenance of such auxiliary airdromes.

9-11. **Advanced landing fields.**—Each detached post and outlying detachment camp should have a field of sufficient size to permit the operation therefrom of scout and combat planes. Many of these fields need have no special facilities, other than the landing area, but certain ones in key locations should be provided with storage facilities for limited amounts of fuel, bombs, and ammunition. It may be desirable to have one or more mechanics stationed at such fields. Necessary protection and assistance in handling airplanes on the ground should be provided by the garrison of the station.

9-12. **Emergency landing fields.**—These are merely possible landing places, located, cleared, and properly marked. Their primary function is to provide disabled or weather-bound aircraft with emergency landing places. They may also be useful in making evacuations of sick and wounded men from isolated patrols, or for facilitating air support in unusual situations. As many as possible of these fields should be provided throughout the area of operations.

9-13. **Specifications of landing fields.**—a. Under normal conditions current types of military airplanes in taking off and landing usually roll on the ground for a distance of from 500 to 700 yards. This distance will be increased by the load carried, by a rough or muddy surface. by hot dry weather, or where the airdrome is situated at high altitudes. Therefore, in order to allow a reasonable factor of safety in operating airplanes under the varying conditions,

SELECTION AND PREPARATION OF BASES

landing fields should have minimum dimensions of from 700 yards for all combat airplanes up to 1,000 yards for transports. The landing fields should be smooth, of firm surface, and without obstructions within or near its boundaries.

b. If obstacles such as hills, trees, or large structures are near the boundary of a landing field, its dimensions must be increased in order that the airplane may clear the obstacles in taking off or landing. Obstacles near the ends of runways must not have a height greater than one-tenth of their distance from the field, i. e., a tree 50 feet high cannot be closer than 500 feet to the end of the runway.

c. Under varying conditions of terrain it will frequently be impossible to locate or construct landing fields which will permit airplanes to land and take off in all directions. Under such conditions the runways or longer dimensions of the landing field should, if possible lie in the direction of the prevailing wind for that locality.

9–14. Minimum size of landing fields.—

Load	Conditions, land and take-off	Transports	All other types
		Yards	*Yards*
Light	No wind	800	700
Do	10 miles per hour	700	600
Military load	No wind	1,000	800
Do	10 miles per hour	900	700

Runways should have a minimum width of 200 yards.

7

SECTION IV

GENERAL CONDUCT OF AIR OPERATIONS

9-15. **Control and command.**—*a.* The senior aviator on duty with a command exercises a dual function similar to that of the force artillery commander. He commands the air force and acts as advisor on air matters to the Force Commander. The air commander will generally have an extensive detailed knowledge of the area in which operations are being conducted—first-hand knowledge which may not be available otherwise—and he should maintain close contact with the Force Commander and staff through the medium of frequent conferences. An aviation liaison officer may be detailed to represent the air commander at headquarters during the absence of the latter on flying mission.

b. Normally, all aviation attached to a small wars expeditionary force will operate from the main airdrome under centralized control. However, when distances are great and weather conditions uncertain, it may become advisable to detach aviation units to subordinate commands, to be operated from auxiliary airdromes.

9-16. **Details of operations.**—*a.* At the close of each day's operations the air commander estimates the situation for the following day, and imparts his decision to his staff and unit commanders. Formal operation orders are seldom written in advance, their substance being posted on the operations board and explained to the pilots concerned. The hour for publishing the daily orders will normally be late enough in the day to permit the commander to analyze the day's reports and receive last-minute instructions from the higher command, but should not be so late as to interfere with the crew's rest. Where possible, the board should be made ready for inspection at a given hour each evening—at 7 or 8 o'clock for example.

b. During daylight hours the airplanes and crews not scheduled for flight should be kept in a condition of readiness to take off within 20 or 30 minutes. Small wars situations often require prompt

GENERAL CONDUCT OF AIR OPERATIONS

action on the part of the supporting air force. Night operations will seldom be required, due to the nature of the support rendered, but should occasion demand, the air units must be equipped to perform night reconnaissance or combat missions. Operations under unfavorable weather conditions will be the rule, rather than the exception, in the average small wars theater. This factor, and the necessity for operating small independent units rather than large formations, requires a large percentage of seasoned and highly trained pilots. At least half of the flight personnel should be in this category.

c. Constant two-way radio communication is desirable between the air patrols and the airdrome operations office. Present equipment will permit such communication within reasonable distances by radio telephone; radio telegraph is available in the same sets for longer range transmission.

9–17. **Reports.**—*a.* Upon the completion of each tactical flight the pilot and observer should compare notes and submit their report on a standard form which will contain a brief chronological record of the flight, including a statement of the mission; time, and place of observation; action taken; comments on the situation; copies of all messages sent or received; weather conditions encountered; ammunition expended; and casualties inflicted or suffered. Reports should be limited to observed facts, and opinions given sparingly. Deductions, except where immediate action is indicated, should be left to the Force staff or appropriate commander. It must be understood, however, that the air observer in small wars operations must be given a greater latitude in estimating a situation on the ground than he would be given in a comparable position in major operations. Often the rapidly moving situation will not permit of delay in the transmission of information to headquarters, but requires immediate positive action on the part of the air patrol commander. In such cases, of course, the written report will eventually be made, with notation of the action taken. In any event, flight reports are submitted immediately upon completion of each mission.

b. In addition to the formal reports submitted upon landing, flight crews may gather information to be dropped to troops in the field, or they may submit fragmentary reports prior to the completion of the flight. Expediency will govern the method of disseminating information, but it is doctrinal for observers to transmit important information without delay to the units most immediately concerned. The airdrome radio station guarding the flight will habitually copy all intercepted messages.

GENERAL CONDUCT OF AIR OPERATIONS

c. The air operations office consolidates the information contained in the individual flight reports into the operations report, which is submitted daily to Force Headquarters. The air force commander is responsible for the accuracy of these reports and for their immediate transmission when urgent action is required. Normally, a brief summary of important or unusual information is telephoned to Force Headquarters immediately, or the air commander calls in in person to discuss the results of important flights. Radio reports received from airplanes in flight should be handled in the same manner, unless Force Headquarters also maintains a radio watch on the aviation frequency. Standard procedure will govern as to the priority of transmission. Formal reports are intended as a summary of the day's operations; vital information should never be withheld pending their preparation.

Section V

EMPLOYMENT OF RECONNAISSANCE AVIATION

9–18. **General considerations.**—*a.* The employment of reconnaissance aviation in small wars situations follows generally the tactics prescribed for major operations. The principal difference lies in the common usage in small wars of the reconnaissance airplane in the dual missions of scouting and attack operations against ground targets. The habitual employment of scouts in pairs or small formation, primarily for mutual protection, favors the dual mission for this type.

b. Reconnaissance may be classified as strategical or tactical as to mission; visual or photographic as to method. Visual reconnaissance will be the principal method of obtaining information in the typical small wars operation. The type of country, unusually densely wooded, and the fleeting nature of the contacts to be expected with hostile forces, will probably limit the use of photographic observation to mapping operations.

c. The effectiveness of air reconnaissance is dependent upon: the nature of the terrain, whether open or densely wooded jungle; the habits of the opposing forces with respect to concealment from aircraft; and, to a greater extent than any other factor, upon the skill and training of the observer. Generally speaking, a trained observer will detect the movement in open country of small groups, while in densely wooded country he will have great difficulty in locating a force the size of a company or larger. However, it will be very difficult for a hostile force of any considerable size to move in daylight without disclosing some indication of its presence, while the mere presence of airplanes in the area will be a deterrent to guerrilla operations. Intensive low altitude reconnaissance over restricted areas will seldom fail to discover the presence of hostile forces, although aviation cannot be expected to always furnish reliable negative information with respect to the hostile occupancy of dense woods, towns, and villages. In small wars, as in major ones, air reconnaissance supplements but does not replace, the normal measures of security.

9-19. **Strategical reconnaissance.**—*a.* Prior to the initiation of the land campaign, the commander should dispatch such reconnaissance aircraft as may be available to make a general air survey of the proposed theater of operations. This mission may include aerial mapping, verification of existing maps, the location and disposition of hostile forces, their methods of operation and supply, location of airdromes and bivouac sites, and the scouting of possible routes of advance into the interior. During this period the flying personnel will familiarize themselves with the terrain and climatic conditions of the country.

b. Strategical reconnaissance may precede the initial landing of troops, if patrol seaplanes, shipbased seaplanes, or carrier-based aircraft are available. Where time is an important factor, much strategical information can be secured in a single flight, although a period of several days may be needed for a comprehensive air survey. Landplanes or amphibians should be used for inland reconnaissance when available, although the urgency of the situation may require the dispatching of seaplanes on such missions. In any event, the importance of a thorough air reconnaissance prior to the advance inland will justify the employment of whatever type of aircraft might be available.

9-20. **Tactical reconnaissance.**—*a.* After a general picture of the situation has been obtained and the ground forces have started their movement inland, reconnaissance becomes more tactical in nature. When contact becomes imminent, reconnaissance aviation maintains a close surveillance over local hostile activities, keeps the ground commanders constantly informed, and furnishes such combat support as may be urgent. The principal task of aviation operating in close support of an advancing column is to supplement the normal security measures taken by the ground forces against the possibility of surprise. Ambush by guerrilla bands is a constant menace. Airplanes should reconnoiter ahead of the ground columns, paying particular attention to those localities recognized by the skilled observer as being dangerous ambush sites. This precaution will protect the ground units from surprise by a *large* force. It must be remembered, however, that detection of *small* forces of irregulars, not in uniform and with no distinctive formation, in heavily wooded country, or in a

14

jumble of mountain boulders, is extremely difficult and largely a matter of luck for even the most skilled air observer. The habitual presence of airplanes in the vicinity of our column will discourage operations of guerrilla forces, even though they escape detection, hence it is advisable to conduct more or less continuous reconnaissance throughout the hours of daylight over the area occupied by our advancing forces. Flights at irregular intervals may accomplish the same purpose with more economy of force.

b. Tactical reconnaissance immediately prior to combat becomes more intensive and is centralized to a definite locality. Detailed information of the hostile positions, strength, movement, and dispositions will be sought out by aircraft and communicated to the friendly ground units without delay. Ground observation will usually be very limited because of the nature of the terrain, and observation of the enemy position from the air may be absolutely essential for the formulation of plans and for the conduct of the action. Airplanes engaged in close reconnaissance missions may participate in combat by employing bombs and machine-gun fire against objectives particularly dangerous to ground troops, especially when requested by the ground commander. It should be borne in mind, however, that combat is secondary to reconnaissance, and attacks which are not coordinated with the ground force action should generally be avoided.

9-21. **Infantry mission.**—*a.* In small wars there does not exist the same line of demarcation between the tactical reconnaissance mission and the infantry mission as is prescribed in air tactical doctrine for major operations. The functions of each merge into the other. Perhaps the best definition of the term "Infantry mission," as understood for small wars, refers to a daily or periodic air patrol which flies over a given area and contacts all the ground patrols and station garrisons located within this area. Tactical reconnaissance is conducted by these air patrols incident to their passage from one ground unit to another, and they are prepared to attack hostile ground forces upon discovery. Their primary mission, however, is to maintain command liaison with detached units of friendly ground forces, and to keep these forces informed of the situation confronting them. The infantry airplanes may be used for the emergency transport of men and supplies, or they may be called upon to assist some ground patrol in a difficult situation by attacking the hostile ground force. In short, the airplanes assigned to the infantry mission, operating habitually in pairs, support the ground forces in whatever manner is expedient, regardless of their normal function in major warfare.

EMPLOYMENT OF RECONNAISSANCE AVIATION

b. Occasions may arise where it is desirable to dispense with air support for some special operation. Considerations of secrecy of movement for some ground unit may justify the responsible commander in making such a decision. Should it be decided that air support will not be furnished a ground patrol, the patrol commander should be so informed, and pilots instructed not to communicate with this unit, nor to disclose its presence in any way. However, to avoid being fired on, the ground patrol should display an identification panel whenever possible. While the infantry airplanes may disclose the position of a ground patrol to the enemy through efforts to establish a contact, it is likewise possible to deceive the enemy as to the true location of our forces by having the airplanes simulate contact with fictitious units in various other places.

c. Contacts between the infantry airplane and ground units are established by means of panels and drop messages, and where open ground is available, by message pick-ups. The use of radio will be more prevalent in the future than has been the case in the past.

9-22. **Special combat missions.**—Airplanes engaged in reconnaissance missions will be prepared to attack hostile ground forces, in order that emergency combat support may be rendered friendly ground units without delay. In small wars operations targets are apt to be fleeting and time may not permit the dispatch of regular attack units. If the enemy is to be struck while he is most vulnerable, he must be attacked immediately by the air patrol which discovers him. When time permits, a contact report should be made, but the patrol leader must make the decision in each case. This doctrine is applicable mainly to jungle warfare, against small groups of irregulars, where the offensive power of a pair of scouting airplanes would be of some avail. In more open country, against larger and better organized forces, search-attack missions by small air units are not generally recommended. In any event, it must be remembered that the primary mission of reconnaissance airplanes is not combat, but the procurement of information, and the mere existence of offensive armament should not encourage their needless diversion to combat tasks.

Section VI

COMBAT SUPPORT

9–23. **General discussion.**—The primary mission of combat aviation in a small war is the direct support of the ground forces. This implies generally that all combat aviation will be used for ground attack. Air opposition will usually be nonexistent or weak, and friendly aviation should be able to operate against hostile ground troops at will. Fighting squadrons, if included in the force, may be employed as light bombers; while the bombing squadrons will find more use for their lighter bombs and offensive machine guns than they will for their major weapon—the heavy demolition bomb. Attack aviation, or its substitute, the dual-purpose scout, is the best type to cope with the targets likely to be encountered in small wars. Troop columns, pack trains, groups of river boats, occupied villages of flimsy construction, mountain strongholds, and hostile bivouac areas are all vulnerable to the weapons of the attack airplane—the light bomb and machine gun. Occasionally, targets of a more substantial nature may require the use of medium demolition bombs. As the type of campaign approaches the proportions of a major conflict, so will the employment of the different types of combat aviation approach that prescribed for major warfare. For the typical jungle country small war, the division of missions between the different types is not so clearly marked.

9–24. **Fighting aviation.**—This class of combat aviation will be included in the small wars air force when there exists a possibility that opposition will be provided with military aircraft. The fighting squadrons should be used to neutralize the hostile air force early in the campaign. Thereafter, the fighting units could be made avail-

COMBAT SUPPORT

able as a part of the general air reserve to be employed for ground attack against particularly favorable targets.

9-25. **Attack aviation.**—The employment of attack aviation (or dual-purpose scouts acting as such) differs little in tactics or technique from the doctrine prescribed for major operations. Such units as are available should be held in central reserve to be dispatched only against definitely located targets. The six-plane division, instead of the squadron, will usually be ample force to employ against the average small wars objective.

9-26. **Bombing aviation.**—The medium dive bomber is a versatile weapon, and although there will probably be little call for the employment of the 1,000-pound bomb against small wars objectives, this type of aircraft can also carry the lighter demolition and fragmentation bombs, and is armed with offensive machine guns. Bombing units may thus be employed against personnel and the lighter material targets usually assigned to attack aviation. Legitimate targets for bombing units include forts, village strongholds, railroad rolling stock, motor trains, and the larger supply boats; secondary targets are troop columns and pack trains. When attack units are available for strafing missions, the bombing squadrons should, like the fighters, be considered as part of the general air reserve, and their use against unsuitable targets avoided.

9-27. **Attacks on troop columns and trains.**—*a.* Troops and animal trains marching in close formations on roads or trails are extremely vulnerable to surprise air attack. Such attacks should be carefully timed to hit columns as they pass through narrow defiles formed by the hills or jungle growth. If the terrain permits, a low altitude strafing attack is preferable, as it favors surprise, and permits a more effective employment of air weapons. An attempt should be made to enfilade the column with machine-gun fire and with fragmentation bombs dropped in trail, repeating the attack as required. Should the hostile column be encountered in very mountainous country it may be necessary to employ the diving attack, each airplane in the column selecting a part of the target, in order to cover the whole effectively on the first assault. Surprise will be more difficult to obtain when the diving approach must be used, although a skilled leader should be able to launch an effective assault without giving the enemy more than a few seconds' warning. Repeated diving assaults are made as required, although the objective may be much less vulnerable after the first surprise attack. In the attack of a long column which cannot be covered in one assault by the air force available, the head of the column should always be chosen as the initial objective, regardless of the method of attack employed. This will ensure the maximum of delay and confusion, and facilitate repeated assaults.

b. The successful attack of a column by an organized air unit is dependent upon the prompt transmission of information by the reconnaissance agency which makes the discovery. Small columns of mobile troops will usually be attacked on the spot when discovered by reconnaissance patrols. If the importance of the target and the nature of the terrain appears to warrant the delay necessary to launch a concentrated attack, the hostile column should be kept under surveillance, *if it can be done without sacrifice of surprise,* and a full report be made by radio to the air commander. Upon the receipt of such a message the air commander should communicate with the Force Commander while airplanes are being prepared, advising him of the contemplated action. Speed of movement and surprise of execution will be the essence of success in the air attack of a column.

9–28. **Support of a marching column.**—*a.* When the size of a column, or the hazardous nature of its advance makes the assignment of combat aviation advisable, two methods of general support are possible. A division of airplanes can be kept continuously in the air over the column; or the column can be contacted at short intervals by a combat patrol of appropriate size. In most cases the latter form of support will suffice, bearing in mind that the column would normally have a pair of infantry planes with it at all times. The reconnaissance airplanes seek out ambushes and enemy positions along the route of march; the air combat units assist the ground forces in routing hostile opposition. Air attacks may be coordinated with the ground attacks if communication facilities and the tactical situation permit, or they may be launched independently to prevent hostile interference with the march of the supported column.

b. Ground commanders supported by aviation should be careful when in action to mark the position of their advanced elements by panels, and where the force is held up by fire from a given locality they should also indicate by the proper panel signal the direction and estimated distance to the enemy position. The ground commander should also indicate, by whatever means is expedient, just when and where he wishes the fire of aviation to be concentrated. In short, he requests fire support in the same manner as he would from artillery. In addition to complying with these requests, the air commander will be constantly on the lookout for the location and move-

COMBAT SUPPORT

ment of any enemy forces in the vicinity, and will be prepared to exploit any success of the ground forces by the immediate pursuit of retreating hostile troops.

9-29. **Attack on hostile positions.**—Combat aviation may be used as a substitute for artillery in the organized attacks of hostile strongholds. As such it provides for the preliminary reduction of the hostile defenses by bombing, for the interdiction of lines of communication and supply, and for the direct close-in support of the attacking infantry by lying down a barrage of machine-gun bullets and fragmentation bombs on the enemy front lines. All these missions cannot of course be performed by one air unit; schedules of fire must be worked out, timed with the infantry advance, and executed by successive waves of aircraft. Details of this form of air support are worked out by the air commander, using such numbers and types of air units as are available and necessary. The ground commander must submit a definite plan if air attack is to be coordinated; otherwise, the air commander on the spot must use his force as opportunity offers. In minor attacks the latter procedure will probably be the rule.

9-30. **Attacks on towns.**—When hostile forces seek the shelter of occupied towns and villages, air combat support cannot be given the attacking troops without endangering the lives of noncombatants. However, it may be feasible to drop warning messages to the inhabitants, and allow them sufficient time to evacuate before initiating an attack. Once the attack is decided upon, aviation again performs the role of artillery. One bomb, penetrating the roof of a small house before exploding will effectively neutralize all occupants; those not being killed or wounded will immediately escape to the streets to become targets for machine guns. Continuous bombing forces the defenders from their shelters and facilitates their capture or defeat by the ground forces. The tactics and technique involved in the air attack of a town do not differ materially from those used against any defended position, except that medium dive bombers may be used here to better advantage than they could be in most small wars situations. Care must be taken not to endanger advancing friendly troops.

9-31. **Aviation as a mobile reserve.**—The employment of aviation as a reserve for infantry in battle is merely an application of the principle of quick concentration of superior force at the decisive point. The mobility and striking power of combat aviation favors such employment in minor operations.

Section VII

Section VII

AIR TRANSPORT

9–32. **General considerations.**—*a.* The transportation of troops and supplies becomes of increasing importance as the ground forces in a small wars campaign work inland, away from the navigable waters and railroads usually found in the coastal regions of tropical countries. Roads for wheeled transport are apt to be poor or non-existent, and dependence for supply of certain units may have to be placed on slow animal transport. As distances from the base of operations increase, this form of supply tends to break down, especially during rainy seasons, and the most advanced of the ground forces may be partially or altogether dependent upon air transport for months at a time. The air force, then, should include a much greater percentage of transport aircraft than is required for the normal needs of the air units themselves.

b. Air transportation is justified only when more economical forms of transport will not serve; it should be considered only as an emergency supplement for land transportation, and its use rigidly controlled by Force Headquarters. Factors which may influence the decision to use air transport are: unfavorable condition of roads and trails; long distances through hostile territory necessitating the provision of strong escorts for land transport; and emergency situations requiring immediate action. When air transport is planned, the air force will usually establish regular schedules for transport airplanes. Force Headquarters will arrange for routine and priority listing of supplies and replacements to be forwarded to outlying stations. Routine evacuation of the sick and wounded is accomplished on the return trips, and only occasionally should the necessity for emergency flights arise. The air force should generally have priority in the use of air transport for its own requirements. Where small air units are maintained and operated on outlying auxiliary fields, the problem of supplying fuel, ammunition, bombs, and other supplies becomes a considerable task.

9-33. **Troop transportation.**—*a.* Possibilities for the transportation of troops in airplanes are limited only by the number of transport aircraft available and the existence of suitable landing fields. In small wars operations, the ability to concentrate forces quickly in any part of the theater, through the medium of air transport, may materially influence the planning of the campaign, and offers a solution to the grave difficulties of moving forces through a country devoid of communication facilities. Small forces, not to exceed a battalion, can be transported and supplied by air everywhere within the operating radius of the aircraft, provided landing facilities are available. The utility squadron of eight transports will carry approximately one rifle company per trip, including combat equipment. While these figures indicate the maximum troop movement possible with the amount of air transport normally provided, they by no means imply that movements on a larger scale are impractical. In the typical campaign of this nature, the movement of a force larger than a company will be exceptional.

b. Troop commanders of units ordered to move by air should be advised in advance of the weight limitations per man, in order that excess equipment may be stored before embarkation. Movement orders should be specific as to time of arrival on the airdrome; details of loading will be supervised by a representative of the air operations officer, who will be guided, insofar as possible, by the principle of tactical unity in the assignment of troop spaces. On outlying airdromes, the senior aviator present is charged with these details and is responsible that safety limitations are observed. While in flight, the regularly assigned pilot of the aircraft exercises command analogous to that of the commander of a surface vessel on which troops are embarked.

c. A general policy classifying persons and articles considered eligible for air transport, with priority ratings, should be adopted and published by Force Headquarters. Permits for air travel should be issued by Force and Area Commanders, and passages coordinated with scheduled or emergency movements of transport airplanes. Requests for special airplanes should be rigidly controlled by Force and Area Commanders.

9-34. **Transportation of supplies.**—*a.* Generally speaking, the transportation of bulky supplies by air is economical only for long hauls in regions of poor communication. Questions of tactical expediency will often outweigh those of economy, however, and where air transport is available it will normally be used to capacity.

b. In order to handle properly the many calls for air transportation of supplies, regulate priority, and expedite the more urgent shipments, a special shipping office, under the control of the air commander, should be maintained at the base airdrome. This agency acts as a regulating depot between the rear echelon and the units in the field. It receives and prepares shipments, loads and unloads the airplanes, and arranges for the storage and delivery of incoming shipments. Adequate storage and transportation facilities should be made available. Shipping agencies should also be provided at the more important auxiliary airdromes if the volume of supplies appears to warrant such installations. Personnel for these regulating stations is supplied by the Force Quartermaster, as requested by the aviation supply officer who is responsible for the preparation and loading of all air shipments. The air operations officer is kept informed at all times regarding amounts and priorities of shipments, and will issue the necessary instructions for the actual loading of the airplanes.

9–35. **Dropping of supplies.**—*a.* Supplies transported by air may be delivered by landing, or by dropping from the airplane while in flight at low altitude. To avoid undue loss by breakage, articles to be dropped must have special packing. Skilled personnel can wrap almost any article so that it will not be injured by contact with the ground after being dropped. Explosives, detonators, liquid medicines, etc., may be swathed in cotton and excelsior and dropped safely; water in half-filled canteens may be dropped from low altitudes with no protection other than the canvas cover; dry beans, rice, sugar, and similar supplies may be dropped by enclosing a half-filled sack in a larger one. The governing principle in packing is to arrange for cushioning the impact and for expansion within the container. Machine guns and similar equipment should be disassembled prior to packing for air drops, although in emergency such loads could be dropped intact by using parachutes. In short, it is possible to drop safely any article of supply provided it is properly packed.

b. The dropping ground should have a clear space at least 100 yards in diameter, with no obstructions which would prevent the airplane from approaching at low altitude and minimum speed. An identification panel should mark the center of the area. Men and animals must be kept clear, or casualties will occur from men being struck by heavy falling articles.

AIR TRANSPORT

c. Emergency supplies of medicines, food, small arms ammunition, clothing, money, and mail are usually transported to detached units in the field by the daily air patrols. The observers stow the articles in their cockpits and drop them when contact is established. The standard scouting airplane will safely handle an overload equivalent to the weight of an extra man, provided room can be found for stowage near the center of gravity of the plane. Unless a landing can be made, however, the load is limited to what the observer can stow in his cockpit.

9–36. **Evacuation of sick and wounded.**—The evacuation by air of the sick and wounded personnel reduces the percentage of permanent casualties, relieves the units in the field of responsibility for their care, and enhances the morale of troops engaged in patrolling or garrisoning remote areas. Air ambulance service should have priority over all utility missions, and should be second only to urgent tactical requirements. The normal flow of sick and slightly wounded personnel are handled on the return trips of regularly scheduled transports, or by smaller airplanes from the more remote districts where no transport fields exist. When it is known in advance that casualties are to be evacuated, a medical attendant should accompany the transport or ambulance plane on its outbound trip in order that medical escort will not have to be provided by the unit in the field. Emergency cases will be handled by the senior aviator present without waiting for formal authority for the flight. Stretcher cases can be moved only by transport or ambulance planes; the patient must be able to sit up if evacuation is to be effected from a small field by a two-seater scout.

○

SMALL WARS MANUAL
UNITED STATES MARINE CORPS
1940

✦

CHAPTER X
RIVER OPERATIONS

RESTRICTED

UNITED STATES
GOVERNMENT PRINTING OFFICE
WASHINGTON : 1940

SMALL WARS MANUAL
UNITED STATES MARINE CORPS

Chapter X

RIVER OPERATIONS

v

Section I

RIVER OPERATIONS IN GENERAL

10–1. **Necessity for river operations.**—*a.* During the estimate of the situation, or after the initiation of the intervention, it may become apparent that navigable inland waterways exist within the theater of operations to such an extent that their use by the intervening force is necessary or advisable.

b. In many countries, water routes are a primary means of transportation and communication, especially if there are few and inadequate railroads, roads, or trails. In some sections of the country, they may be the only avenues of approach to areas occupied by hostile forces. So long as water routes are more economical in time and money than other available means, they will be employed by the local inhabitants and their use must be seriously considered in the plan of campaign of any force entering the country for small war operations. Such river operations as appear practicable should be coordinated with the land operations which are to be conducted simultaneously.

c. In some cases, it may be necessary or advisable to occupy a river valley in order to protect the foreign civilians, of other than United States citizenship, and property located therein against hostile depredations.

d. When offensive operations against the hostile forces interrupt the normal land routes, such forces will turn to navigable rivers as a means of supply and communication, or as an avenue of escape. Adequate and timely preparations should be undertaken by the intervening force to deny these water routes to the enemy.

e. Navigable rivers often form part or all of the boundary between the affected country and an adjacent State. If the hostile forces are receiving assistance and supplies from the neighboring country, river patrols may seriously interfere with, but never entirely suppress, such activities. Amicable agreements should be completed as soon as possible, through the Department of State, for the use of territorial waters by such patrols, and for the pursuit of hostile groups who

RIVER OPERATIONS IN GENERAL

may use the remote districts of the friendly country as a base of operations or place of refuge.

10–2. **General characteristics of rivers.**—*a*. All navigable rivers have certain similar characteristics. Their general profile is best represented as a series of terraces, the levels of which are relatively placid stretches of water of more or less uniform depth and current, and the walls of which are impassible falls or rapids. As one proceeds upstream from the mouth of the river, the depth of water in each successive level is usually less than in the one preceding. This characteristic feature determines the distance that a boat of any given draft can travel and eventually makes the use of any type of boat impossible. The extent of each group of falls and rapids, their relative distance from the mouth of the river, and the length of the intervening stretches of smooth water will vary with every river. For example, the first obstacle in the Congo River in Africa is only a hundred miles from its mouth, although the second level of the river presents no impassable falls for over a thousand miles. The Yangtze River in China is navigable by ocean-going vessels for nearly a thousand miles from its mouth before the Yangtze Gorge is reached. The Coco River in Nicaragua can be traveled for over 200 miles before the first real falls and rapids, extending over 30 miles, are found; the second level is navigable for some 60 miles; and the third level for another 70 miles to the head of navigation.

b. These various levels are customarily the "lower," "middle," and "upper" rivers as one proceeds upstream from the mouth to the head of navigation, and as the depth of water in the succeeding levels necessitates a change in the type and draft of boat which can be used.

c. The condition of the river, the depth and length of the navigable stretches, and the obstacles presented to navigation vary with the seasons of the year. At certain times, the water in the middle and upper rivers may be so low that numerous portages are necessary. When the river is in flood, such obstacles may disappear entirely and the boats normally restricted to the middle river may proceed all the way to the head of navigation, or the lower and middle rivers merge into one. This characteristic will influence the time of year and the ease and practicability of conducting river operations. The probability that supply boats could not reach Poteca, on the Coco River, during the months of April and May, influenced the decision to abandon that outpost in April 1929. In commenting on the Nile Expedition of 1884–85, Callwell says, "And it must be added that the supply difficulties were enormously increased by the lateness of the start, by the unfortunate postponement in deciding on the dispatch of the expedition. A few weeks sufficed to convert the Nile between the second and third cataracts from a great waterway up which the steamers from below Wadi Halfa could have steamed with ease, into a succession of tortuous rapids passable only with difficulty by small boats." ("Small Wars, Their Principles and Practice," by Col. C. E. Callwell, 3d ed., p. 70.)

d. As the river empties into the ocean, the sediment which it carries is deposited to form a bar or shoal. In the case of large rivers, the shoal is usually so deeply submerged that it does not prevent the entrance of ocean going vessels. In those rivers usually found in the theater of small war operations, the bar may be so near the surface of the water that it is a real obstacle and may make the passage of even the ordinary ship's boat a dangerous undertaking, especially if the services of a local pilot are not available.

SECTION II

TYPES AND CHARACTERISTICS OF BOATS

10-3. **General.**—The types and characteristics of boats which are to be used in a particular river operation depend upon several factors, of which the more important are:

(1) Coastwise communications required.

(2) Nature of the river.

(3) Desirable boat characteristics for lower, middle, and upper river use.

(4) Types of boats available.

(5) Method of propulsion.

(6) Influence of tactical principles.

10-4. **Coastwise communications.**—Navy vessels, motor launches, and local coastal schooners, normally will be used for maintaining coastwise communications. Unless a main supply base is located at the mouth of the river on which the operations are being conducted, coastal shipping will be used for the transportation of personnel and replacements, and primarily for the shipment of supplies.

10-5. **Nature of the river.**—The nature of the river, more than any other factor, determines the types of boats which will be used in river operation. The depth of the lower, middle, and upper rivers; the swiftness of the current; the distances between obstacles in the river; the number and length of the portages required; the season of the year; and the probability of securing native boatmen; each of these will have some effect on the decision. Ordinarily, at least three types of boats will be required because of the limitations as to draft in the various river levels. If the lower river is more

5

than 300 miles long, or has a limiting depth of over 8 feet, boats of the coastwise type will be used in addition to the usual river types. On the other hand, if the length of the middle river is quite short, it may be more economical to use only two types of boats, those for the lower and upper rivers only.

10–6. **Lower river boats.**—Boats to be used on the lower river normally should be motor propelled, of 4 feet draft or less, and with a maximum speed of 15 miles or more per hour. Their propellers should be protected to prevent damage from submerged rocks or logs. If they are procured outside of the theater of operations, they should be of such size and weight as to permit them to be transported by Navy transports. They should be provided with .30 or .50 caliber machine guns mounted on swivel mounts at the bow, and light armor provided to protect the gunner, helmsman, and fuel tank.

10–7. **Middle river boats.**—Boats for use on the middle river should be of sufficient size to carry at least one squad and its equipment in addition to the boat crew. Normally these boats should have a draft of 2½ feet or less. The power plant may be an outboard motor or an inboard motor with the propeller protected against damage from rocks and other obstacles. A maximum speed of 20 miles per hour is desirable. These boats should be strongly but lightly built, to facilitate their passage through rapids and rough stretches of water, or their portage around such areas. The .30 caliber machine gun may be mounted forward, either on its regular tripod mount, or on a swivel mount if one has been provided.

10–8. **Upper river boats.**—For the upper rivers, the most suitable boats are those obtained locally from the natives. If these cannot be procured in sufficient quantity, substitutes should be of the light, shallow-draft, canoe-type boat, with fairly wide, flat bottoms and built as strongly as possible commensurate with their light weight. Provision should be made for the attachment of outboard motors, although the normal method of propulsion will be by hand in most situations. They will vary in size from small canoes capable of carrying one half of a squad plus the crew, to cargo canoes capable of carrying 8 to 10 thousand pounds of supplies in addition to the necessary crew. The average upper river boat should be of sufficient size to carry a complete squad with its equipment, in addition to the crew.

10–9. **Types of boats available.**—*a. Local boats.*—Local boats obtained in the theater of operations have been used in the past with a fair degree of success. Unless the operation is planned a considerable length of time before its initiation, local boats will probably be the only ones available. These boats should be purchased outright if they are to be used for combat purposes. If the owners will not agree to sell them, as is sometimes the case, it may be necessary to requisition them. Receipts must be given for such boats. A record should be made of the owner's name, if it can be ascertained, the date and place at which the boat was acquired, its condition, and the estimated value. This information should be forwarded to the area commander or other appropriate commander so that proper adjustment can be made of the owner's claim when it is submitted. If combat boats are rented on a *per diem* basis, the eventual cost for rent, plus the expense of repairs or replacements if the boats are damaged or lost, will be exorbitant. On the other hand, it is usually more economical to rent local boats which are to be used solely for the transportation of supplies after the river has been pacified. Local boats will be nondescript in character. This complicates the repair and upkeep of motor-propelled craft. They have one decided advantage, however, all of them will have been built for use on the river on which the operations are to take place and, in that respect, they probably will be superior to boats imported for the operation.

b. Regular Navy boats will seldom be available in sufficient numbers to meet the needs of the expedition. They may be used for coastwise communications and on the lower river, depending on the depth of the water and the presence of rapids or falls in that section of the river. They are too heavy, draw too much water, and are too slow to answer the helm for use in the middle river.

c. Marine Corps landing boats, especially the smaller types, probably can be used effectively in the lower and middle rivers. Their armament, uniformity of power plant and equipment, protected bottom and propeller, and the fact that trained crews may be available to handle them, are important advantages. Their weight may be a disadvantage for middle river operation if many portages are required.

d. There are numerous boats available in the United States which are suitable for small wars river operations and which can be purchased if the situation makes it necessary. They range in type from the larger shallow draft boats which can be used on the lower rivers, to canoes suitable for employment in the upper river. So far as possible they should have approximately the same characteristics as those found in the local theater of operations. Radical changes in type should be introduced with caution.

e. Rubber boats probably will be used extensively in future small wars river operations.

TYPES AND CHARACTERISTICS OF BOATS

f. Improvements and new developments are constantly taking place in boat design and boat materials. One can never expect to obtain a uniform flotilla of boats for river operations. The difficulty will always be to get enough boats of any description to meet the demands of the situation which are suitable for use in the particular river involved. It is probable that much better boats will be available in the future than have been utilized for such operations in the past.

10–10. **Method of propulsion.**—*a. General.*—Boats used in river operations will be motor propelled, rowed, paddled, poled, or towed, depending upon the type of boat being used, the nature of the river, and the tactical situation.

b. Inboard motor boats.—Inboard motor boats have the following advantages:

(1) Speed.

(2) Usually greater carrying capacity than other types of boats.

(3) Requires small crew.

They have the following disadvantages:

(1) Noise of exhaust, even though muffled, discloses the location of the patrol and gives warning of its approach.

(2) Gasoline and oil must be carried for the period between the initiation of the patrol until the arrival of the first supply boats. This decreases the carrying capacity for troops and rations, which may be offset by the increased speed of the movement.

(3) They draw too much water for use in the upper river, or in some stretches of certain middle rivers.

(4) Their power plant often fails, or propellers are fouled or broken in rapids where power is most essential.

(5) Weight of the boat increases the difficulties of portaging around obstacles in the river.

Inboard motors are especially useful for transporting the main body and supplies of a large patrol, and in the system of supply in the lower and middle rivers.

c. Outboard motors.—(1) Outboard motorboats have the same advantages and disadvantages as inboard motorboats. They are more subject to failure during heavy rains than the inboard type.

(2) Outboard motors can be used with a fair degree of success in the upper river, although the presence of sandbars, rocks, sunken trees, and other debris, and the innumerable rapids normally encountered in this section of the river increase the difficulties of operation.

(3) Outboard motorboats are especially useful for security units with a patrol operating entirely with motorboats; and for liaison and command missions.

(4) Outboard motors purchased for river operations should be of the multiple cylinder type and capable of developing at least 25 horsepower. Motors whose water intake is through the forward end of the propeller housing should not be purchased. They are prone to pick up too much sand, dirt, and other debris in the shallow waters in which they often have to operate.

d. Rowboats.—Rowboats will seldom be used in small war river operations. Disabled navy or large sized motorboats may have to be rowed for comparatively short distances.

e. Paddles.—Paddles are normally used as the means of propulsion with upper river boats which are not equipped with outboard motors. They may be used when moving against the current in quiet stretches of the river, depending upon the strength of the current, and will always be used when going downstream or from one side of the river to the other. They are used as rudders in boats of the canoe type. Because of their reliability under all conditions, they are part of the normal equipment of every middle and upper river boat, whether they are equipped with motors or not.

f. Poles.—In swift water, poles must be used to make headway against the current if the water is too shallow for the operation of motors or if the boat is not equipped with a motor. In many cases, poles can be used to assist a motorboat when passing through rapids and bucking an unusually strong current. They are part of the normal equipment of every middle and upper river boat.

g. Towing.—Towing will have to be resorted to when passing upstream through very bad rapids. Occasionally the overhanging branches close to shore may be grasped to haul the boat along. Before towing a boat through bad stretches of water, it should be unloaded at the foot of the rapid, and the load portaged around it. In some cases, such as falls or extremely bad rapids, the boat will have to be portaged also. In going downstream through dangerous rapids, towlines must be used to ease the boat and keep it under control.

10–11. **Influence of tactical principles.**—Tactical principles will have considerable influence on the type of boats selected for any particular river patrol. Security units should be transported in small, light, easily maneuverable boats, carrying one-half to a complete squad of men in addition to the crews. The command group requires

TYPES AND CHARACTERISTICS OF BOATS

a small, fast boat. Elements of the main body must be transported as units in order to facilitate their entry into action. Supply boats may be of an entirely different type than the combat boats. The necessity for speed will influence the composition of the flotilla. Even in the lower river, these tactical requirements may necessitate the employment of some middle and upper river craft; in the upper river sections, they will influence the size of the boats employed.

Section III

PREPARATIONS FOR RIVER OPERATIONS

10–12. **Introduction.**—*a.* When the decision to seize and occupy a river route has been reached, certain preparatory measures, such as the organization of the force to be employed and the assembling of boats and their crews, must be taken. In many respects these preliminary steps closely resemble the organization of infantry patrols discussed in Chapter VI, "The Infantry Patrol."

b. In the majority of cases, the occupation of a river will proceed from the coastline inland. If the situation requires that the occupation begin near the head of navigation and work downstream, the difficulties of preparation are magnified, especially in the collection of the necessary boats, boat equipment, and native crews. The measures to be taken, however, are similar in either event.

c. If the river to be occupied is not already held by hostile forces, or no opposition is offered, the problem will be relatively simple, provided suitable and sufficient boats to handle the personnel and initial supplies are available. If the mouth of the river is held by the enemy, it must be seized as the first step. This operation does not differ from any landing against opposition, which is completely covered in the "Manual for Landing Operations."

d. River operations are relatively unfamiliar to our forces. They utilize types of transportation whose capabilities and care are comparatively unknown to our personnel. The operations are conducted on routes of travel which are seldom accurately indicated on the available map, and they are executed over waterways of constantly changing characteristics. The condition of the water highways varies with the low or flood stage in the river and with the part of the river in which the boat is operating, whether lower, middle, or upper river. Every opportunity should be given the men to become water-wise and boat-wise, in order to build up their boat-handling ability and their self-confidence. Preparations for river operations should commence, therefore, as far as possible in advance of the date when such operations are expected to begin.

10–13. **Organizing the river patrol.**—Many of the same principles apply to organization of a river patrol as apply to that of an infantry patrol. The size of the patrol is determined by the same factors, except that the number and type of boats available must be taken into consideration. Individual armament, the proportion of supporting weapons to be attached, the necessity for additional officers, cooks, medical personnel and signal personnel, native guides, and interpreters are all considered in the estimate of the situation on the same basis as for a land patrol. The principles to be borne in mind are the same; the difference is that a river is used as the avenue of approach to the hostile area instead of a road or trail, and instead of riding animals or marching, boats are used.

10–14. **Crews.**—a. Whether enlisted or native boat crews, or a proportion of each, are included in the organization of the patrol depends upon the types of boats to be used, the nature of the river, the availability of reliable natives, and the general situation in the theater of operations. Very few natives who are good engineers and mechanics will be found in the usual small wars theater. If the nature of the river is such that only motorboats will be used in the patrol, the crews should consist of enlisted men, with a sufficient number of natives to act as guides in the bow of each boat. Even these can be dispensed with if the patrol is well trained in river work. On the other hand, if the operations are to take place in the middle and upper rivers, where innumerable rapids will be encountered, and boats have to be propelled by hand, natives should comprise the boat crews if they can possibly be obtained. The handling of shallow-draft boats, such as the canoe and sampan, in the upper river is an art not easily or quickly acquired. This is second nature to the native who has been brought up in the upper river country; whereas, only a very few enlisted men will be found who can learn to handle all types of river craft in all kinds of water. Every enlisted man who is detailed as a member of a boat crew depletes the number of effectives in the combat personnel. The procedure relative to hiring native boat crews does not differ from the hiring of native muleteers, and the same principles apply as with land patrols. Every situation must be estimated and decided upon its merits.

b. The crew of a boat powered with an inboard motor should consist of a coxswain, an engineer, and a pilot. An outboard motorboat requires an engineer and a pilot only. Boats which are propelled by hand; that is, by poling when going upstream and paddling when going downstream, require a much larger crew in relation to the size of the boat. The smaller, upper river boats, capable of carrying from a half squad to a squad, should have a poling crew of three or four men at the bow, and a boat captain who handles the steering paddle or rudder at the stern. The larger supply and combat boats may require as many as twelve bowmen and two men at the stern. Smaller crews than these can operate, but the speed of the patrol will be adversely affected, and the dangers of capsizing or losing control of the boat in rough water will be increased.

10–15. **Boat procurement.**—*a.* After a decision has been made to engage in a river operation, the earlier the necessary boats are procured, the better are the chances for success. If such operations can be foreseen when the expedition is organizing in the United States, lower- and middle-river boats, and a few light-draft boats which may be suitable for use in the upper river, should be carried with the initial equipment, as well as a supply of outboard motors.

b. If suitable boats have not been provided, it will be necessary to purchase or charter local boats. If the supply of available craft exceeds the needs of the patrol, only the best of the various types required as determined by the composition of the patrol should be selected for the initial movement. It is advisable, however, to take possession of at least double that number so that they will be immediately available for supply and replacement purposes in the future. Boats should be inspected and inquiries made as to their riverworthiness before they are purchased. In many cases the supply of boats will be less than the required number or type, and the size of the patrol may have to be curtailed, or some compromise effected in the distribution of personnel, equipment, and supplies among the boats.

10–16. **Armament and equipment.**—*a. Organic.*—The organic armament and equipment, and the proportion of attached units, will not differ from that of an infantry patrol of comparable strength. For details, see Chapter II, "Composition of Forces," and Chapter VI, "The Infantry Patrol."

b. Individual.—The same principles apply to the amount of individual equipment carried in a river patrol as on an infantry land patrol. (See ch. VI, "The Infantry Patrol.") Each man, however, should be provided with a rubber, waterproof bag for carrying his personal equipment. The bag should be securely tied at the throat and distended to create the maximum airspace. If the boat capsizes, as is often the case, the bag will float and support the man in his efforts to reach the shore. Mosquito nets must always be included, especially in operations along the lower and middle rivers.

c. Boat.—For the armament of lower- and middle-river boats, see paragraphs 10–6 and 10–7. Each boat should be equipped with the following:

100-foot, stern and bow lines of 1 inch manila rope, in place ready for instant use at all times.

1 paddle for each man required to use it.

1 pole, metal shod, for each man required to use it.

2 long range focusing flashlights.

1 gasoline or makeshift stove for preparing food.

d. Signal.—Patrols operating in the lower river should be equipped with a reliable two-way radio set. Patrols in the middle and upper rivers should carry the light, portable set and establish communication with the base each day. (See ch. II, "Organization.") Panels, message pick-up set, and pyrotechnics should be carried as with infantry patrols. (See ch. VI, "The Infantry Patrol.")

e. Medical.—Medical supplies should be packed and distributed among several boats in the patrol to reduce the possibilities of loss if one or more of the boats should capsize.

f. Ammunition.—The same principles apply as with an infantry patrol, especially as to the amount of ammunition which should be carried on the person. Because of the comparative ease with which it can be transported with a river patrol, it might be advisable to carry at least one complete unit of fire in the train. Like the medical supplies, the ammunition in reserve should be distributed throughout the entire boat flotilla, except the security units.

g. Rations and galley equipment.—See Chapter VI, "The Infantry Patrol."

10–17. Loading boats.—After all other preparations have been made, the boats, in order to facilitate an early start, should be loaded with as much of the patrol equipment and supplies as possible the day before the patrol is to clear its base. Each man should be required to carry his ammunition belt and similar equipment on his person, properly fastened at all times to avoid its loss if the boat is capsized. Arms should be carried within reach at all times. Individual packs are loaded in the boat the man will occupy;

PREPARATIONS FOR RIVER OPERATIONS

they can be used as seats in boats of the upper-river type. Boats assigned to the service of security should be lightly loaded and should carry only the personal gear of the men on that duty. Other boats in the flotilla should carry their proportionate share of the equipment and supplies. Even though a supply train is included in the flotilla, it is necessary to distribute some of the supplies among the other boats as a precautionary measure against their loss if the supply boats are capsized or broken in negotiating rough water. In navigating fast water, boats should be loaded down by the head for work against the current or down by the stern for work with the current, so that the deeper end will always be up-current. As the boat tends to pivot on its deeper end, the current will hold the boat parallel to the flow of the current.

Section IV

OCCUPATION OF A RIVER

10-18. **The mission.**—The missions which determine the necessity for the occupation of a river line have been stated previously: to provide an easier and more economical route of supply to the land forces; to deny the use of the river to the hostile forces; to interfere with enemy lines of communication which are perpendicular to the river line; or to secure an avenue of approach to the hostile area for the establishment of a base from which active land operations can be conducted. Each of these will affect the size and composition of the force employed, and the location of the garrisons established along the river.

10-19. **Similarity to land operations.**—The occupation of a river parallels in every respect the advance of a land patrol from its base, except in the means of transportation. After the initial base at the mouth of the river has been seized, a first objective is selected and patrols are pushed forward until it is captured. Reorganization takes place, supplies and reenforcements are brought forward, and the advance is resumed to the second objective. A third objective is selected and taken in the same way, and so on until the river is brought under control. If opposition is not expected and the mission is to garrison the river more or less equitably throughout its length, as in the case of using it as a route of supply or to deny it to the enemy, the advance may be continuous. The entire river force may

17

leave the original base as a body, provided enough boats are available, and detachments are made as each outpost is established along the route. If opposition is anticipated, or if the supply of boats is not sufficient for the entire patrol, the advance will certainly be made by bounds from objective to objective, and eventually the major portion of the river force will be concentrated at the final objective where it is employed for coordinated land and river operations against the enemy in hostile territory.

10–20. **The day's march.**—As with land patrols, the day's march should begin as soon after dawn as possible. This is facilitated by the fact that most of the supplies and equipment may be loaded into the boats each evening as soon as the rations for the next 24 hours have been removed. Noonday halts should not be made for the purpose of preparing a hot meal. Midday lunches may be prepared and distributed in the morning although usually the ration situation will not permit such action. Unless tactical considerations prevent, the day's movement should be halted at least 2 hours before sundown in order to carry out the necessary security measures, make the camp, and feed the troops and boat crews before dark. The camp should be on fairly level ground, sufficiently above the water level to avoid flooding in the event of a rapid rise in the river during the night. Boats should be secured with a sufficiently long line to prevent their being stranded on dry land because of a sudden drop in the water level, or being pulled under and swamped because of a sudden rise in the river. Boat guards should always be posted over the flotilla.

10–21. **Rate of movement.**—The rate of movement will depend upon the type of boat being used, whether propelled by motor or by hand; the nature and condition of the river, whether in deep comparatively calm water, or in the strong currents and innumerable rapids of the middle and upper river; and the need for careful reconnaissance. A motor flotilla may average between 60 and 100 miles a day under the best conditions; a flotilla moving by hand power will average from 12 to 15 miles per day. The rate of advance will be that of the slowest boat in the column. Regardless of the rate of movement, some word of the approach of the patrol will usually precede it up river, especially if the area is well populated. If the state of the river permits, it may be possible, and in certain situations desirable, to overrun the hostile shore positions by utilizing the speed available to a motorboat flotilla. If the mission of the patrol is to drive the hostile groups out of the river valley, it may be better to advance slowly, sometimes by poling, in order to seek out the enemy by reconnaissance and engage him in combat.

10–22. **Boat formations.**—*a. General.*—Formations for a boat column advancing along a river, either up or down stream, parallels in every respect a march formation for an infantry patrol over land, and the same principles apply. (See "The Infantry Patrol," ch. VI). There should be an advance guard, a command group, a main body, a combat or supply train, and a rear guard. Tactical units, such as half squads (combat teams), squads, and platoons, should be assigned to separate boats so as to maintain freedom of maneuver and yet retain as much control over the various elements of the patrol as possible. The number and type of boats within the formation will depend upon the size and composition of the patrol and the nature of the river in which it is operating. Even in the lower river where no opposition is expected, some security elements should proceed and follow the main body. It would be a mistake to place an entire patrol consisting of a rifle platoon in a single lower river boat, or even to divide it into a point of a half squad in an upper river boat and the remainder of the patrol in a single lower river boat, if opposition is anticipated. To do so might immobilize the entire patrol if the main body should suddenly be fired upon from a concealed hostile position. On the other hand, it would be tactically unsound to employ only upper river boats containing one squad each or less for a patrol consisting of a reenforced rifle company. This would result in an elongated column and a corresponding lack of control. If the nature of the river and the type of boats available make such a disposition necessary, it would be better to employ the split column formation, described in Chapter VI, for large infantry patrols.

b. Type of boats employed.—The elements of the advance guard, the rear guard, and flank security units, as well as the command group should be assigned to small, light, fast boats of the upper-river type. This is especially true of the point, rear point, and command group. This facilitates the movement of the security elements and permits them to adjust the distances in the formation according to the terrain through which they are passing without slowing up unduly the steady progress of the main body. It enables the patrol commander to proceed rapidly to any part of the column where his presence is most urgently required. The remainder of the patrol may be assigned to types of boats which are best suited to the tactical requirements and the nature of the river.

OCCUPATION OF A RIVER

c. Distances in formation.—The distances between the elements of the column will vary with the terrain, the size of the flotilla, and the speed of movement. The principles involved are analogous to land operations in which the troops are proceeding along a fairly wide, open road. The leading elements should never be out of sight of the next element in rear for more than a minute or so at a time. Where the river is straight and wide, distances between the various parts of the column should be great enough to prevent the main body coming under machine-gun fire before the hostile position has been disclosed by the security detachments. Where the river is winding and tortuous the distance between groups should be shortened. If the distance between elements is too great each unit may be defeated in detail before the next succeeding unit can be brought up, disembarked, and engaged with the enemy.

d. Location of patrol commander in column.—The patrol commander's boat will usually move at or near the head of the main body.

e. Location of supply boats.—Normally the supply boat or boats should be located at the tail of the main body. The rear echelon of the command group acts as train guard. In the event of combat, the train guard assembles the boats of the flotilla and the crews, and moves the train forward to maintain liaison with the main body as the attack progresses. If the rear guard is committed to action, the train guard assumes its functions to protect the column from an attack from the rear. If the train is unusually long, as may be the case when a patrol is to establish an advance base at the end of its river movement, it may be advisable to detach the majority of the supply boats from the main column and form it into a convoy, following the combat part of the patrol at a designated distance.

10–23. **Reconnaissance and security.**—*a. Methods of reconnaissance.*—A river patrol employs the same methods of reconnaissance as an infantry patrol ashore. (See "The Infantry Patrol," ch. VI.) Since the route of advance is limited to the river, it is often necessary to halt the movement temporarily while small land patrols reconnoiter suspicious localities some distance from the river banks.

b. The advance guard.—The advance guard may consist of a point boat only, or it may be broken into a point, advance party, support, and reserve, depending upon the strength of the patrol. As in operations on land, the function of the point is primarily reconnaissance, to uncover and disclose hostile positions in front of the advancing column before the main body comes within effective range of the enemy's weapons. The upper-river type boat is best suited for this purpose; it can be handled easily and does not expose too many men to the surprise fire of an ambush laid along the shore lines. The elements of the advance guard should increase in strength from front to rear so that increasing pressure is applied as succeeding units engage the hostile position. If the river is wide, the advance guard should employ a broad front, with at least one boat near each bank. The main body should proceed near the center of the river to reduce the effects of hostile fire from either bank.

c. Flank security.—(1) It is almost impossible for men in boats to discover a well-laid ambush. When operating in hostile territory, or when there are indications that combat is imminent, shore patrols should precede or move abreast of the advance guard boats on each bank of the river. Although this will slow the rate of travel, it is an essential precaution unless speed is the most important factor in the mission of the patrol.

(2) Navigable tributaries entering the route of advance should be reconnoitered and secured by some small boat element of the patrol while the column is passing them.

d. March outposts.—March outposts should be established at every temporary halt. This is accomplished by reconnaissance to the front and rear by the point and rear point, resepectively, and by the establishment of flank boat or shore patrols as necessary.

e. Security at rest.—Immediately upon arriving at a temporary or permanent camp site, boat reconnaissance patrols are sent up and down river for a distance of one or more miles depending upon the nature of the river. Trails and roads leading into the camp site and suspicious localities in the vicinity of the site are reconnoitered by land patrols. Other precautionary measures are taken as prescribed for infantry patrols. (See "The Infantry Patrol," ch. VI.)

10–24. **Initial contact with the enemy.**—The initial contact with the enemy in river operations may be in the nature of a meeting engagement, with all the elements of surprise for both forces found in such contacts, or, as is more often the case, it consists of uncovering his outpost positions. In either event, once contact has been made, the choice of position and the time of future engagements will pass to the hostile force attempting to prevent the further advance of the patrol. In most small war operations, these engagements will be in the nature of an ambuscade.

10–25. **A typical ambush.**—The typical hostile ambush will resemble those found in land operations. It will be located at a bend in the river in order to provide suitable locations for automatic weapons to enfilade the advancing column of boats. The nature of the river will be such that the boats will be forced close to one bank to negotiate the current. Along this bank will be located the main hostile position so sited that rifle and automatic weapon fire can be directed at the column from the flank. The terrain will be heavily wooded to afford cover and concealment. Under these conditions, the possibilities that the ambush can be detected by men in boats will be very slight. Portages, rapids, and canyons may also be selected as ambush positions in order to engage the patrol when it is widely dispersed and out of control of its commander.

10–26. **The attack.**—Men in boats present a concentrated, vulnerable target to a hostile force ashore. The hostile fire should be returned by the weapons carried on the boats as normal armament. A few riflemen may be in such a position that they can open fire without endangering the crew or other members of the boat. However, any fire delivered from a moving boat will be erratic and comparatively ineffectual. The full power of the attacking force cannot be developed until the troops are on shore and deployed for the fight. If the attack occurs in a wide, deep stretch of river in which the boats can be maneuvered, it may be possible to run past the hostile fire and land the troops above the ambush to take the attack in the rear and cut off the enemy's prearranged line of retreat. Usually the ambush will be so located that this is impossible. In that event, the leading boats should be beached toward the hostile position. Disembarking, the men in these boats take up the fire and hold the hostile force in its position. Those boats not under the initial burst of fire should be brought upstream as close to the hostile position as possible, the troops disembarked, and the attack launched from the flank to envelop the ambush, overrun the position, and intercept the hostile line of withdrawal. Ordinarily the patrol should land on one side of the river only. In some situations it may be desirable or necessary to land on both banks, especially if the hostile force is deployed on both sides of the river. This action increases the difficulties of control, and may result in inflicting casualties among friendly personnel. Once the troops are ashore, the tactics are similar to those employed by regular infantry patrols.

10–27. **Garrisoning the river.**—*a.* The location of the various posts to be established along the river is determined by: foreign settlements and investments which require protection; junctions of important river-ways; location of intersecting roads and trails; supply dumps and reshipping points between the lower and middle rivers or the middle and upper rivers; and the strength, aggressiveness, and disposition of the hostile forces.

b. The strength of each post will depend upon its mission and the hostile forces in the area. The largest forces should be located at those points on the river which are most vulnerable to attack, or from which combat patrols can operate to best advantage against hostile forces.

c. The distance between posts on the river is determined by the existing situation. If the hostile force is active and aggressive in the area, the posts should be within supporting distance, not over 1 day's travel upstream, from each other. If the hostile force is weak, unaggressive, or nonexistant, a distance of 150 miles between posts may not be too great a dispersion of force.

d. In some situations, it may be necessary to establish outposts on navigable tributaries to the main river in order to protect the line of communications. This is especially important if the tributary leads from hostile areas or if trails used by the hostile force cross its course.

10–28. **Defensive measures.**—*a.* Each garrison along the river must be prepared for all-around defense. Wire entanglements or other obstacles should be erected, machine guns and other defensive weapons should be supplied, and normal defensive measures taken. Active patrolling by land and water should be maintained. Communications by radio or other means should be established with the area headquarters and with other outposts along the river. Boats should be supplied to each outpost for reconnaissance, liaison, and local supply purposes, and as a means for evacuation down river in case of necessity.

10–29. **Passage of obstacles.**—Obstacles in the river, such as narrows, gorges, bad rapids, and falls, whether they can be navigated or require a portage, are similar to defiles in ordinary warfare and similar protective measure must be taken. A combat patrol should proceed to the head of the obstacle, and flank security patrols should reconnoiter both banks of the river and dangerous commanding localities, in order to secure the safe passage of the main body through the obstacle.

10–30. **Night operations.**—Night operations may be conducted:
(1) To make a reconnaissance.
(2) To make a search.
(3) To secrete small detachments and picket boats.

OCCUPATION OF A RIVER

(4) To send out a patrol.

(5) To change the location of a post.

(6) To avoid aimed fire from shore and to avoid combat. Night operations must be conducted by poling or paddling, never by motor, if secrecy is to be attained. Movements upstream against the current at night are extremely slow, difficult, and fatiguing to crew and combat force alike. They should be avoided except in the most urgent situations. They have all the attendant difficulties of a night march by an infantry patrol. See "The Infantry Patrol," Ch. VI.) On the other hand, night movements by boat downstream with the current can be silently and easily executed if the night is clear and if the river is free of dangerous obstacles. Such night movements are often profitably employed in river operations.

10–31. **Supporting forces.**—*a. Infantry patrols.*—River operations often can be coordinated with the operations of infantry patrols if the trail net is satisfactory and such supporting troops are available in the area. Such coordinated efforts should be employed whenever possible to effect the seizure of important towns or localities along the river, or to increase the probability of inflicting a decisive defeat upon the hostile forces.

b. Aviation.—Aviation support is fully as important for the successful conclusion of river operations as for the corresponding land operations. For details, see Chapter IX, "Aviation."

SMALL WARS MANUAL
UNITED STATES MARINE CORPS
1940

✦

CHAPTER XI

DISARMAMENT OF POPULATION

RESTRICTED

UNITED STATES
GOVERNMENT PRINTING OFFICE
WASHINGTON : 1940

TABLE OF CONTENTS

The Small Wars Manual, U. S. Marine Corps, 1940, is published in 15 chapters as follows:

SMALL WARS MANUAL

UNITED STATES MARINE CORPS

CHAPTER XI

DISARMAMENT OF POPULATION

11–1. General.—*a.* Due to the unsettled conditions ordinarily prevailing in a country requiring a neutral intervention, and the existence of many arms in the hands of the inhabitants, the disarming of the general population of that country is not only extremely important as a part of the operation of the intervening forces but also to the interests of the inhabitants themselves. It is customary in many undeveloped or unsettled communities for all of the male population upon reaching maturity, to be habitually armed, notwithstanding that such possession is generally illegal. There is a logical reason for the large number of weapons in the hands of the inhabitants. The arbitrary political methods which frequently result in revolution, and the lawlessness practiced by a large proportion of the population, is responsible for this state of affairs. The professional politicians and the revolutionary or bandit leaders, as well as their numerous cohorts, are habitually armed. Legal institutions cannot prevail against this distressing condition; persons and property are left at the mercy of unscrupulous despots, until in self-

DISARMAMENT OF POPULATION

preservation the peaceful and law abiding inhabitants are forced to arm themselves.

b. If it has not been done previously by the intervening forces, the disarming of the people should be initiated upon the formal declaration of military government, and must be regarded as the most vital step in the restoration of tranquility. The disarming of the native population of a country in which military occupation has taken place is an imperative necessity.

c. One of the initial steps of an intervention is the disarming of the native factions opposing each other. If this action is successful, serious subsequent results may be averted. To be effective, this action must be timely, and the full cooperation of native leaders must be secured through the proper psychological approach. The disarmament can be effected only through the greatest tact and diplomacy. It is only one of several successive steps in the settlement of the local controversy, and any agreement effected must insure not only ultimate justice but immediate satisfaction to all contending parties. To secure this concession, the arbiter must have the confidence of the natives and must be ready, willing, and able to insure the provisions of the agreement. This involves the responsibility to provide security not only for the natives who have been disarmed but for the individuals depending upon them for protection. This implies the presence of the arbiter's forces in sufficient numbers to guarantee safety.

d. Peaceful inhabitants, voluntarily surrendering their arms, should be guaranteed protection by those forces charged with the restoration and maintenance of peace and order. Were it possible to disarm completely the whole population, the military features of small wars would resolve themselves into simple police duties of a routine nature. Obviously, considering the size of the population, the extent of territory, and the limited number of available troops, any measures adopted will not be 100 percent effective. However, if properly executed, the native military organizations and a large proportion of the populace may be disarmed voluntarily; many others will be disarmed by military or police measures designed to locate and confiscate arms held clandestinely. These measures will limit the outstanding arms to those held by a few individuals who will seek to hide them. In many instances, these hidden arms will be exposed to the elements or to deterioration which in time will accomplish the same end as surrender or confiscation. Although complete disarmament may not be attained, yet the enforcement of any ordinance restricting the possession of arms will result in the illegal possession of such arms only by opposing native forces, outlaws or bandits, and a few inhabitants who will evade this ordinance as they would attempt to do with any unpopular legislation. Comparatively few of this latter class will use their weapons except in self-defense. Thus the inhabitants are partially segregated at the outset of the negotiations. The disarming order will probably not influence the professional guerrilla fighters to give up their weapons but such source of supply and replenishment of weapons and ammunition within the country will be practically eliminated.

11-2. **Estimate and Plans.**—*a.* Prior to the issuance of any order or decree disarming the inhabitants, it is necessary to make an estimate of the situation and analyze all features of the undertaking, the powers and limitations, the advantages and disadvantages, and then make plans accordingly. The plans should include the following provisions:

(1) The measures necessary to strengthen the local laws.

(2) The civil or military authority issuing the disarming order, or decree.

(3) The forces necessary to enforce the order or decree.

(4) The form of the order or decree.

(5) The method of promulgating the order or decree.

(6) The measures and supplementary instructions to place the order in effect.

(7) The designation and preparation of depots, buildings, and magazines in convenient places for the storage of the arms, ammunition, and explosives.

(8) The disposition of the munitions collected.

(9) The method of accountability for such munitions, including the preparation of the necessary forms, receipts, tags, and permits, to be used in this system.

(10) The arrangements for the funds necessary to execute the disarmament.

(11) The designation of the types and classes of munitions to be turned in.

(12) The exceptions to the order or decree, definitely and plainly stated for the information of subordinates. (Special permits to individuals.)

(13) The agencies (civil officials or military commanders), who will collect, guard, and transport the material.

DISARMAMENT OF POPULATION

(14) The supplementary instructions for the guidance of the agencies charged with the execution of the order or decree.

(15) The instructions governing the manufacture and importation of munitions.

(16) The instructions governing the sale and distribution of munitions manufactured or imported.

(17) The time limit for compliance and penalties assigned thereafter.

b. Small wars take place generally in countries containing primitive areas where many of the inhabitants depend on game for their fresh meat. The peasants in the outlying districts accordingly are armed with shotguns for hunting, as well as for self-protection. Many demands for the retention of such arms will be made on this score and they should be satisfied in accordance with the seriousness of the situation, the justice of the request, and the character of the individual making it.

c. A feature of the disarming of the inhabitants which is a source of difficulty and misunderstanding is the question of retaining their machetes, cutachas, knives, and stilettos. Machetes in these countries are of two general types; one is for work and the other for fighting. The working machete is practically the only implement found on the farms or in the forest; it is used for clearing and cultivating land as well as harvesting the crops. It would be obviously unfair to deprive the natives of this general utility tool. It is distinguished by its heavy weight, the blade being broader and slightly curved at the end away from the handle, and without a guard or hilt. The fighting machete or cutacha has a hilt and is narrow, light, and sharp. Sometimes working machetes are ground down into fighting weapons but these are readily distinguished. Directions issued for the collection of arms should contain instructions so that subordinates may be informed of the difference in order to insure the collection of these dangerous weapons, and to avoid depriving the peasants of their implements which mean their very livelihood. Similarly, one finds that the natives are almost always armed with some kind of knife. They are used when packing animals and for all kinds of light work; they are often the only implements used in eating; they are used in butchering, in trimming the hoofs of their animals, and for many other chores. Certain weapons are obviously for fighting only and these are banned without question; these are the stiletto or narrow blade, dagger type of weapon. They have little or no cutting qualities but they are deadly.

4

DISARMAMENT OF POPULATION

d. The disarming order, or supplementary instructions thereto, should describe these weapons sufficiently to properly guide the subordinates who will execute the order. They should provide that cutachas will not be permitted to be carried at any time; agricultural machetes will not be permitted on the public roads or in public gatherings; stilettos will not be permitted at any time or place.

e. Care should be taken to allow sufficient time for all inhabitants to turn in their arms, and opportunity to turn in arms must be assured. If sufficient time is allowed, or if instructions to turn in arms are not widely published, a number of inhabitants may have arms in their possession though willing to turn them in. They will be fearful of the consequences, and through their ignorance will constitute a ready field for recruiting for bandit ranks. This is particularly true in remote areas. It is therefore most important that time, notice, and opportunity be given all concerned.

11–3. Laws, Decrees, Orders, and Instructions.—*a.* In most countries, there are statutes restricting the possession of arms and explosives. As a rule these laws are not enforced rigidly and even at best are not sufficiently comprehensive to meet the immediate requirements. The laws and their enforcement agencies must be strengthened by appropriate measures to insure the effective execution of the measures intended.

b. The first step in disarming the population is to issue a disarming order forbidding all inhabitants to have in their possession firearms, ammunition, or explosives, except under special circumstances to be determined by a specific authority. This order is directed to the authority who will be responsible for its execution. It specifies that the prohibited articles will be turned in to the proper officers of the forces of occupation, who will receipt and care for such as are voluntarily surrendered, but that such articles as are not voluntarily surrendered will be confiscated. It will further stipulate that after a certain date the illegal possession of arms, ammunition, or explosives will render the person apprehended liable to punishment. The details of carrying out this order are properly left to the discretion of appropriate military authority.

c. The official who has the authority to issue the disarming order will be indicated by the nature of the intervention. In a simple intervention where the civil authorities are still in charge, a decree might be issued by the Chief Executive, or a law might be enacted in proper form and sufficiently forceful to fit the situation. Such decrees have been issued in emergencies in the past and have proven effective. In case a military government is established, the military governor would issue the decree or order. Under some circumstances the commander of the military or naval forces might issue the disarming order.

d. To give the order the force and character of a public document it should be published in appropriate official publications of the government for the information and guidance of the citizens of the country. This method not only gives the order an official character but insures its prompt and legal distribution throughout the country. The order should be published in the native language and, as necessary, in the language of the intervening forces. Circumstances will determine the time limit in which the prohibited munitions must be surrendered; after which date their possession will be illegal. This will depend upon the ability of the natives to comply before a given date, or the availability of the forces to make it effective. The necessity for explosives required for the routine peaceful vocation of some inhabitants should not be overlooked. Prohibitory restrictions against their possession or use would materially interfere with industrial and commercial enterprise and development. So, in keeping with the policy of fair and liberal treatment of the natives, provision must be made for these special cases. Before incorporating in the disarming order any exception thereto, the military authorities should consider first, the conditions which might result under legalized use of firearms and explosives by certain favored individuals (civil officials, land owners, etc.,), and second, the extent and character of supervision that will be required to control their use and their sources of supply. Once these points have been determined, the order should be prepared to incorporate the necessary provisions. The disarmament of that portion of the native population living in remote and lawless districts should only be undertaken with a full appreciation of the responsibilities involved. Ranch overseers, mine superintendents, paymasters and local civil authorities, should be given special consideration in the matter of arms permits. There is such a thing as being over-zealous in the matter of disarmament, and it is often advisable to make certain concessions to responsible parties in order to secure their full cooperation in the enforcement of the laws.

11-4. **Manner of Collecting Arms.**—*a.* When opposing native forces are operating in the field, the intervening forces, if acting as an arbiter, should institute measures to secure the arms of all the opposing forces by organizations prior to their disbanding. Every endeavor should be made to have the full cooperation of the leaders and to prevent the escape or departure of any subordinate leader and his followers with arms in their possession. Disarming such organizations involves disbanding them and providing for their return home. With the twofold purpose of insuring the turning in of their arms and the return of the natives to their homes, a price is often paid to individuals for their weapons in accordance with a schedule fixing the rates for the various types of firearms and ammunition. This is a reasonable charge against the native government and the money from this source must be assured before proceeding. This procedure may be a source of chicanery and fraud to deceive the authorities and get money dishonestly. Every precaution should be taken to see that money is paid only for the arms of men regularly serving with the units at the time of the agreement. Precautions must be taken that the armories and magazines are not raided after the agreement is in effect and that the same individuals do not repeatedly return with such rifles for payment. On the other hand, ready money in sufficient quantity from the local government must be available at the time and place of payment agreed upon, or where the forces are found.

b. If part of the native forces remain armed, the full benefits of disarmament are not obtained, and serious consequences may later develop. When this occurs, some of these small armed groups may take the field and continue their operations not only against the local government but also against the intervening forces. This would place the intervening forces in an embarrassing position. After having disarmed the forces which might have been capable of controlling the movement, the intervening forces may be required either to halt the disarming negotiations and again rearm those forces or send out its own troops to take the field against these armed groups. In other words, to be fully effective, disarmament must be practically complete.

11-5. **Collecting Agencies.**—*a.* The following agencies may be employed to collect firearms, ammunition, explosives, etc.:

Provincial governors and local police authorities, particularly the communal chiefs, the chiefs of police, and the rural policemen.

The military forces of occupation.

Special agents or operators of the Force Intelligence office, or the Provost Department.

The native constabulary forces.

If military government has been instituted, the Provost Department may very appropriately be assigned the task of collecting the

DISARMAMENT OF POPULATION

munitions, the responsibility for the storage and custody of same, keeping records and submitting necessary reports. In other situations where a native constabulary has been organized, there may be advantages in assigning this duty to that organization.

b. (1) When employed as a collection agency, the civil authorities are issued supplementary instructions at the time the disarming order is promulgated, stating explicitly the manner in which firearms, ammunition, and explosives will be collected and stored or turned over to the military forces. These instructions may be amplified where necessary by field commanders who will visit the various communities and issue instructions to the local officials, imposing such restrictions as to the time and place the prohibited articles will be surrendered. The civil officials may be required to make personal delivery of the collected articles to the military forces or to make report of same and the material collected periodically by designated agencies.

(2) The success attained through employment of civil officials depends upon the spirit and conscientious effort which they display. Some who have been thoroughly indoctrinated with the advantages of the idea will have remarkable success; others who consider the disarming order an unjust imposition will perform their duties in a perfunctory manner, and still others will use the order to promote dishonest practices, disarming some of the people and permitting others to retain their arms, for personal, political, or monetary reasons.

(3) The disarming of the inhabitants through the intervening instrumentality of the civil officials possesses many redeeming features over the utilization of the armed forces for the same purpose. It is the most peaceful means of accomplishing the desired object, less provocative, and the least likely to engender antagonism or create friction. It gives the peaceful, law-abiding citizens, who are worn out by the constant political abuse of the past, the opportunity to hand over gracefully their weapons without being subjected to what they might consider the indignity of making a personal surrender to the military authorities. Misunderstanding will thus be avoided that might otherwise occur if the armed forces are employed, because of a difference of language and custom. Moreover, it relieves the armed forces of the unpleasant responsibility and eliminates the factor of personal contact at a time when the population views the intentions of the forces of occupation with doubt and suspicion.

8

DISARMAMENT OF POPULATION

c. It is not to be assumed that an order as exacting and far reaching in its effect as this disarming order will meet with willing and universal compliance. As a consequence, it may be necessary to resort to more stringent enforcement in order to compel the recalcitrants to surrender their weapons. The civil officials may be directed to secure the prohibited articles, or the armed forces may conduct a house-to-house search for concealed weapons. Both means may be employed simultaneously. Stringent measures may be unavoidable and wholly justifiable, in an effort to promote an early return to peace and order.

d. (1) Special agents or operators of the Force Intelligence office or the Provost Department may trace, or make collections of weapons. Their action is taken on what is considered reliable information and generally applies to comparatively large quantities of firearms and ammunition held by certain prominent individuals. The success of these operations depends upon the skill and courage of the agents who have to rely in a great measure upon their own initiative and resources.

(2) The Intelligence Service through special operatives may be employed to trace imports of arms and ammunition during a period of several years preceding the occupation. Government permits and correspondence, custom house files, and other records will aid in identifying receipts of these munitions, and a search for their subsequent disposition may be undertaken. Deliveries of rifles, special weapons, automatics, machine guns, howitzers, artillery pieces, ammunition, and explosives are noted and compared with issues, sales, and expenditures.

e. (1) Upon the establishment of a native constabulary, this organization may assist the military forces in the collection and confiscation of firearms. These troops may perform valuable service in this connection through their knowledge of the country and their familiarity with the habits of the people.

(2) After a reasonable time has elapsed, or when it appears that the civil officials have exhausted their usefulness in the collection of arms, the military authorities may issue an order to the effect that after a given date the military forces will be responsible for the collection of arms and the gathering of evidence for the conviction of persons involved.

11-6. Custody of Arms.—*a.* Included in the plans for disarming the population must be the designation of personnel necessary to receive and care for the material turned in. Buildings and storerooms

DISARMAMENT OF POPULATION

suitable for the safekeeping of the weapons, munitions, and explosives must be provided prior to the actual receipt of the material; the volume of this material may assume unwieldy proportions by the large increments arriving during the early days of the disarmament. An accurate system must be devised to keep a complete record of everything received, and the material tagged and stored in such manner as to identify it easily; the place should be of such construction as to preserve the material and also to make it secure. When material is received in condition which makes its keeping dangerous, authority should be requested to destroy it or to dispose of it otherwise. A frequent inventory and inspection of all such material in custody should be made not only by the custodian official but by inspecting officers.

b. Receipts should not be given for weapons delivered upon payment of money, nor for arms and material confiscated. There will be, however, a number of reputable citizens including merchants authorized previously to deal in these stocks, who wish to comply with the latest order and turn their stocks in to the custody of the military forces. The latter are obliged to accept this material and must be prepared to deliver it when a legitimate demand is made for its return.

c. Instructions should be issued designating and limiting the agencies which will accept the material and give receipts for same. There have been instances in the past, where sufficient time has not been allowed for proper organization and preparation for the methodical receipts of arms; in the avalanche of arms turned in simultaneously at many places, junior officers, in good faith, have accepted the material and have given personal receipts for it without having a proper place for its safe keeping. In the rush of official business, they did not demand a receipt from other officers to whom they delivered the material collected. No records were made of the ultimate disposition of the material. When proper authorities subsequently requested information concerning the final disposition of special material, the information was furnished only after a most difficult search. In many cases, the material could not be located or its disposition determined due to the lack of records.

d. If only certain officers are designated to issue receipts on the prescribed forms, and if the material is assembled by areas or districts, confusion may be avoided. Such confusion may arise from junior officers giving personal receipts in several different districts in which they may serve during the disarming period with no record

DISARMAMENT OF POPULATION

being made of the receipts. The receipts should be in standard form, and should indicate the name and residence of the owner of the weapon, and the date and the place of issue of the receipt. The material or weapon should be properly identified and other appropriate remarks should be added. This receipt should be signed by the officer authorized to receive the material.

e. District Commanders should be required to submit monthly reports of all arms and ammunition collected within their respective districts. The larger part of the weapons collected will be obsolete and in such poor condition as to render them of little or no practical value; those which have been paid for or confiscated may be destroyed by burning or dumping at sea. Those of better type and condition may be retained or issued to the native constabulary troops. The collection of arms cannot be said to be terminated at any given time; it is a process which continues throughout the occupation.

11–7. **Disposition.**—*a.* When arms have been received from various sources, they are classified as follows:

(1) Material for which a receipt has been issued.

(2) Material confiscated, collected upon payment of money, or otherwise received.

b. The custody of material under "class one" implies responsibility to guard and preserve it for return to the rightful owner when law or decree permits.

c. The material under "class two" is further divided into serviceable, unserviceable, and dangerous material. The serviceable material may be of a type, caliber, and condition suitable for reissue to native troops, local police, special agents or others whom it is desired to arm. The question of uniformity, adaptability, and ammunition supply is involved. The unserviceable material, or that whose keeping is hazardous, is disposed of as directed; firearms are burned, the metal parts being used in reinforcing concrete, or disposed of in other effective ways to preclude any possible future use as a weapon. Sometimes material is dumped in deep water beyond recovery. Dangerous material, such as explosives, should be stored in special places apart from other material.

d. Whenever any material is disposed of in any manner, permanent records should be made of the transaction. Receipts should be demanded for that which is reissued or transferred no matter in what manner. When material is destroyed or otherwise disposed of, a certificate should be made, attested to by witnesses, which voucher

DISARMAMENT OF POPULATION

should set forth in sufficient detail by name, mark, and quantity, the identity of the material disposed of. This will prove a valuable aid if and when information is ever demanded at a later date.

e. In general, records are made and subscribed to by witnesses whenever material is destroyed or disposed of otherwise. Appropriate receipts should be demanded whenever material is issued in accordance with orders from higher authority. Care must be taken that no weapon or material is issued or otherwise disposed of except in an authorized manner.

f. Great care should be exercised in keeping material which has been confiscated, or whose ownership is transferred to the government, segregated from that material which the government simply holds in custody. The latter is not subject to destruction nor available for issue. Beware of the ever present souvenir hunters of all ranks who wish to get possession of articles of unusual design, of historical interest, or of special value. All such unusual weapons or articles create a peculiar interest whenever they come into our custody. They arouse the attention and interest of enlisted men or civilian workers who assist around the magazines or storerooms in the receipt or storage of such material. These unusual articles which are in greatest demand as souvenirs, and the disposition of which is most closely watched and remembered by subordinates, are the very articles that the original owners wish to have returned sooner or later. Minor indiscretions in the disposition of material received assume serious proportions in the minds of the natives which are not at all in keeping with their actual importance.

11-8. Permits.—*a.* The military authorities determine who shall be empowered to issue arms permits, to whom they may issue them, and any other pertinent restrictions. Under certain circumstances District Commanders and District Provost Marshals may be the designated agencies. In any event the process must be coordinated to prevent conflict or overlapping authority. Certain civil officials, such as provincial governors, judges, and others exercising police functions, may be authorized to carry arms. Certain permits are issued which are honored throughout the country; some are issued which are good for more than one district or jurisdiction but not for the whole country; in either case the higher authority approving same notifies the responsible officers in the several subordinate jurisdictions concerned. When permits are requested, information is furnished concerning the nationality, character, commercial and political affiliations, occupation, and address of the applicant and the necessity for the granting of the permit. Officers issuing permits must exercise great care to the end that permits be issued only where real necessity exists; any application which has the appearance of being made simply to enhance the prestige of the individual making it, as often happens, should be promptly refused.

b. Permits should be issued on a standard form with a description of the person to whom issued, together with the character and serial number of the firearm, the purpose for which it is to be used, and the locality in which it is to be carried. These permits should be nontransferable and should be renewed each year or the firearms must be turned in to the authorized agency. Holders of permits should be warned that the unauthorized use of their firearms will result in certain disciplinary measures in keeping with the gravity of the offense and the punitive authority of the official; this may include revocation of the permit and confiscation of the firearm, and even fine or imprisonment or both.

c. Permits should be issued only for the possession of pistols, revolvers, and shotguns. The privilege of possessing rifles should be refused consistently. It should be exceedingly difficult to secure any kind of permit.

d. In order to maintain a strict account of all arms permits in effect, all issuing officers should be directed to keep a record of all permits issued by them, copies of which should be forwarded to the district commander. The district commanders in turn should submit to Force Headquarters, annually or semiannually, a list in duplicate of all permits issued within their respective districts. In addition to this annual or semiannual report, they should also render a monthly change sheet in duplicate, containing a list of permits issued and cancelled during the month.

11-9. Control of Sources of Supply.—*a.* As the military force is charged with the preparation and execution of regulations concerning the possession and use of firearms, ammunition, and explosives, it is only proper that it should exercise similar supervision over the sources of supply.

b. The military forces should control the entire legal supply of arms and ammunition. This control may be exercised either (1) by requiring purchases to be made from official sources by the Provost Marshal General and turning this ammunition over upon requisition to the District Commanders, who may distribute it to their provost marshals for sale in limited and necessary quantities to persons having permits, or (2) certain merchants may be authorized to sell

DISARMAMENT OF POPULATION

munitions. If there are munitions manufacturing plants in the country, they must be controlled; in addition, the introduction of munitions into the country must be restricted vigorously.

c. Any person, or representative of a business or firm desiring to import these articles should make written application for permission for each separate shipment of arms, ammunition, or explosives, in which application should appear in detail, the quantity and character of the stores to be imported, the use for which supplies are contemplated, the name of the firm from which the stores are to be purchased, and the port from which they will be exported. All applications should be forwarded through local Provost Marshals or other designated authorities who should endorse the request with such information or recommendation as will establish the character and identity of the applicant. In case there is a legal restriction on the importation of arms, the approved application should be forwarded to the office of the Minister of Foreign Relations for request on proper authorities through diplomatic channels.

d. There have been two methods used by marines in the past to control the sale of munitions in the occupied territory. Either of these two methods shown below apears to be effective.

(1) Immediately upon the arrival of the arms, ammunition, or explosives at the port of entry, the customs officials should notify the local Provost Marshal, who receives the shipment and deposits it in the provost storeroom or other suitable place. The articles may then be drawn by the consignee in such quantities or under such conditions as the Provost Marshal may indicate. Except for an exceptional shipment of explosives for some engineering project, the shipments will ordinarily be small.

(2) Immediately upon the arrival of the approved munitions shipment at the port of entry, the customs officials should notify the local Provost Marshal. This officer should notify the consignee, and with him check the shipment for its contents and amount. An enumerated record of the contents and the amounts should be prepared by the Provost Marshal, one copy given to the consignee and the other retained by the Provost Marshal. The shipment is then turned over to the consignee, after payment of all duties and with the written approval of the Provost Marshal. A monthly check should then be made by the Provost Marshal of all munitions stores in the hands of the approved sales agency; the sales agency making a monthly report of all sales (on individual approved permits) to the Provost Marshal.

11–10. Measures Following Disarmament.—*a.* Even after the population has been effectually disarmed, energetic measures must be taken to discourage or prevent rearming. Some plan must be evolved without delay to make it impracticable or dangerous to procure firearms illegally, either from within or without the country. If the existing laws of the country prohibiting possession of arms are sufficient in themselves, measures should be taken to make them effective. To the extent that authority is delegated or assumed, additional or new laws should be put into effect restricting the possession of firearms. This latter method can be applied only if military government is established. In issuing these laws one must bear in mind the responsibility assumed by the military forces in enforcing the laws and guaranteeing the security of life and property. If there be remote sections where law enforcement is difficult due to the limited number of the military forces, certain concessions may have to be made in order to permit the local inhabitants to protect themselves against the lawless element. On the other hand if the lawless elements remain in the field in numbers greatly in excess of the military forces, special considerations may make it advisable to provide means for arming a certain proportion of the reliable and responsible natives, to compensate in a degree for that inferiority in numbers; this should not be prejudicial to the other law-abiding elements. Sometimes this action will greatly discourage the lawless factions.

b. The military forces, or native constabulary, in conjunction with the customs officials, should be particularly alert along the coast and frontiers of the occupied country to prevent illegal entry of munitions. Where a native constabulary exists or is later established, a portion of such organization should be constituted a coast guard, equipped with fast boats, to prevent such arms being smuggled along the coast and rivers. Until such a constabulary is constituted, a unit of our own military forces, adequate in size and equipment, should be established as soon as active intervention takes place.

○

SMALL WARS MANUAL
UNITED STATES MARINE CORPS
1940

CHAPTER XII

ARMED NATIVE ORGANIZATIONS

RESTRICTED

UNITED STATES
GOVERNMENT PRINTING OFFICE
WASHINGTON : 1940

TABLE OF CONTENTS

The Small Wars Manual, U. S. Marine Corps, 1940, is published in 15 chapters as follows:

SMALL WARS MANUAL
UNITED STATES MARINE CORPS

―――――――

CHAPTER XII

ARMED NATIVE ORGANIZATIONS

v

SECTION I

GENERAL

12–1. Local armed forces.—In most sovereign states, the executive authority is enforced by the national military forces, national forces, and organized reserves under the control of the state. In addition, there may be an organized militia and police forces under the control of political subdivisions of the state. Police forces are normally maintained by municipalities. These armed forces represent the national-defense forces of the state and the armed forces employed to preserve peace and order within its borders.

12–2. United States intervention.—*a.* When the domestic situation of a foreign country is such that it is necessary for the United States Government to intervene, the national and local armed forces of the country concerned are usually powerless to suppress the domestic disorder or enforce the laws. At the time of intervention, the armed forces of the country will probably have disintegrated due to defeat by insurgent forces or because of desertions. In some cases, the armed forces may be engaged in action against an insurgent force, whose operations have created havoc and destruction throughout the country. Due to the magnitude of the domestic disturbance, the local police authorities are usually ineffective in the suppression of lawlessness, and may even have ceased to function entirely. Upon arrival within the foreign country, the armed forces of the United States Government immediately become responsible for the protection of the life and property of all the inhabitants of the foreign country. In order to discharge this responsibility, it may become necessary for the United States forces to assume the functions of the national armed forces of the foreign country in addition to the duties of the local and municipal police.

b. In assisting any country to restore peace and order, it is not the policy of the United States Government to accept permanent responsibility for the preservation of governmental stability by sta-

1

tioning its armed forces indefinitely in the foreign country for that purpose. The United States forces seek to restore domestic tranquility as soon as possible and to return the normal functions of government to the country concerned. To accomplish this, the United States Government will usually insist upon the establishment of an efficient and well-trained armed native force, free from political influence and distatorial control.

12-3. **Restoration of authority to local government.**—Having assumed the obligation for the restoration of domestic tranquility within the foreign country concerned, the obligation is fulfilled by the use of United States forces. There is also present the obligation to restore to the foreign country its organic native defensive and law-enforcement powers as soon as tranquility has been secured. The organization of an adequate armed native organization is an effective method to prevent further domestic disturbances after the intervention has ended, and is one of the most important functions of the intervention since the United States armed forces may have superseded or usurped the functions of armed forces of the country concerned at the beginning of the intervention. It is obvious that such armed forces must be restored prior to withdrawal.

12-4. **Formation of a constabulary.**—*a.* In the case of smaller countries whose national and international affairs are of limited magnitude and whose finances support only a small budget, the defense functions of the country and the police functions within the country can usually be combined and assigned to one armed force. Such a force is termed a "constabulary." The constabulary is a nonpartisan armed force patterned along the line of the military forces of the United States, with modifications to suit local conditions. The legal authority or approval for the formation of such an armed native organization must emanate from some person or body empowered with such sovereign right.

b. The authority for formation of a constabulary may be a decree of the *de jure* or *de facto* Chief Executive of the country in cases where a legislative agency does not exist. In such cases, the authority for any law enactment rests with the Chief Executive alone, who legally has the authority to issue a decree for the establishment of armed forces for his government. Provision is made for the appropriation of the necessary funds from the national budget for maintenance of the constabulary.

c. Authority for the formation of a constabulary may be granted by legislation initiated by the legislative body. In such cases, the existent armed forces of the country concerned are legally disbanded and the new constabulary force lawfully created by modification of the organic law of the country. Provision is made for the appropriation of the necessary funds from the national budget for its maintenance.

d. Authority for the formation of a constabulary may be the result of a treaty between the United States Government and the country concerned, providing for creation of such a constabulary. The treaty normally outlines the powers and limitations of the organization and provides funds for its maintenance. Often a treaty between the two governments will already exist, granting authority to the United States Government to intervene in the domestic affairs of the country concerned whenever the latter is unable to control domestic disorder within its boundaries. In such cases, this treaty is usually the basis or the authority for the creation of new armed forces within the country concerned, either through the executive or legislative agencies of the State, or through the powers of a military government set up within the country concerned by United States forces.

e. Authority for the formation of the constabulary may be the result of a decree of the military commander of United States forces in cases where a military government has been established to supplant the local government. In such cases, the maintenance of the constabulary is provided by means of appropriation of local revenues under control of the military government.

Section II

ORGANIZATION OF A CONSTABULARY

12–5. **Planning agency.**—*a.* The establishing of a constabulary is preceded by the appointment of a planning group to draft the necessary plans for its formation. The initiative in the creation of the constabulary devolves upon the United States forces, since it has assumed the obligation to restore law enforcement and defense forces to the country concerned prior to withdrawal of United States forces.

b. The planning group, or the majority of the members of such a group, are usually drawn from the military and naval forces of the United States Government within the country concerned. The selection of the planning group from among officers of the United States forces then in the country is advisable since such officers will normally be more familiar with existing political, economic, geographical, and psychological conditions. In addition to such members, it may be advisable to select officers who have had prior experience in constabulary duty in other countries.

12–6. **Approval of plans.**—After the planning group has completed its plans for the organization of the constabulary, the plans must first be approved by the proper officials before the constabulary may be considered existent. Among the officials who approve the plans are the Chief Executive of the local state, the diplomatic representative of the United States accredited to the foreign country concerned, the senior naval officer in command of the United States forces operating within the foreign country concerned, the Secretary of the Navy, the Secretary of State, the Congress of the United States, and the President of the United States. When a legislative body exists, the approval of the legislature of the foreign country concerned is also secured. When a military government has been established, only the approval of the United States executive, legislative, and departmental agencies is required.

12–7. **Local creative law.**—In order that the constabulary may be the constituted military instrument of the local government, it must be legally established and provided with the legal power to execute its functions. In the law or decree establishing the constabulary, there should be definite provisions setting forth the authority and responsibility of the commander of the constabulary in order that the constabulary may be entirely free from autocratic or political control within the country concerned. The law or decree should state definitely the specific duties that the constabulary is legally empowered to perform.

12–8. **United States creative laws.**—*a.* The plans for the establishment of a constabulary will invariably contain certain provisions relative to the employment of members of the United States armed forces as officers or directing heads of the proposed constabulary upon its initial formation. The Constitution of the United States, Article I, Section 9 (8), states: "No title of nobility shall be granted by the United States; and no person holding any office of profit or trust under them shall, without the consent of the Congress, accept of any present, emolument, office, or title of any kind whatever from any king, prince, or foreign state." In order that members of the United States forces may accept office, including emolument for such office, from the foreign country concerned, it is necessary that the Congress of the United States grant specific authority by law. The necessary law for service with the constabulary is drawn up by the planning group, and, after approval, is presented to the Congress for enactment and subsequent approval by the President of the United States. Such authority may be included in a treaty between the United States and the country concerned.

b. Since all treaties of the United States are ratified only by the United States Senate, without action by the House of Representatives, it becomes necessary to enact a separate law approved by both Houses of Congress, even though authority for members of the United States

ORGANIZATION OF A CONSTABULARY

forces to serve in the constabulary may be included in the treaty. When a general law has been already enacted by the Congress of the United States permitting members of the United States forces to serve in the armed forces of the foreign country concerned, no specific law is required.

12-9. **Composition.**—*a*. Initially, the officers of the constabulary are selected officers and enlisted men (usually qualified noncommissioned officers) of the United States military and naval forces. In time, as the domestic situation becomes tranquil and the native members of the constabulary become proficient in their duties, the United States officers of the constabulary are replaced by native officers. Officers and enlisted men of the United States forces appointed as officers of the constabulary should be acceptable to the local government and have the qualities considered essential for a position of similar importance in the United States forces. They must be physically fit to withstand arduous duty in the field and should be proficient in the language of the country concerned. A general knowledge of local conditions is an important requirement. They should be known for their tactful relationships, and should be in sympathy with the aspirations of the inhabitants of the country concerned in their desire to become a stable sovereign people. They should be educationally and professionally equipped to execute the varied functions that they will be called upon to perform.

b. Native troops make up the enlisted personnel of the constabulary. Service is not compulsory. Recruiting is carried on throughout the country, and the desired personnel is acquired by enlisting only those volunteers who possess the requisite qualifications.

c. Plans are made for the operation of recruit depots. Schools in academic and governmental subjects are conducted for enlisted personnel. Consideration must be given to the formation of a medical department. In some cases, a coast guard may be required. The medical department and the coast guard are included in the constabulary organization. Early establishment of a school for training candidates for commission should receive much thought and consideration. The establishment of such a school will provide orderly replacement of the personnel of the United States forces utilized initially to officer the constabulary. It also indicates to the local government the altruistic motives of the United States Government and indicates its intention to turn over the control of the constabulary to the local government at the earliest possible moment.

ORGANIZATION OF A CONSTABULARY

12–10. **Duties and powers.**—*a.* The police duties formerly performed by the organic military and naval forces of the country concerned are assumed by the organized constabulary. The constabulary is the national-defense force of the country concerned and also performs police duties and civil functions.

b. The military duties of the constabulary consist of the defense of the country against outside aggression and the suppression of domestic disorder when local police in the territorial subdivisions of the country are ineffective in the maintenance of law and order.

c. Among the police duties of the constabulary are the prevention of smuggling and the control of the importation, sale, and custody of arms, ammunition, and explosives. It is also empowered to arrest offenders for infractions of local laws, not only of the state, but also of the territorial subdivisions and municipalities. It is charged with the protection of persons and property, the control of prisons, and the issuance of travel permits and vehicular licenses. The constabulary provides guards for voting places and electoral records, and exerts plenary control during natural disasters, such as floods and earthquakes.

d. The civil duties of the constabulary include the distribution of funds for the payment of civil employees in outlying areas and the distribution of executive, legislative, and judicial notices. When required, the constabulary operates the lighthouse and lifesaving service by means of a coast guard. Members of the constabulary may act as communal advisors to municipalities. The constabulary may be assigned the task of supervision of the construction of roads and bridges. Census compilation, supervision of local sanitation, and operation and control of telephone and telegraphic systems, including air and radio communication may also be included among the civil duties of the constabulary. Other civil duties are the supervision of weights and measures, the enforcement of harbor and docking regulations, compilation of reports on the use of public lands, supervision of the occupancy of public lands, and periodic reports of agricultural conditions.

12–11. **Size of force.**—In determining the strength of the constabulary force, it is necessary to consider carefully the domestic situation in each territorial division of the country concerned, particularly the situation in the principal cities and seaports. The strength of the constabulary detachment required for one locality may be entirely inadequate or excessive in another locality. Factors that enter into the determination of the strength of the constabulary are the organic

ORGANIZATION OF A CONSTABULARY

strength of the military forces employed prior to the intervention, the organic strength of the civil police forces in territorial divisions and municipalities, the normal domestic situation relative to law observance and law enforcement in the territorial subdivisions, and the relative importance of the larger cities within the state. The political, economic, and geographical importance of the various territorial subdivisions should also be considered. The constabulary should be large enough to suppress active rebellion, as well as to repel outside aggression. The original estimate of the strength required is based upon the normal domestic situation in all territorial subdivisions of the country concerned. Local conditions in a particular section may call for a material increase in the strength that would be normally required for that section. Such conditions should be taken into consideration in order that the initial strength may be adequate to meet all situations that may require the employment of the constabulary. Although the foregoing considerations may dictate the necessity for a larger force, restrictions on the strength of the constabulary may be imposed by the limited finances of the local government, as well as the financial requirements of other governmental activities.

12-12. **Administrative organization.**—The constabulary is organized administratively in the following manner: the headquarters, consisting of the commander and his staff; the administrative, technical, and supply departments or groups; the operating forces, organized as administrative or tactical units and stationed in tactical localities or at posts and stations in conjunction with other governmental activities. The geographical divisions of the state are normally the determining factor in the formation of "groupments" or territorial commands.

12-13. **Supply and equipment.**—a. Any estimate that is made to determine the required strength of a constabulary must naturally include provisions for the supply and equipment for such troops. Among the items of equipment are weapons and military uniforms or distinctive dress for the troops and, in some cases, vehicular transportation. The confidence and loyalty of the native troops is promoted by careful supervision of their material needs. More often than not, they will have been accustomed to meager salaries irregularly paid, scant food carelessly provided, as well as indifferent shelter, clothing, and equipment. When they are regularly paid in full on the date due, when fed adequately as provided by the allowance, and when good shelter, clothing, and equipment are provided, native troops will usually respond in the quality of service rendered.

9

ORGANIZATION OF A CONSTABULARY

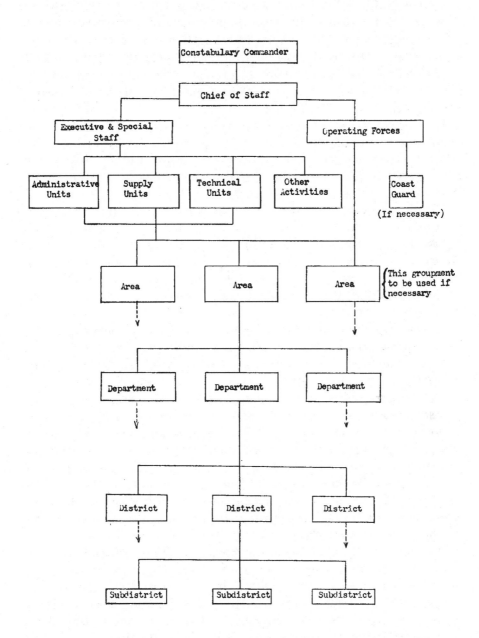

ORGANIZATION OF CONSTABULARY

ORGANIZATION OF A CONSTABULARY

In establishing and maintaining an organization of native troops, attempts should be made to provide better clothing and shelter and particularly better food than native civilians of the same social class enjoy. This is decidedly an important morale factor. The equipment of the constabulary is governed by the type of service required. Often different types of equipment are employed in various localities.

b. In many countries, the distinctive uniform or dress of native troops prior to intervention by United States forces is likely to be of a type more adapted to purely peaceful military display or ceremony than to combat. In some cases, the uniform is of a type that cannot be termed a "distinctive dress" within the meaning of the Rules of Land Warfare. There is a natural inclination on the part of United States forces when organizing a constabulary to outfit the troops with a uniform similar to that of the United States forces, with slight modifications in the distinctive ornaments, texture of clothing, and design. Any uniform adopted for the constabulary should be suitable for the combat and climatic conditions likely to be encountered.

c. The organic armed forces of the country may have been only indifferently armed. Such weapons as they have are likely to be in a poor condition, due to carelessness in upkeep. However, modern weapons are becoming more accessible to all countries, due to the lowered costs as a result of modern mass production. Such arms as are in good condition are retained and reissued to the constabulary after the disarmament of government and insurgent forces. Plans for arming the constabulary should take into consideration all probable tasks that may be assigned, as well as the capabilities of the troops in the employment of the various types of weapons. It may be advisable to arm the constabulary with weapons of different types, make, and in different proportions from the organic armament of the United States forces.

d. There are three methods for subsisting the constabulary. The first method is by the organization of general messes at those points where sufficient troops are quartered together to make the method feasible. The second method is to permit individuals to subsist themselves upon the payment of an adequate subsistence allowance in addition to their normal pay. The third method is the subsistence of personnel by contract with civilian contractors. The ration allowance should be announced in orders. The psychology of making the ration allowance the same for general mess, subsistence allowance, and contract mess is self-evident, since the troops will feel that under each system they are receiving the same treatment in regard to their food. The organized mess is practicable only at those posts and stations having a sufficient number to make this method economical. Recruit depots, officers' schools, and commands of over 20 men are normally fed in general messes. However, even in commands of over 20 men, activities in the field may dictate that a subsistence allowance is more practicable. In small detached posts of only a few men, it is usually more practicable to furnish food by contract messing, or to pay the troops a subsistence allowance. In outlying posts and stations, troops normally ration themselves on the cash allowance. As a general rule, a cash allowance should not be granted if a general mess can be organized or if contract messing is practicable. When a cash allowance is paid, there is a tendency to squander the cash allowance and to contract indebtedness for food, with no assurance that troops are subsisted on a well-balanced ration. When food is procured under contract, the contractor may be so interested in making a profit, that troops will not receive the proper amount or quality of food. The commanding officers of detached posts should continually check on the quantity and quality of food served when troops are subsisted in a contract mess. In the conduct of a general mess, no attempt should be made to supply foreign food products. The ration component should be confined to local staples and garden products, since it is this type of food to which the troops are accustomed.

e. Estimates should be made covering the type and quantity of miscellaneous supplies required by the constabulary. Many materials may be purchased locally. To facilitate the acquisition of supplies not obtainable locally, they are normally procured from the continental United States.

12–14. Records and reports.—*a.* The records and reports in general use by military organizations are used by the constabulary. In addition, periodic reports may be required covering local economic and political conditions, reports of arrests and disposition of such cases, reports of military activities of the various units, and such special reports as may be required by higher authority.

b. Reports and records should always be in the language of the country concerned. It is unreasonable to require natives to learn the English language simply because that is the language of the United States forces. In the preparation of texts to be used in the training of troops and in the preparation of instructions for handling of legal cases, the language of the country concerned should always be employed.

12–15. **Finances.**—*a.* When planning the creation of a constabulary, the financial status of the country concerned is naturally a feature that will influence the strength of the constabulary as well as the acquisition of supplies for such a force. When the necessary funds have been estimated, it is imperative that such funds be allotted from the national treasury by presidential decree or by the legal sanction of the legislature of the country concerned. Funds are allotted from the revenues by the military government in those cases where the constabulary is organized during the tenure of a military government. Appropriations for the establishment and maintenance of the constabulary may be difficult to obtain, not only because administrative authority is required for such allotment, but also due to the fact that, in most instances, the country will have few funds available for such a purpose. The scarcity of funds is likely to be the consequence of unstable economic conditions due to widespread disorder, the despoilation of the treasury by individuals or groups, and the lack of an efficient system for the collection and control of taxes and custom duties.

b. Initially, a large part of the revenue of the country concerned will necessarily be devoted to the financing of the constabulary. After the initial allotment of funds has been authorized for the establishment of the constabulary, it is necessary to assure that the annual or other periodic allotments are continued, and that these allotments are given the highest priority in the national budget. This is insisted upon at all times, and efforts to decrease or subordinate this allotment for the constabulary should be resisted energetically.

c. The pay of officers and enlisted personnel forms a large part of the expenditures of the constabulary. Consideration should be given to the standards of living within the country in computing the rates of pay. The rates of pay should be such as to attract the best type of natives to join the constabulary. By making the rates of pay attractive, natives of the highest type will be encouraged to make the constabulary a career. This feature is particularly desirable since it will tend to promote tranquility throughout the country after the withdrawal of the United States forces, if the majority of the officers and men have served in the constabulary for a number of years. Initially, all the officers of the constabulary are members of the United States forces. The rates of pay granted

ORGANIZATION OF A CONSTABULARY

them as officers of the constabulary are in addition to pay and allowances received from the United States Government.

12–16. **Recruiting.**—*a.* Age and height limits are established for recruits. Physical requirements are decided upon for regular enlistments, but these may be relaxed somewhat in case of emergency. If recruit depots are maintained, a definite period for recruit training is assigned. In some cases, it may be advisable to refuse enlistment of men from disturbed sections of the country. In many cases, the political affiliations of applicants must be considered. This matter may be adjusted satisfactorily by the enlistment of recruits of different political beliefs in proportion to the voting strength of the principal political parties. In some cases, it may be deemed advisable to refuse enlistment to members of former military forces of the country. In some countries, the best method of obtaining recruits may be to enlist troops from one locality to serve in that locality under their own noncommissioned officers after a period of training at a recruit depot. In accordance with the plan of organization of the constabulary into a chain of command through departmental control, it may be advisable to distribute the officers of the constabulary to their respective posts and stations in order that the recruiting of enlisted personnel may be accomplished under their direction and control. The officers of the constabulary act as recruiting officers in addition to their other duties. To assist in recruiting, notices are published in the official gazette of the local government, advertisements are inserted in local newspapers, and notices are furnished local civil officials for publication to the populace. Itinerant recruiting parties may be employed in thickly settled areas. Medical units are attached to garrisons in the more important towns and villages in order that applicants may receive prompt medical examination. In some cases, it may be necessary to utilize the services of contract physicians for the initial examination of recruits, further examination to be conducted later by medical personnel of the constabulary.

b. Before accepting applicants for service, the recruiting officer should assure himself of the proper qualifications of applicants. In addition to an oral examination, recommendations from reputable citizens of the home locality of the applicants are usually required. In many instances, the recommendations of local civil officials are invaluable in the selection of applicants.

12–17. **Housing and shelter.**—When the organic armed forces of the country have been disbanded upon the formation of the constabulary, it will be found that many public buildings are available to house the constabulary. These public buildings will consist of barracks, offices, forts, prisons, camps, police stations and, in some cases, naval craft. Public buildings are within the eminent domain of the local government and as such can be lawfully employed by proper authority to house and shelter the constabulary. When such housing does not exist, it may be necessary to rent suitable buildings or to erect permanent buildings. Prison labor may be used in such construction and every effort should be made to use construction materials obtainable locally. Warehouses may have to be leased for the storage of supplies when such space is not available in old arsenals, forts, or former military warehouses.

12–18. **Military courts.**—The system of military courts-martial set up by the constabulary must have the legal sanction of the local government. Usually, the constitution of any sovereign state will provide for military tribunals. In such cases, it is necessary only to secure legislative approval for the system of courts martial applicable to the constabulary. A modification of the courts-martial system employed by the United States forces, adapted to local conditions and the basic laws of the country concerned, will usually be acceptable. The system of courts-martial set up within the constabulary does not usurp any of the judicial functions of the civil courts. Members of the constabulary, who commit civil offenses, should be brought before civil courts for trial and punishment. (See par. 12–27.) Trial by courts-martial is reserved for military and for criminal offenses, when civil jurisdiction is lacking in the latter case.

Section III

OPERATIONS AND TRAINING

12–19. **Recruits.**—The methods adopted for the training of recruits are dependent upon the military situation at the time of enlistment. Normally, recruits are sent to a central recruit depot for a stated period of training. Several recruit depots may be conducted in different sections of the country. In some cases, recruits may be retained at the local station or military post, trained at that station, and later assigned to duty in that locality or sent to another military post that may have been unsuccessful in obtaining the required number of recruits. The adoption of a single method may be practicable in some areas while in others a combination of training methods may be necessary in order to meet local conditions. Some troops are more effective when serving in their own community, while others will be found to operate more effectively in other localities due to changes in climate, environment, and food. In some situations, it is better to employ troops away from their home localities to prevent the use of their authority improperly against personal enemies or for the benefit of friends. Recruiting officers should be supplied with uniforms and equipment sufficient to outfit the number of recruits desired from the various sections of the country. The training of the recruit has two distinct objects in view, namely, training as a member of a military combat organization and training for police duties. The military instruction of a recruit covers the basic individual training of a soldier including target practice and drill. A recruit training textbook in the language of the country concerned will be found extremely useful. Instruction of the recruit in police duties includes instruction in the constitution of the country, civil and criminal laws, powers and limitations in making investigations and arrests, and the assistance the constabulary is to render local civil officials. A handbook in the language of the country concerned, covering these police duties will materially aid in presenting this instruction and will also provide a useful guide to all members of the constabulary. For the larger cities, it may be advisable to train units for the primary duties of municipal police with only secondary instruction in military duties. The early training of competent police forces for the larger cities is one of the most effective methods to strengthen the local government and secure the good will of the better class of inhabitants. Medical enlisted personnel is obtained by enlistment of qualified individuals for duty with the medical service.

12–20. **Unit training.**—Unit training is carried out by individual units of the constabulary as a part of their routine training in order to maintain their military and police efficiency. This training embraces unit combat training, target practice, field firing, specialist training, instruction in law enforcement and, in some cases, instruction in elementary academic subjects. Instruction schedules are so arranged that training does not interfere with the normal military and police duties of the unit. In preparation for special operations, units may be more effectively trained at a central point prior to engaging in such operations.

12–21. **Officers.**—As soon as practicable after the formation of the constabulary, a school for the training of native candidates for commission should be organized. The staff of this school is composed of officers of the United States forces, who are specially qualified for this work. Rigid physical qualifications are adopted to cover the admittance of candidates. All candidates should have sufficient scholastic qualifications to insure their ability to absorb the military instruction. The period of instruction for such a school is 1 year. At the end of this period, the candidate is given a probationary commission that is confirmed after 1 year of service with the troops. This method of instruction provides a steady supply of native officers to replace the members of the United States forces. The gradual replacement of commissioned medical personnel of the United States forces is effected by commissioning native physicians as vacancies occur. This is usually commenced just prior to withdrawal of all United States forces.

12–22. **Field operations.**—a. Each race of people has its peculiar characteristics and customs. These may be modified somewhat under influence, but cannot be entirely destroyed or supplanted. These characteristics and customs should always be recognized and considered when dealing with persons of different races.

SWM 12–23

OPERATIONS AND TRAINING

b. In the organization of the constabulary, consideration should be given the form of warfare to which the troops are accustomed. No attempt should be made to impose entirely new forms of tactics unless a long period of training and indoctrination is available. In emergencies, or when only a limited time is available for training, it may be better to organize the troops according to native methods. Different types of organizations, equipment, and tactics will often be required in various localities.

12–23. Troop leading.—*a.* Strict justice exerts a marked influence on the discipline of native troops. A few lessons suffice, as a rule, to impress upon them that orders are to be obeyed. When this idea has been implanted in their minds, they generally become amenable to discipline.

b. During the earlier field operations of the constabulary, it is usually advisable to employ mixed units composed of members of the United States forces and the constabulary. Later, the United States forces are used only as a reserve available to support the constabulary in emergencies. The constabulary gradually assumes full responsibility for the maintenance of law and order. In active operations, the officers of the constabulary should be models of leadership, inspiration, and an example to their troops. Members of the United States forces serving with the constabulary must possess good judgment and extreme patience, coupled with tact, firmness, justice, and control. Firmness without adequate means of support may degenerate into bluff. Tact alone may be interpreted as weakness.

SECTION IV

AUXILIARY FORCES

12–24. Urban and rural agents.—Small detachments of varying size are stationed throughout the country in towns, cities, and villages. Each detachment is assigned the task of restoring and maintaining law and order within a given area. To assist the detachments in the performance of their duties, urban and rural agents are employed as part of the constabulary. These agents are selected from among the inhabitants of communities and outlying sections. Only men of high standing in the community are selected for this duty. These agents are, in reality, the rural police of the constabulary. They are appointed or commissioned by the constabulary and are paid as a separate budgetary unit of the constabulary. They are granted powers similar to those granted a sheriff in the continental United States. They are not given any distinctive uniform, but are provided with a badge of office, together with a special police permit to bear arms in the execution of their duties. These agents are under the direct command of the local constabulary commander. Employment of such agents is invaluable, since they are thoroughly familiar with their section or community and know all the individuals residing in the vicinity, thus making the apprehension of any resident malefactor a comparatively easy task. They keep the local constabulary commander informed of the domestic situation within their respective sections, thereby forestalling any organized attempt at insurrection or rebellion against the local government.

12–25. Special agents.—In addition to urban and rural agents, individuals may be armed and endowed with police powers. These special agents are employed by owners of large estates, plantations, mines, ranches, banks, and other large financial and commercial houses. They act as guards for the protection of life and property from marauders, bandits, and robbers. They are paid by the estate or firm employing them and are legally empowered by the constabulary to make arrests of trespassers as agents of that force. They are given a distinctive badge of office and are issued a special police permit to bear arms in the performance of their duties. The appointment of special agents should be made only after a careful investigation by the local constabulary commander. Under no circumstances, should the practice of appointing special agents be permitted to grow to such an extent that any large land owner has a considerable number of armed men in his employ and under his control. The hiring of additional special agents should be strenuously opposed when sufficient personnel of the constabulary is present in the vicinity to provide protection.

12–26. **Auxiliary units.**—When an organized rebellion or insurrection develops, or when banditry assumes such proportions that the local units of the constabulary are unable to combat such domestic disorders successfully, volunteer units under the direction of the constabulary may be organized from the inhabitants to assist in quelling such disorders. These auxiliary units are composed of inhabitants who are armed and rationed by the constabulary. Auxiliary units are temporarily armed forces, and are employed only for the duration of the emergency. During their period of service, they are governed and controlled in the same manner as regular members of the constabulary.

Section V

CIVIL AND MILITARY RELATIONSHIP

12–27. **Relation to civil power.**—*a.* The constabulary represents the power of the executive branch of the government and its territorial subdivisions. Unlawful acts committed by members of the constabulary are usually found to be in contravention of the regulations of the constabulary or the civil or criminal laws of the country. In the former class are military misdemeanors and crimes that are within the jurisdiction of the military power; that is, the constabulary courts-martial system. In the latter class are those crimes and felonies that are set forth in the penal code of the country. Generally, any infraction of constabulary regulations by a member of the constabulary should be tried by the constabulary itself, either by the member's immediate commanding officer or by court martial. Likewise, members of the constabulary charged with conspiracy against the local government should be tried by court martial and the punishment executed by the constabulary after confirmation of the sentence by the Chief Executive. Alleged civil offenses are first investigated by the constabulary. If an offense is found to be sufficiently proved by evidence as to its commission, the member should be discharged from the constabulary and delivered into the custody of the civil authorities for trial and punishment as a civilian. If, however, after investigation of the offense by the constabulary, the evidence indicates that the member is guiltless, he should under no circumstances be delivered to the civil authorities for trial and punishment until such authorization has been secured from the commanding officer of the constabulary.

b. It is to be expected that some animosity and jealousy will be prevalent during the establishment of the constabulary by officers of the United States forces. Attempts may be made to interfere with or embarrass the constabulary in its operations indirectly by civil-court actions and by noncooperation on the part of minor officials. Complaints against members of the constabulary should

CIVIL AND MILITARY RELATIONSHIP

be thoroughly investigated. When warranted, a just trial should be immediately conducted, with prompt punishment of guilty individuals, thus indicating to the populace that the constabulary enforces the law among its own members and that they receive no preferential treatment not granted civil violators of the law. The chief of the constabulary is responsible directly to the Chief Executive of the country, who is the commander in chief of all the armed forces of the country.

c. In their contacts with civil officials, members of the constabulary must be courteous, firm in the execution of their duties, and just in dealing with any and all classes of inhabitants, regardless of rank, title, creed, or social position. Tact is one of the most necessary attributes which may be possessed by members of the constabulary. Fair and just operation of the constabulary must always be tempered with tact. Brutality in making investigations and arrests should be firmly and promptly suppressed. The inhabitants should be encouraged to regard the constabulary as an honest, impartial, and just law enforcement agency, friendly toward the law-abiding population. In times of emergency during fires, floods, and earthquakes, the constabulary should be quick to render aid to the distressed.

12-28. Relation to United States forces.—*a.* The line of demarcation between the execution of the military power of the United States forces and the constabulary should be definite. When it has attained full strength, the constabulary should have sole responsibility for the preservation of law and order. Since the United States forces have set up this military instrumentality for the local state and endowed it with a certain strength, the constabulary should have unhampered opportunity in its conduct of operations as the armed force of the country. Interference by United States forces not only seriously decreases the prestige of the constabulary, but also denies to the local state the ability to utilize freely the force that has been created to increase its power and prestige. The constabulary assumes its functions gradually, as it recruits to full strength, and takes over the police functions of the country under the guidance and observation of the United States forces. When the constabulary has demonstrated its competence to perform its duties, the United States forces relinquish control and command, and are withdrawn and concentrated at central points where they are available to be employed as reinforcements in case of unexpected emergencies. During the organization of the constabulary, the assignment of detachments of the constabulary to operate with elements of United

CIVIL AND MILITARY RELATIONSHIP

States forces in joint action against hostile forces may be advisable. In this manner, the constabulary, as well as the native population, will feel that the local situation is being handled by their own governmental agency and not by a foreign power. Unlawful acts committed by members of the constabulary or by civilians against the United States forces are legally under the jurisdiction of the United States forces and may be punishable by an exceptional military court martial. Whenever possible, every effort is made to have the offenders tried by the constabulary courts-martial system or by the local civil judicial agency in order that such unlawful acts may be punished by agencies of the country and not by agencies of the United States forces.

b. When joint operations are conducted by United States forces and the constabulary, the principle of seniority according to rank of members of United States forces present should be retained. Thus, if the senior constabulary officer present, who is also an officer of the United States forces, is senior to the officer in command of the United States forces present, the senior constabulary officer assumes command of the joint forces. When the officer in command of the United States forces present is senior, he assumes command of both organizations.

O

SMALL WARS MANUAL
UNITED STATES MARINE CORPS
1940

✦

CHAPTER XIII

MILITARY GOVERNMENT

RESTRICTED

UNITED STATES
GOVERNMENT PRINTING OFFICE
WASHINGTON : 1940

TABLE OF CONTENTS

The Small Wars Manual, U. S. Marine Corps, 1940, is published in 15 chapters as follows:

SMALL WARS MANUAL
UNITED STATES MARINE CORPS

CHAPTER XIII

MILITARY GOVERNMENT

v

SECTION I

GENERAL

13-1. **Scope of chapter.**—*a.* The features of the subject of military government herein discussed relate to the powers, duties, and needs of an officer detailed to command a force on a mission involving intervention into the affairs of a foreign country under conditions which are deemed to warrant the establishment of complete military control over the area occupied by the intervening force. While the form of military control known as military government is designed principally to meet the conditions arising during a state of war, it has been resorted to, by the United States in numerous instances, where the inhabitants of the country were not characterized as enemies and where war was neither declared nor contemplated.

b. Military government being founded on the laws of war, many questions arise with regard to the method to be used in the application of these laws in situations requiring the establishment of such a government where no state of war exists. It is the purpose of this chapter to outline the general principles involved in the exercise of authority and functions of military government and to indicate how those principles are applied in the various situations with which the marine or naval officer may be confronted.

13-2. **Definitions.**—*a. Military government.*—Military government is the exercise of military jurisdiction by a military commander, under the direction of the President, with the express or implied sanction of Congress, superseding as far as may be deemed expedient, the local law. This form of jurisdiction ordinarily exists only in time of war, and not only applies to the occupied territory of a foreign enemy but likewise to the territory of the United States in cases of insurrection or rebellion of such magnitude that the rebels are treated as belligerents.

1

GENERAL

b. Martial law.—Martial law is that form of military rule called into action by Congress, or temporarily by the President when the action of Congress cannot be invited, in the case of justifying or excusing peril, in time of insurrection or invasion, or of civil or foreign war, within districts or localities whose ordinary law no longer adequately secures public safety and private rights.

c. Distinctions.—The most important distinction between military government and martial law is, that the former is a real government exercised for a more or less extended period by a military commander over the belligerents or other inhabitants of an enemy's country in war, foreign or civil; martial law, on the other hand, is military authority called into action, when and to the extent that public danger requires it, in localities or districts of the home country which still maintain adhesion to the general government. The subjects of military government are the belligerents or other inhabitants of occupied territory, those of martial law are the inhabitants of our own territory who, though perhaps disaffected or in sympathy with a public enemy, are not themselves belligerents or enemies. The occasion for military government is usually war; that for martial law is simply public exigency which, though more commonly growing out of pending war, may nevertheless be invoked in time of peace in great calamities such as earthquakes and mob uprisings at home.

13–3. Authority for exercise of military government.—Military government usually applies to territory over which the Constitution and laws of the United States have no operation. Its exercise is sanctioned because the powers of sovereignty have passed into the hands of the commander of the occupying forces and the local authority is unable to maintain order and protect life and property in the immediate theater of military operations. The duty of such protection passes to the occupying forces, they having deprived the people of the protection which the former government afforded. It is decidedly to the military advantage of the occupying forces to establish a strong and just government, such as will preserve order and, as far as possible, pacify the inhabitants.

13–4. Functions of military government in general.—As to its function, military government founded on actual occupation is an exercise of sovereignty, and as such dominates the country which is its theater in all branches of administration whether administered by officers of the occupying forces or by civilians left in office. It is the government of and for all the inhabitants, native or foreign, wholly superseding the local law and civil authority except insofar as the same may be permitted to exist. Civil functionaries who are retained will be protected in the performance of their duties. The local laws and ordinances may be left in force, and in general should be subject, however, to their being in whole or in part suspended and others substituted in their stead, in the discretion of the governing authority.

13–5. **By whom exercised.**—Military government may be said to be exercised by the military commander, under the direction of the President, with the express or implied sanction of Congress. The President cannot, of course, personally administer all the details, so he is regarded as having delegated to the commander of the occupying forces the requisite authority. Such commander may legally do whatever the President might do if he were personally present. It follows that the commander of the occupying force is the representative of his country and should be guided in his actions by its foreign policy, the sense of justice inherent in its people, and the principles of justice as recognized by civilized nations. A single misuse of power, even in a matter that seems of little importance, may injure his country and its citizens. Foreign, official, commercial, and social relations depend in a great measure upon the friendliness of other countries and their people. Acts of injustice by a force commander jeopardize this friendliness, especially in neighboring countries, or in those whose people have racial or other ties in common with the people of the occupied country.

13–6. **How proclaimed.**—In a strict legal sense no proclamation of military occupation is necessary. Military government proclaims itself; a formal proclamation, although not required, is invariably issued and is essential in a practical way as announcing to the people that military government has been established and advising them in general as to the conduct that is expected of them. It should be remembered that the inhabitants do not owe the military government allegiance; but they do owe it obedience. A sample form for a proclamation may be found in the current issue of Naval Courts and Boards.

Section II

ESTABLISHMENT AND ADMINISTRATION OF MILITARY GOVERNMENT

13–7. Importance of organization.—The efficient administration of a military government requires that the officers chosen for the administration of various departments should be particularly qualified for such office.

13–8. Plans.—*a.* Whenever it becomes known or can be foreseen that territory is to be occupied, the commander of the military forces that are to occupy it will no doubt be called upon to formulate beforehand his plans for administering the military government. These detailed plans are prepared by him under the policies prescribed by higher authority pursuant to such general plans or policies as may previously have been prepared, announced, or approved by the Navy Department. They will always depend upon the military situation, and will be influenced by the political, economic, and psychological factors which may prevail in the area to be governed.

b. The actual preparation of the main plan or plans is primarily a function of that section of the commander's staff which will later take part in the administration of the military government. The commander of the occupying forces should ordinarily organize a separate

5

MILITARY GOVERNMENT

and additional staff for the administration of civil affairs. However, the plan as determined upon by the commander of the occupying force requires coordinated study and assistance of the several staff officers. F–1 provides the data as to personnel; F–2 the data as to the situation in the territory to be occupied; and F–3 the data for coordination between the tactical plan and the military government plan.

c. Too much emphasis cannot be placed on the necessity for having the military government under a separate staff section, thus avoiding the interference with the military functions of the usual staff sections, and yet coordinating the whole under the supervision of the force commander. The chief of this separate staff is designated as Officer in Charge of Civil Affairs. An outline of the organization of this staff showing the various subdivisions, along with the duties assigned to each is set forth in the following outline which is intended as a guide only, there being no hard and fast rule prescribed for organizing this staff:

The military governor and civil affairs staff

Rank	Title	Outline of duties
Major general	Military governor	Acts under the authority and by the direction of the President of the United States. Exercises military law applicable to the occupation. Issues proclamations and supplemental regulations. Assumes the duties of the President and Congress of the occupied territory. Continues in effect the laws not conflicting with the objects of, or regulations issued by, the occupation. Supervises and controls officials. Supervises the collection of revenue and controls its expenditure. Establishes military tribunals. Respects personal and property rights.
Colonel	Officer in charge of civil affairs.	Acts as officer in charge of civil affairs. Promulgates the orders of the Military Governor and supervises their execution. Assumes the duties of the Secretary of State (Foreign Relations).
Commander (CEC), U. S. Navy or lieutenant colonel, U. S. Marine Corps.	Officer in charge of public works and utilities.	Assumes the duties pertaining to the Department of Public Works. Conducts the public business of this department in accordance with the laws and Constitution of the occupied territory. Supervises public works of all kinds, public utilities, public building, mining, agriculture, forestry, and fisheries.
Lieutenant Colonel (PM or QM).	Officer in charge of fiscal affairs.	Assumes the duties pertaining to the Department of the Treasury and Public Credit. Conducts the public business of this department in accordance with the laws and Constitution of the occupied territory. Has general supervision of public finances, taxes, excises, banking, postal service, state insurance, foreign commerce, and customs service.
Commander (M. C.), United States Navy.	Officer in charge of sanitation and public health.	As indicated by name. Includes quarantine service.
Major	Officer in charge of schools and charitable institutions.	Assumes the duties pertaining to the Department of Education, as indicated by name. Includes religious societies and activities.
Major	Officer in charge of the legal department.	Assumes the duties pertaining to the Department of Justice, supervises courts, prisons, provost courts, and military commissions. Gives legal advice and opinions.
Colonel or Lieutenant Colonel.	Officer in charge of the constabulary.	Assumes the duties pertaining to law enforcement through a native force organized as a constabulary.

MILITARY GOVERNMENT

The Department of War and Navy will be abolished during the occupation.

d. The plan of the commander for the administration of the military government should give expression to his decisions and instructions on the following points:

(1) The distribution and territorial assignment of his military forces in the occupied territory.

(2) The immediate changes, if any, to be made in the local governmental system.

(3) The extent to which the more important local civil officials are to be displaced and officers appointed to fill their places.

(4) The relationship which is to exist between the civil and military administrations, especially the extent to which tactical subdivisions are to be used as units of control of the civil administration.

e. The following form illustrates a guide which might be utilized by the commander of the forces in drawing up the general instructions set forth above. Annexes to this plan will be prepared showing the proclamation to be issued, the supplemental regulations to be published at the beginning, and proposed staff organization for administering the military government.

MILITARY GOVERNMENT PLAN

Distribution and territorial assignment of the occupying forces: General distribution.

(1) Number of areas into which the occupied territory is to be divided for administrative purposes.

(2) General policy of the commander in respect to the distribution and administration of the military government.

Territorial assignment of units.

Unit	Headquarters	Area
List task groups	Designate city	Give political area to be occupied by the task group noting any extraordinary mission.

NOTE.—Within their respective areas, commanders will assign districts and subdistricts conforming to political subdivisions when practicable.

Immediate changes to be made in the governmental system:

(1) Powers to be exercised by the Military Governor in his administration of the occupied territory.

(2) Immediate changes to be made in the existing governmental system as a whole.

MILITARY GOVERNMENT

Extent to which more important civil officials are to be displaced:

(1) Status of Chief Executive and his cabinet under the military government.

(2) Status of the Congress under the military government.

(3) Status of the Governors and other civil officers of the several provinces.

(4) Status of the customs and tax collectors.

Relationship between the civil and military administrations:

(1) Military commanders assigned to various areas—will they perform civil administrative duties or act purely in an advisory capacity?

(2) Local laws and ordinances—will they be adopted, amended, or abrogated?

(3) Internal judicial system—to what extent will it function?

(4) Will a strict or liberal policy in relations with officials and inhabitants be pursued?

13–9. **The proclamation.**—*a.* The proclamation of the commander to the inhabitants of the occupied territory should be prepared beforehand, should above all be brief, and should cover the following points:

(1) Announcement as to the exact territory occupied.

(2) The extent to which the local laws and governmental system are to be continued in force, including a statement that the local criminal courts have no jurisdiction in cases of offenses committed by or against members of the occupying forces.

(3) Warning that strict obedience of the orders of the commander of the occupying forces is to be expected of all, and that those who disobey such orders or regulations, or commit acts of hostility against the occupying forces, will be severely punished; but those who cheerfully accept the new sovereignty and abide by its orders will be protected.

(4) A statement that the occupying forces come not to make war upon the inhabitants but to help them reestablish themselves in the ways of peace and to enable them to resume their ordinary occupations.

(5) In conclusion, the proclamation should make reference to supplemental regulations to be issued containing more detailed instructions.

b. The proclamation should be published in English and in the national language of the occupied territory. The conditions which might call for such proclamations are varied, and in each case the

MILITARY GOVERNMENT

particular circumstances must control. (For form for a proclamation see current edition of Naval Courts and Boards.)

13–10. **Supplemental regulations.**—*a*. The military government, being supreme, can lawfully demand the absolute obedience of the inhabitants of the area over which it is exercised. There should, therefore, always be prepared and ready for issue contemporaneously with the proclamation, or as soon thereafter as practicable, a supplementary order giving definite expression to regulations and detailed instructions on a variety of subjects in order that the inhabitants may be fully informed from the first as to the conduct that is expected of them.

b. The drafting of these regulations, usually at a headquarters far removed from the theater of operations, is by no means an easy task. If they are more harsh than is necessary for the preservation of order and the proper decorum and respect, the force commander and his government are bound to stand in disrepute before the civilized world.

c. One of the principal aims should be to so administer the military government that upon conclusion of the occupation, the transition to the new state of affairs may be accomplished without radical change in the mode of life of the inhabitants or undue strain in the return to, or setting in motion of, the machinery of their own laws and institutions. Yet, restrictions must be placed upon assemblages, notwithstanding that the people, looking to the future, will want to gather together and discuss platforms of political parties or campaigns for supremacy in their national affairs. Parades and gatherings in celebration of national holidays, and even religious processions on church holidays, may have to be restricted. The problem of reconciling these conflicting features is one of the most difficult and delicate with which the military government will have to deal.

d. The principal restrictions included in the supplemental order relate to unlawful assembly, circulation, identification, possession of arms and ammunition, policy as to manufacture and sale of alcoholic beverages, and offenses in general against the personnel, establishments, installations, and material of the forces of occupation.

e. Consideration should be given to the following matters:

(1) The force and effect of the instructions, rules, and regulations contained in the order.

(2) The fact that existing civil laws shall remain in effect, and be enforced by the local officials except those laws of political nature, and except that the civil laws shall not apply to members of the occupying force.

MILITARY GOVERNMENT

(3) A list of additional rules and regulations imposed by the military authority, and to be enforced by military tribunals, declaring it to be unlawful:

To act as a spy or to supply information to the opposing forces.

To cause damage to railway property; war materials, and other public utility.

To impair sources of water supply.

To destroy, damage, or secrete any kinds of supplies or materials useful to the occupying forces.

To aid prisoners to escape, or to willing assist the opposing forces.

To harm or injure members of the occupying forces.

To attempt to influence members of the occupying forces to fail or be derelict in the performance of their duties.

To damage or alter military signs or notices.

To circulate propaganda against the interests of the occupying forces.

To recruit troops, or to cause desertion by members of the occupying forces.

To commit any act of war, treason, or to violate the laws of war.

To utter seditious language.

To spread alarmist reports.

To overcharge for merchandise sold to members of the occupying forces.

To interfere with troops in formation.

To commit arson or to unlawfully convert property to the injury of the occupying forces.

To circulate newspapers or publications of a seditious nature.

To signal or communicate with the opposing forces by any means.

To sketch or photograph places or materials used by the occupying forces.

To escape or attempt escape from imprisonment.

To swear falsely.

To forge, alter, or tamper with passes or other documents issued by the occupying forces.

To interfere with or refuse to comply with requisitions.

To perform any act in substantial obstruction to the military government.

To show disrespect to the flag or colors of the United States.

To print, post, circulate, or publish anything antagonistic or detrimental to the Military Government or the Forces of Occupation. (Publications may be suspended or censored for cause.)

To violate any proclamation or regulation issued by the occupying forces.

To conspire, attempt to do so, or aid and abet anyone violating the foregoing regulations.

f. It is important to have beforehand a thorough knowledge of the customs of the country to be occupied, for the enforcement of regulations which run counter to long-established customs is always extremely difficult. It is not likely that much difficulty will be encountered in the enforcement of purely military regulations, but where the customary daily life of the civilian population is circum-

MILITARY GOVERNMENT

scribed by many restrictions and inconveniences, the tendency is towards frequent or continual violations. Desirable as such restrictions may seem from an idealistic standpoint, they will not be conducive to success unless they are so framed as to harmonize to the fullest possible extent with the psychology of the population which they are expected to govern.

13-11. **Digest of information.**—In addition to the study of the theater of operations, the commander should be furnished with a digest, prepared by the Law Officer, utilizing all information at the disposition of the Second Section, and such other pertinent information which would be of value to the commander in administering the military government. A sample form for such a digest follows:

DIGEST OF MILITARY, POLITICAL, ECONOMIC, AND PSYCHOLOGIC INFORMATION

1. MILITARY SITUATION (omitted).
2. POLITICAL SITUATION.
    a. National government.
        (1) Executive power.
            (*a*) In whom vested.
            (*b*) Method of accession.
            (*c*) Term of office.
            (*d*) Cabinet and advisers.
        (2) Legislative power.
            (*a*) Composition of legislative body.
            (*b*) How chosen.
            (*c*) Term of office.
            (*d*) Legislative procedure.
        (3) Judicial department.
            (*a*) Existing system.
            (*b*) Efficiency of existing courts.
    b. Local government.
        (1) Description of political divisions of country.
        (2) Administration of political subdivisions.
        (3) Administration of municipalities.
    c. Political parties.
        (1) Principal parties.
        (2) Leaders.
        (3) Sphere of influence.
        (4) Political tenets.
        (5) Political background prior to establishment of military government.
    d. Treaties and conventions.
        (1) Existing and pending.
    e. Franchise.
        (1) To whom granted.
        (2) How exercised.

MILITARY GOVERNMENT

3. ECONOMIC SITUATION.
    a. Geography.
        (1) Area.
        (2) Climate and rainfall.
    b. Population.
        (1) Entire country.
        (2) Important cities and ports.
        (3) Distribution of population.
        (4) Percentage and distribution of foreigners.
    c. Production and industry.
        (1) Chief industries and resources.
        (2) Location.
        (3) Exports and imports.
        (4) Ships and shipping.
        (5) Mines and quarries.
    d. Finance.
        (1) Monetary system.
        (2) Financial condition of country.
        (3) Sources of revenue.
        (4) Customs administration.
    e. Communications.
        (1) Railroads.
            (*a*) Extent and condition.
            (*b*) Ownership.
        (2) Roads and trails—extent and condition.
        (3) Waterways and harbors—extent and navigability.
        (4) Telephone, telegraph, radio and cables.
            (*a*) Extent, equipment, and possibilities.
            (*b*) Ownership.
        (5) Air transportation.
            (*a*) Extent, equipment.
            (*b*) Ownership.
        (6) Postal service.
    f. Public utilities.
        (1) Extent.
        (2) Control and supervision.
    g. Labor conditions.
        (1) Unemployment situation.
        (2) Wages, and hours.
        (3) Presence and effect of labor organizations.
        (4) Social conditions of laboring class.
    h. Sanitation.
4. PSYCHOLOGIC SITUATION.
    a. General racial characteristics.
        (1) Type: superstitious—vacillating—susceptible to propaganda—excitable.
        (2) Degree of corruption in politics.
        (3) Fighting ability.
        (4) Language and dialects.

MILITARY GOVERNMENT

4. Psychologic Situation—Continued.
   b. *Education.*
      (1) Percentage of illiteracy.
      (2) Compulsory or voluntary.
      (3) Outline of school system.
      (4) Location of important universities.
   c. *Religion.*
      (1) Prevailing form.
      (2) Effect of religion on life of people.
      (3) Location of religious centers.
   d. *Attitude toward other peoples.*
      (1) Foreigners in general.
      (2) Members of the occupation.

13-12. Attitude toward local officials and inhabitants.—*a.* Considering the data obtained with regard to the political situation, decision must be made as to immediate changes to be effected in the local government. Civil control must be subordinated to military control. All the functions of the government—executive, legislative, or administrative—whether of a general, provincial, or local character, cease under military occupation, or continue only with the sanction, or if deemed necessary, the participation of the occupier.

b. The functions of the collectors of customs at all important ports should be assumed, and officers of the naval service appointed to fill their places. No other civil officials should be displaced except as may be necessary by way of removal on account of incompetency or misconduct in office. The policy should be to retain the latter in their official positions and hold them responsible to the military officers in charge of the various areas within which their jurisdiction lies; the idea of this responsibility should be emphasized from the beginning of the occupation.

c. The following general rules should guide the commander of the occupying force in his dealings with the local government machinery to the extent that the latter is functioning:

(1) Acts of the legislature should not become effective until approved by the military governor.

(2) The acts of city and minor councils should likewise not become effective until approved by the military commanders having immediate jurisdiction of the political subdivisions concerned.

(3) In general, a liberal policy should be preserved in all relations with the inhabitants and the greatest latitude permitted in public and private affairs, consistent with the rights and security of the military forces and the termination of the occupation.

MILITARY GOVERNMENT

(4) More specifically, all local laws should be permitted to remain in full force and effect, except as specifically provided by the military governor.

(5) All local civil officials, except those duly removed or suspended from office by the military governor or by the military commander having immediate jurisdiction over said officals, should be encouraged to remain at their posts and be protected in the performance of their official duties. They should be required to take an oath to faithfully perform their duties. This oath is not an oath of allegiance.

(6) Vacancies among local civil officials by death, flight, or removal from office should be filled as follows:

(a) Where the local law provides for their selection by the President or by the head of a department, or for their popular election—by the military governor.

(b) Where the local law provides for their selection by a subordinate civil official or minor legislative body—by the military commander having immediate jurisdiction over the said official or legislative body.

(7) An official of the hostile government who has accepted service under the occupant should be permitted to resign and should not be punished for exercising such privilege. Such official should not be forced to exercise his functions against his will.

(8) Any civil official found guilty of acts subversive of the occupying power should be subject to trial and punishment by military commission.

13–13. **Law enforcement agencies and public services.**—The proclamation of the commander of the force announces the extent to which the local law and governmental system are to be continued. It should request the inhabitants to resume their usual occupations. Public services and utilities should continue or resume operations under the direction and control of military authorities. The administration of justice should be given special attention. All courts, unless specifically excepted by the commander of the force, should be permitted to function and their decisions enforced except that:

(1) No person in the service of the naval forces and subject to naval law will be subject to any process of the local courts. However, writs of subpoena may be served with permission from the local commanding officer.

(2) Persons charged with violations of military orders, or with offenses against persons or property of members of the occupying forces, or against the laws of war are to be tried by military tribunal.

MILITARY GOVERNMENT

(3) Persons employed by or in the service of the occupying forces should be subject exclusively to the military law and jurisdiction of such forces.

13–14. **Exceptional military courts.**—Since a naval court martial is a court of limited jurisdiction restricted by law to the trial of officers and men of the naval service, it is apparent that, in order to exercise the power conferred upon the force commander when his duty is such as to place under him a wider jurisdiction in accordance with the principles of this chapter, it is necessary to employ tribunals other than those used in connection with the administration of naval law. Such tribunals have been referred to by the Navy Department as exceptional military courts, and include the military commission, the superior provost court, and the provost court. At such time as the proclamation and supplemental regulations are issued, an order establishing military tribunals and defining their jurisdiction and procedure should be published. For a discussion of these courts and their procedure, see Naval Courts and Boards.

13–15. **Control of civil and military administration.**—*a.* The greatest efficiency of government will be acquired by centralization of policy and decentralization of execution. In order to accomplish this, it is necessary that the actual administration of the military government be decentralized by means of a special organization of military personnel. This special organization should be designed to facilitate the military supervision necessary within the territorial subdistricts into which the occupied area has been divided for the purpose of governmental control.

b. In subdividing the area for the purpose of administering the military government, the preexisting political subdivisions, such as counties, townships, municipalities, etc., should be considered, and overlapping and mixture of these subdivisions should be avoided as far as possible. The subdivision should also be made so as to lend itself to geographic unity; that is, no district or area should be separated from one of its parts by a range of mountains; each should have adequate means of communication and a fair share thereof; each should have a reasonable proportion of ports of entry and egress; and, each should have a reasonable proportion of the population, industries, etc. The principal feature of the organization should be that each territorial district or subdistrict will be placed under the control of a tactical commander, and that each tactical commander charged with duties pertaining to the supervision of civil affairs will

MILITARY GOVERNMENT

have his staff increased by personnel to be organized as a staff section similar to that previously referred to.

c. The principle of making the military commands coextensive with the political subdivisions of the occupied territory tends to subordinate tactical considerations to the necessities of civil administration. However, the relationship between the civil and military administrations should be such that should it become necessary for the military forces to move on to a continuation or renewal of hostilities, the civil affairs section of the staff may, with a minimum of interference with the military administration, remain in the area and be capable of extending its sphere of activity to include additional territory that may be occupied. So long as a tactical unit remains in a particular subdistrict, its commander will exercise the usual functions of command through the agencies normally at his disposal. He will exercise his special functions relative to civil affairs through the staff which has been organized especially for that purpose. The Officer in Charge of Civil Affairs in each area or district, together with his staff, should be subject to supervision and coordination in technical and routine matters by the Officer in Charge of Civil Affairs next higher in the hierarchy of military government, but all orders involving announcements or changes in policy, or affecting personnel should come through the military commander in the usual way.

13–16. Public utilities.—*a.* Municipal water works, light and power plants should be permitted to remain open and function as in normal times, but should be supervised by the officer on the civil affairs staff having jurisdiction over public works. Payments for public services should be made in the usual manner, while appeals in the matter of rates, wages, etc., should be referred to the military governor. Wilful damage to, or interference with, any public utility should be considered as an offense against the occupying force. Railways, bus lines, and other public carriers needed for military purposes may be seized and operated by the public works officer of the military governor's staff. This officer is also responsible for the upkeep of the highways.

b. All telegraph and telephone lines, cable terminals, and radio stations, together with their equipment, may be taken over and conducted under the supervision of the force communication officer. He should prescribe which will be operated by the occupying forces, which closed, and which will continue to be operated by civilian companies; and, in case of the latter, may requisition the services of the personnel, whether individually or as an organization, as may be

MILITARY GOVERNMENT

deemed advisable. The use by civilians of telegraph, telephone, and cable lines and of radio systems should be permitted only under regulations issued by the force commander.

13–17. **Trade relationship.**—*a.* Ships and shipping should be treated according to the rules of war. The normal port service should be interfered with as little as possible. The port regulations in effect should so remain as long as they are consistent with the orders issued by the military governor. The occupying force may take over any ferry or water transportation regarded as necessary to supplement other transportation lines.

b. The general policy of the military government should be to encourage, foster, and protect all citizens of the occupied territory in the energetic pursuit of legitimate interior and exterior trade relationship. It should be unlawful for inhabitants of the occupied territory to engage in any form of traffic with the enemies of the occupying forces, or to engage in commerce with foreign states in contraband articles of war, or to export money, gold, silver, jewelry, or other similar valuables for the time being.

13–18. **Mines and quarries.**—Mines and quarries should be permitted to operate as in peace except that all explosives on hand should be reported to the local military representative. The manager of a mine or quarry should be held responsible and the explosives are used only for proper purposes. Where the stock is large, a guard therefor should be furnished by the local military commander. Requisitions of local military commanders should be given priority over all orders at mines and quarries. Quarries of road material in areas or districts may be exploited by the local military commander.

13–19. **Public revenues.**—Generally, the policy of the military government should be to divert no public revenues from their normal uses except to defray the legitimate expenses of the military government. Taxes, excises, and custom duties collected at the current legal rate by agencies already operating, or otherwise provided for in orders issued by the military governor, should be turned over to the military government for accounting and disbursement according to law. Supplies for the forces of occupation, and the carriers of same while employed as such, should be exempt from taxes or other public revenue charges of any nature.

13–20. **Requisitions and contributions.**—*a.* Requisitions for supplies required by the occupying forces should be issued under the supervision of the military commander. They should be made upon the officials of the locality rather than upon individuals; they must

MILITARY GOVERNMENT

be reasonable in proportion to the resources of the country so as to avoid unnecessary distress among the inhabitants. They should be paid for in cash, if possible. Otherwise, receipts should be given.

b. Contributions may lawfully be levied against the inhabitants by authority of the military governor or the commander of the occupying force (but not by a subordinate), for the following purposes:

(1) To pay the cost of the military government during the occupation.

(2) Compensation for the protection of life and property, and the preservation of order under difficult circumstances.

(3) As a fine imposed upon the community as a whole for acts injurious to the occupying force.

c. Contributions should be apportioned like taxes, and receipted for. One method of exacting contributions is to take over the customs houses, thus controlling the revenues from import receipts.

13–21. **Public and private property.**—Public buildings and public property of the occupied country, except charitable institutions and those devoted to religious, literary, educational, and sanitary purposes, may be seized and used by the forces in the manner of leaseholder. Title does not pass to the occupying sovereignty. Other buildings are not to be used except in case of emergency. Private property must be respected.

13–22. **Employment of inhabitants.**—*a.* Services of the inhabitants of occupied territory may be requisitioned for the needs of the occupying force. These will include the services of professional men and tradesmen, such as surgeons, carpenters, butchers, barbers, etc., employees of gas, electric light, and water works and other public utilities. The officials and employees of railways, canals, river or coastal steamship companies, telegraph, telephone, postal and similar services may be requisitioned to perform their duties so long as the duties do not directly concern operations of war against their own country, and thereby violate the Rules of Land Warfare as recognized by the United States.

b. The prohibition against forcing the inhabitants to take part in operations of war against their own country precludes requisitioning their services upon works directly promoting the ends of the war, such as the construction of forts, fortifications, and entrenchments; but there is no objection to their being employed on such work voluntarily for pay, except the military reason of preventing information concerning such work from falling into the hands of the enemy.

MILITARY GOVERNMENT

13–23. Police and elections.—The civil police force may be continued in operation in conjunction with the military forces and the members thereof may be required to shoulder the burden of enforcing certain additional police regulations imposed by the various military commanders. Elections may be suspended or held at the discretion of the military governor; and he may regulate such elections to avoid fraud, disorder, and intimidation.

SECTION III

APPLICATIONS OF PRINCIPLES TO SITUATIONS SHORT OF WAR

13–24. **General considerations.**—The history of our Government indicates that we have occupied territory and established complete military control over such territory in time of peace as well as in time of war. If war has been declared, the establishment of a military government over the territory of the enemy occupied by our forces gives rise to very few questions. However, when military government is established over a foreign sovereign state or a portion thereof, without Congress declaring war, that is, when the native inhabitants of the country occupied are not considered enemies, it brings up the question of whether or not the laws of war can be legally applied. Such situations have presented themselves in the past and will probably present themselves in the future. They arise from the policies of our Government which dictate what our attitude should be toward assisting our neighbors in maintaining peace and order and in protecting the personal and property rights of our own and other foreign nationals. When we intervene in such cases, our action will always appear to many, especially those of the country concerned, as a quasi-hostile act and they will be ready to protest and criticize the conduct of the military government in all its functions. If, as we are taught, a military governor, even in time of war, should be careful to make his government humane, liberal, and just, because of the undesirability of making a return to peace difficult, how much more this principle must apply when there is no war.

13–25. **What laws apply.**—If the commander of the force of occupation establishes a military government and there is no war, what laws can he apply? He cannot apply the laws of our own country in the occupied territory and he cannot accept and enforce on the laws of the occupied territory. Our own constitution cannot be made to apply to a foreign territory, and the existing laws in the occupied territory manifestly will contain no provisions which will guarantee

SITUATIONS SHORT OF WAR

the security of the forces of occupation, their installations and material. The fact remains that the commander must govern and he must utilize a military form of control. Therefore, he will be justified in adopting any reasonable measures necessary to carry out the task or mission that has been assigned him. Whether his government has declared war is no concern of his—that is a diplomatic and international move over which he has no control. The very nature of his mission demands that he must have absolute power—War Power. However, as a matter of policy, the more rigorous war powers should not be enforced. Contributions, requisitions, treatment of war traitors, spies, etc., should not be more rigorous than absolute necessity demands for self-protection. The commander's policy should be to enforce the laws of war but only to such extent as is absolutely necessary to accomplish his task.

O

SMALL WARS MANUAL
UNITED STATES MARINE CORPS
1940

✦

CHAPTER XIV

SUPERVISION OF ELECTIONS

RESTRICTED

UNITED STATES
GOVERNMENT PRINTING OFFICE
WASHINGTON : 1940

TABLE OF CONTENTS

The Small Wars Manual, U. S. Marine Corps, 1940, is published in 15 chapters as follows:

III

SMALL WARS MANUAL

UNITED STATES MARINE CORPS

Chapter XIV

SUPERVISION OF ELECTIONS

v

SECTION I

GENERAL

14–1. **Introduction.**—*a.* The Government of the United States has supervised the presidential or congressional elections of neighboring republics on 12 different occasions. By accepting the responsibility for such supervision, the Government of the United States has settled serious political disturbances and assisted in the reestablishment of law and order. Sanguinary revolutions were stopped and countries rescued from a state of civil war. Assistance rendered by the Government of the United States was, in most cases, the direct result of requests of the conflicting political elements. In some instances, the aid was given in order to preserve peace and to settle controversies in accordance with existing treaties.

b. The supervision of an election is perhaps the most effective peaceful means of exerting an impartial influence upon the turbulent affairs of sovereign states. Such supervision frequently plays a prominent role in the diplomatic endeavors that are so closely associated with small war activities. Due to the peaceful features of electoral supervision, there will probably be more of this form of aid rendered neighboring republics in the future. Such action is in keeping with the popular revulsion against armed intervention in the internal affairs of other countries, and supports the principles of self-determination and majority rule.

c. Whenever the Government of the United States assumes the responsibility of supervising the elections of another sovereign state, it compromises its foreign political prestige as effectively as by any other act of intervention or interposition. There is, perhaps, no other service which may be rendered a friendly state, the motive of which is more actively challenged or criticized, as an endeavor to control the internal affairs or the political destiny of that state.

1

GENERAL

The duty of electoral supervision is normally performed by military and naval personnel. In addition, the electoral supervision will often be conducted under the protection afforded by United States military or naval forces assigned for that specific duty. A knowledge of the subject of electoral supervision will be found useful to all personnel engaged in small wars operations.

14–2. **Request for supervision.**—The supervision of elections within a sovereign State is normally undertaken only after a formal request for such supervision has been made to the President of the United States by the government of the foreign state, or by responsible factions within the foreign state, provided no de facto government exists. The formal request is usually accompanied by statements from the principal officials of recognized political parties to the effect that they desire the electoral supervision, that they pledge their active aid and support in cooperation with the proposed electoral mission, and that they agree to exercise their influence over all their followers to the end that a peaceful election may be held.

14–3. **Definitions.**—_a._ The group of individuals representing the Government of the United States that proceeds to the foreign country concerned to supervise a particular election in accordance with agreements between the Government of the United States and the foreign government is known as the ELECTORAL MISSION TO_____ (E. M.). It is normally composed of officers and enlisted personnel of the military and naval services of the United States, augmented by certain qualified civilian assistants.

_b._ The committee that directs and controls the national electoral machinery and electoral procedure of the country concerned is known as the NATIONAL BOARD OF ELECTIONS (N. B. E.). It is governed by the existing electoral laws of the country concerned, and is normally composed of citizens of the country. During the supervision of a particular election by the Government of the United States, a member of the Electoral Mission serves as a member of the National Board of Elections in lieu of one of the members who is a citizen of the foreign country. In the past, it has been customary for the Chairman of the Electoral Mission to serve also as President of the National Board of Elections.

_c._ The committee that directs and controls the electoral machinery and electoral procedure within a Department is known as a DEPARTMENTAL BOARD OF ELECTIONS. A member of the Electoral Mission serves on this board in lieu of a citizen member. The Department is a political subdivision of the country analogous to a State in the

GENERAL

United States. These political subdivisions are sometimes known as Provinces.

d. Depending upon the further political subdivision of the country concerned, minor boards of election are set up. Such boards may be known as CANTONAL BOARDS OF ELECTIONS, DISTRICT BOARDS OF ELECTIONS, or COMMUNAL BOARDS OF ELECTIONS. In each case, the political subdivisions of the country concerned will be the governing factor in the organization of such minor boards. Normally, a District (Arrondissement) is a political subdivision of a Department (Province) analagous to a County in the United States. A further political subdivision of the District (Arrondissement) is known as a Canton (Commune) and is analogous to a Ward or Township in the United States.

14–4. **Responsibilities of an electoral mission.**—*a.* A "free and fair" election implies the unrestrained popular choice of the whole people expressed at the polls by all who are lawfully entitled to suffrage. There must be no restraint or reservation, either physical or mental, exerted upon any aspirant to office or upon any of his supporters, except those normal restrictions required for the preservation of law and order. The fear of restraint may be real and with sufficient reason, or it may be imaginary and without cause. In either case, every effort must be made to remove the fear of restraint. It is only by a studied impartiality on the part of the Electoral Mission that charges of favoritism can be avoided.

b. It is well to consider the internal conditions that make the electoral supervision necessary. The electoral laws, the economic conditions, and the educational problems of the country concerned will often be found to be factors. The Electoral Mission can actually institute few permanent electoral reforms during the limited time that it is present in the country. It can, however, demonstrate a method of conducting elections that may serve as a model to the citizens for future elections. A free, fair, and impartial election cannot be held in a country torn by civil strife. Before such an election can be held, the individual must be made to feel safe in his everyday life. The presence of United States military and naval forces is often necessary to furnish that guarantee.

c. The Electoral Mission can assume the responsibility for the conduct of a "free and fair" election only within the definite limitations of the authority granted it and the facilities made available for carrying out its mission. There may be political and legal restrictions over which the Electoral Mission has no control. To guaran-

3

tee a "free and fair" election, the Electoral Mission should have the necessary authority over the executive, legislative, and judicial departments of the government to make effective its legal decisions. It must also have the active cooperation, if not the actual control, of military and police forces sufficient to enforce its rulings.

14–5. **Intimidation of voters.**—The employment of military and police forces for the protection of voters will often be a vital factor in the conduct of a "free and fair" election. The selfish personal and political partisanship of individuals, groups, or political parties may induce them to use various and sundry methods, including force, in an attempt to influence the outcome of the election. Guerrilla elements may be encountered, whose announced purpose is the prevention of an impartial election. These guerrilla elements may be banded together on their own initiative for this announced purpose, or they may be in the hire of some political group or party. Military and police forces are employed to prevent violence to personnel conducting the elections at voting booths, to prevent the destruction or seizure of ballots and electoral records, and for general protection of the populace from guerrilla activities. Protection is furnished the inhabitants in towns, in cities, and along lines of communication in order that registrants and voters may not be prevented from registering or voting due to threats of bodily violence while proceeding to and from registration and polling places. In some cases the homes of voters may be threatened, and the safety of their lives, families, and property may be endangered as a result of their announced desire to exercise their right of suffrage. This is particularly true in the case of members of campaigning (propaganda) parties normally employed in countries that do not enjoy good communication facilities. The foregoing measures of violence may be attempted by individuals, small groups, or large bands of guerrillas. A large organized group may make raids into quiet sections of the country in order to frighten the peaceful inhabitants and disturb the peace to influence the elections in the locality, attack isolated posts, ambush patrols, and conduct other operations of such nature as to demand the employment of a comparatively large military force for protection of the inhabitants.

14–6. **Military and police measures.**—*a.* It is essential that the Chairman of the Electoral Mission have military and police forces available in sufficient number to insure peace and order during the election campaign, the registration period, and the voting period.

4

GENERAL

Without such forces, it may be highly impracticable to assure the electorate of a free, fair, and impartial election.

b. During the electoral period, United States naval and military forces already stationed in the country may be augmented temporarily by troop detachments from the continental United States, by detachments from ships of the United States Fleet, or others armed forces of the United States. In order that such units may be readily available for recall or return to their normal stations for duty and to save time and transportation, it will usually be advisable to employ such temporarily attached troops in or near the larger bases or along lines of communication (railroads, automobile roads, lake, and river boats). Veteran troops that are accustomed to the country and inhabitants are employed in the more exposed districts. Such assignment of troops will promote more efficient performance of duty.

c. In some countries, there may be a native constabulary or similar organization under the command of United States or native officers. Whenever practicable, the larger portion of the military and police duties required to guarantee an impartial election should be provided by the native military organization. This force should be employed to its maximum capacity before employing United States forces. The display of United States armed forces at or near the polling places is kept to a minimum in order to avoid the charge that the Government of the United States has influenced the election, or placed favored candidates in office by the employment of military forces. However, the safety of Electoral Mission personnel must be considered at all times. The use of the native military organization places the responsibility for law and order where it properly belongs. It also tends to give the electorate the impression that the election is being conducted under the control of their own country. Care must be exercised to prevent the native military organization and individuals composing that organization from exhibiting any partiality. There cannot be a "free and fair" election if the use of the native constabulary degenerates into a partisan display of force. If the organization is not under the immediate command of United States officers, it becomes even more necessary to supervise its conduct during this period.

d. Local police are generally political appointees and, as a rule, cannot be depended upon to support a "free and fair" election. Their local, political, and personal interests will often result in prejudices and injustices, which will compromise efforts for impartial control. If they are not federalized, nonpartisan, and under

GENERAL

neutral superior authority, it is better to confine the duties of the local police agencies to their normal functions of preserving the peace in their localities, thus furnishing indirect support toward the conduct of a "free and fair" election. Their actions should be observed for any sign of partiality or improper activities that may tend to influence the outcome of the election. In some instances, it may be deemed advisable to suspend the civil police and similar organizations during the period of registration and election. This may be done by decree or other legal means. Their duties are then temporarily assigned to the native military organization. Armed guards from the constabulary force may be stationed at polling places to assist the regular civil police force in the maintenance of order. When so employed, the members of the constabulary force are given civil police power and may make arrests for ordinary civil offenses.

e. It is sometimes desirable to station an armed member of the United States forces inside each polling place to protect the electoral personnel, to guard the electoral records and ballots, and to preserve order within the building, thus relieving the Chairman of the local Board of Elections (usually a member of the Electoral Mission) of those responsibilities. The latter can then devote his entire attention to his electoral duties. At times, it may be necessary to assign detachments of United States troops to protect electoral personnel and records at polling places and while traveling between polling places and departmental capitals.

f. During the electoral period, and particularly on registration days and the day of election, aviation is employed to patrol polling places in outlying sections. This action is particularly valuable in that it gives tangible evidence to the voters that they are receiving protection in the exercise of their civil rights. Airplane patrols also furnish an excellent means of communication with polling places. They are a constant threat to any organized attempt to foment disorder.

14–7. **Unethical practices.**—*a.* In addition to the military or police features which may materially influence the ability of the Electoral Mission to guarantee a "free and fair" election, there are other elements that may operate to prevent that desirable result. These elements may properly be grouped under the heading of "political pressure" practices. Political pressure exerts a powerful influence in the conduct of elections since it reaches and touches every voter, whether he resides in the capital or a remote district of the interior. This political pressure consists of practices of long standing in some coun-

6

GENERAL

tries, is extremely difficult to uncover, and requires tactful and pains-taking effort to circumvent.

b. The incumbent Chief Executive may find it politically expedient to declare martial law in sections of the country at the beginning of the election period, giving as a reason, the preservation of law and order. The action may have no relation to the military situation at the time, and may possibly be taken in spite of recommendation against it by the military authorities concerned. As a consequence of such executive decree, the duly elected civil officials are automatically ousted from office and replaced by presidential appointees. By carrying out the process to its logical conclusion down to and including game wardens scattered throughout the province or provinces, political henchmen, willing and anxious to use every kind of pressure on any voter who might be opposed to the national administration, are in a position to interfere radically with an impartial election. This is a most un-healthful condition under which to attempt to conduct a free and fair election. By appealing to the sense of fair play of the executive, and through other diplomatic endeavors, it may be possible to have the decree rescinded. Unless the civil officials that have been appointed by executive action are removed from office, however, the effect of having such individuals continue in authority is likely to be deleterious to the conduct of an impartial election in the sections affected.

c. Public lands may be distributed to citizens with a tacit under-standing that they will vote for the candidates of the party controlling the distribution of the land, an act which is clearly contrary to the laws of the country. By this process, a political victory for the party dis-pensing the land is practically assured in a district whose inhabitants are known to be about equally divided between two political parties. Investigation and exposure of this practice, coupled with diplomatic efforts on the part of the Electoral Mission, will serve to put a stop to this activity, but it is likely to be too late to prevent full political profit from being derived.

d. In some countries, it is an established custom during electoral periods to arrest numerous citizens of the party, not in power, for old offenses, for charges of minor infringement of law, for honest political activities, and even upon charges that have absolutely no foundation whatsoever. In accordance with the law of the country, such citizens are automatically disfranchised, due to their having been arrested within a given period prior to the date elections are to be held. This action gives the party in power a powerful weapon in influencing the result of election. It is also not uncommon for the Chief Executive to

GENERAL

banish prominent political opponents from the country, thereby ridding the party in power of some of its intelligent opposition. Another pernicious custom is the employment of the "warrant for arrest" as a means of depriving citizens of their constitutional right of suffrage. By this means, citizens may be prevented from voting or holding office during the time the charges are pending against them, even though no arrest may be made. Charges may cover real or imaginary offenses or crimes alleged to have been committed on a date many years before. Through dilatory procedure on the part of the civil courts, trial of such cases may be delayed beyond the registration period, thus effectively disfranchising the alleged offender. This method is employed principally against members of the party not in power, since it is a difficult matter to swear out a "warrant for arrest" against members of the party in power due to the partisanship of the civil officials charged with this duty. By exerting diplomatic pressure, it may be found practicable to have the national laws amended by the insertion of a statute of limitations providing that "warrants for arrest" for civil offenses expire automatically after 2 years and those for criminal offenses after a period of 5 years, provided the civil authorities have taken no steps to bring the case to trial before the expiration of such periods.

e. Public utilities and other agencies owned or controlled by the government may be used in a discriminatory manner for the benefit of the party in power. Campaign (propaganda) parties, voters, workers, and others who may be of assistance to the party in power may be found to have free passage granted them on railroads, river and lake steamships, airplane lines and suburban street car lines owned or controlled by the government. The party in power may employ government trucks to carry voters to registration and voting places, denying such transportation to members of the party not in power. Provided the government owns or operates the telephone, telegraph, radio, or postal service, the party in power may be found to have full and free use of these public utilities, while the opponents of the party in power do not. Telegrams sent and paid for by the party not in power may be garbled en route, and delayed in delivery. When delivered, the message may be so changed from the original that it contains an entirely different meaning from that intended. Members of the party not in power may be subjected to delays in telephone connections, in the transmission of telegraphic and radio messages, as well as delay in the delivery of personal mail. A tactful appeal to

GENERAL

the sense of fair play of the government officials concerned is generally productive in terminating such practices.

f. Just prior to elections, public works projects may be undertaken in districts of doubtful political complexion or in those districts where the party not in power is known to have only a slight majority. Workers of the party in power may be imported into such districts to work on the projects in order that they may vote in that district, such workers normally being transported from districts where a majority for the party in power is definitely assured. Since the worker resides in the doubtful district at the time of registration and election, he is entitled to vote there, and thus may gain a clear majority for the party in power. This situation may be met by imposing a residential qualification for voters. For example, 6 months residence in a given district just prior to registration may be required as a qualification to vote in that district. This qualification may be a part of the electoral law of the country, or may be imposed as an interpretation of the law by decision of the National Board of Elections.

g. In some countries, the government has a monopoly on the manufacture and distribution of distilled liquors. This places a strong weapon in the hands of the party in power during the electoral period. The government party may dispense free liquor at political rallies in order to influence the opinion of the voters. This practice may be continued through the registration and voting period. Adherents to the party in power may give liquor to voters of the opposite party on election day, and then attempt to have them disqualified due to their intoxicated condition when they appear at the polls to vote. By restricting the distribution and sale of distilled liquor to normal amounts, this situation may be alleviated. Distilleries are padlocked, and an amount of liquor withdrawn for legitimate sale to authorized dealers. The amount withdrawn is equal to the average withdrawn over a reasonable period as shown by official records. To prevent the further use of intoxicants during this crucial period, stores and cafes dispensing them at retail are closed or prohibited from selling intoxicants on registration and election days.

SECTION II

PERSONNEL

14–8. **Chairman.**—The Chairman of the Electoral Mission is designated by the President of the United States. Usually, he is a flag officer or general officer, and holds the title of Envoy Extraordinary and Minister Plenipotentiary during the period covered by his assignment to the Electoral Mission. In order to carry out his duties in connection with the elections, he holds the appointive position of President of the National Board of Elections of the foreign country concerned.

14–9. **Electoral mission staff.**—The Electoral Mission staff consists of such officers as are required to carry out the mission in a particular case. Although the size and requirements may vary to some extent, the following tabulation covers the usual staff requirements:

> Chairman.
> Vice Chairman.
> Executive Officer.
> Secretary of Electoral Mission.
> Secretary of National Board of Elections.
> Assistant Secretary of National Board of Elections.
> Inspector.
> Intelligence and Press Relations Officer.
> Assistant Intelligence and Press Relations Officer.
> Legal Advisor.
> Assistant Legal Advisor.
> Communications Officer.
> Disbursing and Supply Officer.
> Assistant Disbursing and Supply Officer.
> Medical Officer.
> Aides.

PERSONNEL

14–10. **Commissioned officers.**—*a.* The officer personnel of the Electoral Mission should have rank commensurate with the importance of their duties. Whenever practicable, officers who have had prior duty with electoral missions should be selected for the more important positions. A knowledge of the language of the country concerned is one of the most important qualifications.

b. When United States forces are present in the country concerned for the purpose of restoring law and order, the officers assigned to duty with the Electoral Mission should not be taken from the United States forces unless suitable replacements are immediately available. In normal small wars situations, a proportionately large number of officers are required, and the United States forces present cannot be expected to have extra officers available. In general, the qualifications for officers assigned the Electoral Mission are identical with the qualifications required in the case of officers serving with the United States forces. The selection of a few of the latter officers for duty with the Electoral Mission may be practicable. The replacement of a large number of the officers serving with the United States forces by officers who are unacquainted with the local situation appears to be inadvisable during this critical period.

14–11. **Enlisted personnel.**—*a.* The most important qualification of enlisted personnel selected for duty with an Electoral Mission is fitness for independent duty requiring a large measure of responsibility. They should be able to speak, read, and write the language of the country concerned. Men selected should have a scholastic background of at least 2 years of high school, and preferably should be high school graduates. A clear record is an important qualification together with a reputation for tact, good judgment, and patience. For duty in remote areas, in districts known for their unhealthful living conditions, and in sections of the country where it is known guerilla bands operate, it is a decided advantage to assign men who have previously served in those areas to electoral duty with the boards of minor political subdivisions. Their prior experience with the military and police situation, in combating unhealthful conditions, and in the procurement of food and shelter, will enable them to assume their electoral duties with less difficulty than men who are unaccustomed to their surroundings.

b. In the event that United States forces have been in the country concerned for some length of time to preserve law and order, the majority of the enlisted personnel for electoral duty should be taken from among those forces. Since the employment of a number of men

12

on electoral duty will tend to reduce the activities of the forces engaged in purely military pursuits, it is obvious that they must be replaced by an equal number of troops from the continental United States or other source. Inasmuch as the electoral period is one which requires a maximum effort in maintaining law and order, it may be neccessary to increase the military forces during the electoral period.

14–12. **Civilian personnel.**—A number of United States civilians should be included in the personnel of the Electoral Mission. This is done to reduce the likelihood of the charge that the elections are being controlled by the military, a charge to which an enterprise of this nature is peculiarly susceptible. Since employment of any considerable number of qualified civilians will generally be impracticable due to the expense involved, it will be possible to employ only a small number of expert legal advisors and technical men, particularly individuals who have made the study of elections and government their life work.

14–13. **Instruction of personnel.**—As stated in paragraph 14–11 *a*, one of the most important qualifications of enlisted personnel selected for electoral duty is a knowledge of the language of the country concerned. In order to improve that knowledge, a language course is included in the instruction received prior to taking over their electoral duties. The language course is confined to the essentials of the language, with particular emphasis placed upon vocabulary adapted to the particular requirements of electoral duty. In addition, they receive instruction in the electoral law and procedure. The electoral law is studied by sections. Each section is discussed in connection with its historical background, its applications, and weaknesses that may have been disclosed in prior elections. The course covered by such a school depends primarily upon the time available and the need for the instruction. One month's instruction may be considered the minimum time required. A longer course of instruction will probably be found to be advantageous.

14–14. **Replacements.**—The personnel of the Electoral Mission will have the same losses, due to sickness and other casualties, as other United States forces serving in the same country. For the proper execution of electoral procedure, certain positions of the various electoral boards are filled by members of the Electoral Mission. In order that their work may continue, trained replacements must be available in sufficient number to take care of the estimated number of losses due to sickness and other casualties. Losses among

PERSONNEL

United States forces who are already in the country may be used as a basis for computing the radio of replacements required.

14–15. **Pay and allowances.**—*a.* The government of the country in which the elections are being conducted normally should provide a monthly money allowance for officers of the United States forces performing duty with the Electoral Mission. This allowance is to cover extra expenses incurred in the performance of electoral duties.

b. When performing electoral duty in cities and towns garrisoned by United States forces, enlisted personnel of the Electoral Mission may be subsisted with those forces. If this is impracticable, they should be furnished a per diem cash allowance for subsistence and lodging. This allowance should be ample to provide them with suitable subsistence and lodging, and should be uniform for all enlisted personnel serving on electoral duty throughout the country.

PERSONNEL

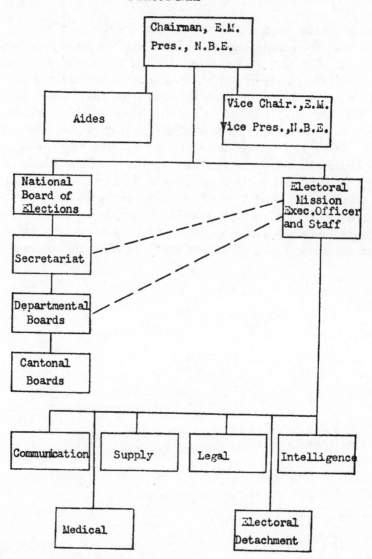

LEGEND:

——————— Authority and Communication

----------- Communication but no authority

FUNCTIONS OF ELECTORAL MISSION AND

NATIONAL BOARD OF ELECTIONS

Section III

ELECTORAL MISSION

14-16. **Chairman.**—*a.* After his designation by the President of the United States, the Chairman of the Electoral Mission visits the State Department in Washington, D. C., and acquaints himself with the history and the existing status of the situation in the foreign country concerned. He is also informed of the policy of the Government of the United States relative to the responsibility of the Electoral Mission, and the doctrine and procedure to be followed. He receives definite assurances from the State Department defining his authority, particularly in relation to other diplomatic representatives of the Government of the United States who may be present in the foreign country.

b. The Chairman then visits the foreign country, accompanied by such expert advisors as may be deemed appropriate, in order to make a personal survey of the situation including the legal, personnel, financial, and material requirements of the Electoral Mission. He assures himself that all necessary preparations are made, by personal contact with the appropriate authorities. The preparation and adoption of an electoral law to meet the requirements may be involved. It will probably be necessary to secure agreements establishing the legal status of the United States electoral personnel and the obligations of the foreign government concerned in relation thereto, particularly those obligations having to do with salaries, allowances, expenses, shelter, supplies, and the right of "free entry."

ELECTORAL MISSION

c. The Chairman issues instructions to United States naval, military, and civil authorities in the country concerned, directing their full cooperation in the successful accomplishment of his mission. Such instructions may involve questions having to do with reassignment of personnel, shelter, rationing, transportation, protection, disciplinary procedure, and replacements. Plans are prepared for instituting necessary military and police protection to maintain order during the registration and electoral period. If available, the native constabulary or similar organization is generally charged with the principal effort with relation to military and police problems arising during the electoral period. This force may fail to cooperate fully with the Electoral Mission and may even attempt to defeat the purpose of the Electoral Mission, of left under the command or control of an officer of the country concerned. For its proper control, it is obvious that such military force should be placed under the authority, if not the actual control, of the Chairman of the Electoral Mission during the electoral period. The Chairman of the Electoral Mission need have no administrative authority over the United States forces in the country. He does, however, expect full support from such forces. The military protection required for the Electoral Mission will include guards at polling places and in the towns where polling places are situated. It may be necessary to increase the number of combat patrols during the period, particularly in sensitive areas. Escorts may be required to convoy groups of citizens to and from places of registration and voting. A general redistribution of military forces may be required to support the Electoral Mission during the critical period.

d. Upon his arrival in the country concerned, the Chairman of the Electoral Mission should have absolute local United States authority on all questions pertaining to the election. He communicates with the State Department, direct, concerning all electoral matters. The State Department representative regularly assigned to represent the Government of the United States in the foreign country concerned forwards reports or other correspondence concerning or affecting the election through the Chairman of the Electoral Mission. The absence of a definite delineation of authority may lead to confusion and disagreement.

e. Direct contact with the Chief Executive of the country places the Chairman of the Electoral Mission in a position to exert a favorable influence, most important in its effect, toward the successful accomplishment of his mission. The political party in power may attempt

ELECTORAL MISSION

to adopt unethical measures that are tolerated by local civil law and accepted by customary usage. Such unethical measures may affect the conduct of a free, fair, and impartial election. Control over such matters generally rests with the Chief Executive of the country. Insofar as practicable, such unethical practices should be controlled by restrictive measures in the election laws themselves. An understanding concerning these practices should be reached by the Government of the United States and the Chief Executive of the country concerned prior to the assumption of responsibility by the Government of the United States for the proper conduct of the elections. In the event problems arise that are beyond his control, the Chairman of the Electoral Mission communicates the circumstances to the State Department. When the counsel or instructions of the latter have been received, he visits the Chief Executive of the foreign country and, with his cooperation, makes definite arrangements to remedy a situation which may become intolerable if permitted to continue. The Chairman makes an estimate of the funds that will be required to cover the expenses of the Electoral Mission, submits such estimate to the proper authority of the country concerned, and arranges for its appropriation. He arranges for the United States funds necessary for salaries, allowances, and travel expenses to and from the country concerned.

f. The Chairman selects his staff and organizes the Electoral Mission. Through his Vice Chairman, he arranges to have the staff undertake studies and the preparation of detailed plans.

g. Prior to the departure of the other members of the Electoral Mission from the United States, the Chairman assists in drawing up a definite agreement between the two governments on the following matters:

(1) The portion of the expenses of the Electoral Mission to be borne by the foreign government concerned, including the appropriation and the deposit of such funds to the credit of the Electoral Mission.

(2) The strength of the constabulary or other military organization of the country concerned, and the police force estimated to be necessary to insure peace and order in the country during the electoral period, and during the 30 days immediately preceding that period. He receives definite assurances that the military and police forces will be maintained at the estimated required strength.

(3) The laws and decrees of the government concerned that are considered necessary in order that the Electoral Mission may accom-

ELECTORAL MISSION

plish its task. Such laws and decrees should be passed and promulgated prior to the assumption of its functions by the Electoral Mission in order that the appearance of pressure on the part of the Electoral Mission may be avoided.

h. The Chairman of the Electoral Mission should be installed in a properly furnished residence by the foreign government concerned. He arranges for the government concerned to provide the Electoral Mission with sufficient office space where complete privacy is assured.

i. After his appointment as President of the National Board of Elections, the Chairman of the Electoral Mission organizes that Board and trains the personnel necessary for its proper functioning. He calls meetings of the National Board of Elections to discuss the electoral law and its procedure, and to make such decisions as are required for the proper interpretation and execution of that law. In some instances, the candidates for office, their followers, and even the political parties themselves may require emphatic instruction to guide them in their conduct. Political parties and their leaders are required to assume certain definite responsibilities, and are charged with maintaining discipline within their respective organizations.

14-17. **Vice chairman.**—The Vice Chairman of the Electoral Mission is the assistant to the Chairman of the Electoral Mission. He is also Vice President of the National Board of Elections. In general, his duties are comparable to those performed by a chief of staff. As Vice President of the National Board of Elections, he attends all meetings of that Board and is prepared to assume the office of the President in the event of the latter's inability to serve.

14-18. **Executive officer.**—*a.* The duties of the Executive Officer include the administration. training, and supervision of United States electoral personnel in their activities of a nonpolitical nature. He controls the expenditure of United States funds, and arranges for the procurement of necessary United States supplies. He is the liaison officer between the Electoral Mission and United States forces. The activities of the Executive Officer are confined to Electoral Mission duty only, as differentiated from duties pertaining to the National Board of Elections. This tends to keep the administration of the Electoral Mission and the training of its personnel separated from the electoral functions of the Electoral Mission.

b. The Executive Officer may be assisted in the administration of the personnel of the Electoral Mission by the assignment of one or more officers as commanding officers of units of the electoral detachment. When the administration of their detachments does not require their full time, and when there is a need for their services, the latter may be assigned duty as assistants to the staff officers of the Electoral Mission.

14-19. **Secretary.**—*a.* The duties of the Secretary include the operation, direction, and supervision of the office personnel of the Electoral Mission. He records the assignments of personnel, including transfers and other items of importance. He is charged with the safeguarding and filing of all correspondence of the Electoral Mission. He keeps the daily history of the Mission. The Secretary arranges for the reception of officials and other visitors. The information and reception desks operate under his supervision. He supervises the work of the nonclerical civilian employees of the Electoral Mission. He is not assigned the dual function as Secretary of the National Board of Elections. An additional officer is required for that duty.

b. The Chief Clerk operates under the supervision of the Secretary. His duties include the management of the clerical force attached to the Executive Officer's office, and the proper handling of incoming and outgoing correspondence, files, and orders issued by the Executive Officer. He is charged with the duty of posting guards to safeguard the office and correspondence therein. The Electoral Mission bulletin board is under his direct supervision.

14-20. **Inspector.**—The Inspector of the Electoral Mission should be thoroughly familiar with the electoral law and all interpretations, decisions, and instructions of the National Board of Elections. He should be qualified to conduct investigations and interviews in the language of the country. His duties include the investigation of complaints of a serious nature. He keeps the President of the National Board of Elections informed of the operations of the electoral law and procedure by observation of their operation throughout the country. He visits the various outlying departments, and studies conditions which may have a direct and important bearing on the elections.

14-21. **Intelligence and press relations officer.**—*a.* The Intelligence and Press Relations Officer collects, evaluates, and distributes all intelligence information of interest to the Electoral Mission. He is in constant touch with the military, economic, social, and political situations and developments insofar as they may affect the elections. A periodic intelligence report, covering the political, economic, and military situation, is made regularly. The period covered may be biweekly or monthly. He prepares and submits to the Chairman of the Electoral Mission, periodically or upon call, general estimates covering the military or political situation, and may be directed to prepare special studies of particular localities or activities. A personnel file of all important characters in the country is kept up to date. This file has a complete history of each individual, including his military and political affiliations. The Intelligence and Press Relations Officer is the liaison officer of the Electoral Mission with intelligence sections of United States military forces, native military and police forces, and with the local and foreign press. He prepares and releases information to the press. A clipping bureau is maintained and brief résumés are prepared of all articles appearing in the press concerning the Electoral Mission or the National Board of Elections. The latter are prepared daily and submitted to the Chairman of the Electoral Mission. If deemed advisable, a résumé of such information may be mimeographed and distributed to personnel of the Electoral Mission. Prior to leaving the United States and after arrival in the country concerned, the Intelligence and Press Relations Officer collects books, texts, and articles of a nonfictional nature pertaining to the country or region concerned, and maintains such material for the use of the personnel of the Electoral Mission. Photographs are taken of various subjects having to do with the activities of the Electoral Mission and are later included in the final report made to the United States State Department.

b. Assistants to the Intelligence and Press Relations Officer are assigned to relieve him of the details of compiling briefs of articles appearing in the press, and to assist in the preparation of press releases. These assistants also make special investigations and reports, and assist in the instruction of Electoral Mission personnel during the period immediately preceding the registration of voters.

c. The office force will generally require a minimum of three proficient stenographers and typists, at least one of whom should be thoroughly conversant with the language of the country concerned. It will often be convenient to employ one or more native interpreters to assist the Intelligence Officer. They should be men of education, good bearing, and neutrality in politics. If it is impossible to find nonpartisan interpreters, they should be selected equally from among partisans of both political parties.

14–22. **Law officer.**—*a*. The Law Officer furnishes information to the President of the National Board of Elections on legal matters relating to that Board. In the discharge of such duties, he takes into consideration the current electoral law, the constitution of the country concerned, various local laws and decrees, and, as a matter of precedent, the rulings and decisions of former Electoral Missions and National Boards of Elections. When required, he renders opinions on complaints submitted to the National Board of Elections for a ruling. He prepares the initial draft of the electoral law, during the period of survey by the Chairman of the Electoral Mission, provided such a document has not already been furnished through the United States State Department. Upon the completion of its duties by the Electoral Mission, the Law Officer prepares a report showing any weaknesses in the legal features of the electoral law or procedure that have been disclosed in the election just completed. This report is prepared in the form of proposals for laws, decrees, and instructions, and is delivered to officials of the country concerned as a suggestion for the improvement of the electoral law and procedure.

b. The Law Officer may be assigned one or more assistants, including United States civilian experts in electoral law. Such clerical assistance as may be necessary is provided. Two stenographers and typists may be considered the minimum requirement for this office. If practicable, they should have had prior experience in legal research and reports.

14–23. **Communications officer.**—*a.* The Communications Officer prepares plans and recommendations for the maintenance of efficient communications for the Electoral Mission and National Board of Elections, by the utilization of existing means of communication, and through the supply of such additional channels as are necessary and practicable. Methods of communication will generally include telegraph, radio, telephone, airplane panel and pick-up, and messenger. The Communications Officer is responsible for the coding, decoding, routing, and filing of dispatches. He maintains a record of communication facilities throughout the country.

b. All the usual agencies of communication in the country should be available to the Electoral Mission and the National Board of Elections. In addition, the communication facilities of United States military and naval forces may be placed at the disposal of the Electoral Mission, provided this can be done without interference with the normal communications of the latter. Such forces will often have only extremely limited facilities for communications, and these will be heavily burdened with necessary traffic. In order that these facilities may not be further burdened, the Electoral Mission should limit its requests for transmission of messages to cases of urgent

ELECTORAL MISSION

necessity, or after the failure or proved inadequacy of other means of communication. Communication with detachments at polling places may sometimes be difficult and slow. It may be necessary to depend chiefly upon the airplane for communication with remote places lacking other communication facilities, and for communication with places with which other means of communication is unreliable. For short messages, advantage can be taken of planes that patrol remote registration and voting places.

14-24. **Disbursing and supply officer.**—The Disbursing and Supply Officer is responsible for the custody and expenditure of all funds of the country concerned, placed to the credit of the Electoral Mission. He prepares estimates for funds required by the Electoral Mission, and presents them to the Chairman of the Electoral Mission for requisition on the foreign government concerned. He drafts all correspondence pertaining to funds of the Electoral Mission placed to its credit by the foreign government. When required, he submits vouchers and requisitions requiring expenditures to the Executive Officer for approval. He submits monthly itemized reports of receipts and expenditures, together with the necessary vouchers, to the Minister of Finance of the government concerned, via the Chairman of the Electoral Mission. The Disbursing and Supply Officer visits the Minister of Finance and secures his approval of the wording and arrangement of vouchers in order that the latter may conform to the current governmental practice. Electoral Mission funds are kept on deposit in an approved bank in the name of the Electoral Mission. The Disbursing and Supply Officer keeps the books, records, vouchers, and reports pertaining to such funds, according to approved methods. The transportation of personnel and freight by rail, airplane, motor, etc., and the coordination of the employment of such agencies is a function of the Disbursing and Supply Officer. Prior to the departure of the Electoral Mission from the United States, the Disbursing and Supply Officer prepares a budget estimate of the elections in considerable detail. To this estimate should be added the item, "Unestimated, Underestimated, and Unforeseen Items." Insofar as can be foreseen, the budget estimate should contain all contingent items, such as per diem allowances and rentals. Some items which are included in the budget estimate, may be found to be unnecessary later, but are included as a precautionary measure to insure an adequacy of funds. It is far easier to reach an agreement upon a definite sum initially, even though this sum is somewhat large, than it is to procure supplementary funds from time to time in order to make up for a budget estimate that was originally inadequate. A favorable impression is created when unexpended credits are turned back to the credit of the country concerned by the Electoral Mission. The funds appropriated by the country concerned should be placed to the credit of the Electoral Mission for its use immediately upon arrival at the beginning of the electoral period, in order to provide funds for expenses, supplies, and rents, that will be needed immediately. Before his departure from the United States, the Disbursing and Supply Officer should be furnished a statement by the State Department indicating the exact United States funds and funds of the country concerned that will be available for the conduct of the elections.

14-25. Medical officer.—The Medical Officer is charged with caring for the health of the personnel of the Electoral Mission. He instructs such personnel in hygiene, sanitation, and related subjects that are peculiar to the country concerned. When the bulk of the personnel of the Electoral Mission is distributed in outlying regions, the Medical Officer makes inspection trips to the various departments to investigate living conditions, health, hygiene, and sanitation.

14-26. Aides.—Aides are assigned to the Chairman of the Electoral Mission to perform such duties of an official, or personal nature as the Chairman may direct. It may be found practicable to assign one aide additional duty as morale officer. As such, he is responsible for the recreational activities and equipment of the Electoral Mission personnel.

14-27. Departmental board personnel.—a. The Chairmen of Departmental Boards are commissioned officers of the United States forces. They are directly responsible to the President of the National Board of Elections for the proper conduct and operation of the electoral procedure within their respective departments. The necessary facilities to carry out these duties are placed at their disposal. They are inducted into office by the National Board of Elections, and proceed to their respective departments about 6 or 8 weeks prior to the first day designated for registration. Additional commissioned officers are designated as Vice Chairmen of Departmental Boards and serve as assistants to the Departmental Chairmen. Upon arrival at the capital of his department, the Chairman makes contact with the local civilian officials, and organizes the Departmental Board of Elections. He surveys the departmental political organization and studies any changes recommended by the political members of his board. Armed with this information, he makes a personal reconnaissance of his department to establish contacts in the various cantons (districts) and to determine if any rearrangement of proposed polling places is advisable. He surveys the political and military situation throughout his department, ascertains the housing and rationing facilities available to the Electoral Mission personnel, and determines the number of guards necessary for their protection and for the maintenance of order at the polling places. The Chairman then returns to the capital of the country concerned, where he makes a detailed report to the President of the National Board of Elections. Included in this report are recommendations made as a result of a survey of the department.

b. The Departmental Board of Elections is organized in a manner similar to the National Board of Elections. The Board consists of the Chairman, who is a commissioned officer of the United States forces, and two political members, one representing each of the two political parties. The political members are appointed by the National Board of Elections after nomination by the representatives of their respective political parties, who are members of the National Board of Elections. Provision is made for the appointment of substitutes to act in case of incapacity of regular political members.

c. The Chairman in each department is authorized to appoint a secretary for the Departmental Board of Elections. It may be advisable to restrict such appointments to commissioned officers of the United States forces. The Secretary may be a United States civilian or a civilian of the country concerned. He takes no part in the deliberations or decisions of the Board.

d. The Departmental Board of Elections has general supervision of the election in its own department, and deals directly with Cantonal (District) Boards of Election. The Departmental Chairman is frequently called upon to reconcile the opposing views of the political members of the Board. Every attempt is made to dispose of complaints, appeals, and petitions by action of the Departmental Board, permitting only the more important complaints, appeals, and petitions to go to the National Board of Elections for decision.

14–28. **Cantonal board personnel.**—*a.* The Cantonal (District) Boards of Election are similar in composition to the Departmental Boards of Election. Each Cantonal (District) Board of Elections has a Chairman, who is usually an enlisted man of the United States naval or military forces, and two political members, one from each of the two political parties. The political members of a Cantonal (District) Board of Elections are appointed by the Departmental Board of Elections in a manner similar to the appointment of the political members of the latter by the National Board of Elections. Provision is made for the appointment of substitutes to act in case of incapacity of regular political members. In some cases, the Chairman, as well as the other two members of the Board, is a citizen of the country concerned. Care should be exercised that the number of Chairmen assigned from among citizens of the country concerned are drawn equally from both political parties, in order to avoid charges of partisanship.

b. The Cantonal (District) Board of Elections exercises direct supervision over the registration and voting of the individual voter. The Cantonal (District) Board is responsible for the enforcement of provisions of the electoral regulations to insure a "free and fair" election. This responsibility rests primarily upon the Chairman, as the representative of the United States Government, who is in direct contact with the voters themselves. He is placed in a position of responsibility and authority, and his relations with the political members of the Board and the military guards will require a maximum of tact and good judgment. The Cantonal (District) Chairman and his guards should arrive at the location of their polling places at least one week prior to the first day designated for registration, in order that they may be established and ready for the transaction of official business on the opening day. In each case, circumstances will determine whether or not the Cantonal (District) Chairman and his guards will be withdrawn to the nearest garrison during the period between the close of the registration and the time it will be necssary to return to the polling places for the election. This decision will be influenced by the challenges, complaints, and other official business to be transacted, and by the travel time required to make the trip. The final decision is made by the Departmental Chairman or higher authority after consultation with the military commander concerned.

c. Many of the cantons (districts) may be situated in remote and outlying places where United States members may be forced to undergo some hardships. It may be necessary for them to live in uncomfortable and unhealthful surroundings without immediate medical aid. Airplane drop and pick-up may be the only method of communication in some cases.

d. When the Electoral Mission personnel available is limited in number, it may not be feasible to assign a Chairman to *each* Cantonal (District) Board. In such cases, it may be necessary to use supervisors. A supervisor is an enlisted member of the Electoral Mission who acts as Chairman of *two* or *more* Cantonal (District) Boards of Election.

Section IV

NATIONAL BOARD OF ELECTIONS

14-29. **Members and staff.**—*a.* The membership of the National Board of Elections includes the following:

1. **President.**—The Chairman of the Electoral Mission. He is designated as President of the National Board of Elections by the executive of the foreign country concerned. He is legally inducted into office in accordance with the laws of that country, as are also the two political members.

2. **Member.**—This member is designated by one of the major political parties and represents that political party on the National Board of Elections.

3. **Member.**—This member is designated by the other major political party and represents that party on the National Board of Elections.

b. The Staff of the National Board of Elections consists of a secretary, assistant secretary, translators, and clerks. The number of translators and clerks is governed by the need for their services, and will vary with the size of the country, the number of voters, and the electoral laws of the country.

c. In order that the work of the National Board of Elections may continue without interruption, substitutes for all members are provided to take the place of any members who are temporarily unable to serve during meetings. The substitutes are designated by the major political parties of the country concerned, and are legally inducted into office in accordance with the laws of the country in the same manner as provided for regular members. The Vice Chairman of the Electoral Mission is designated as Vice President of the National Board of Elections. Although only one secretary and one assistant secretary are normally required to carry on the work of the

NATIONAL BOARD OF ELECTIONS

National Board of Elections, it is convenient to have substitutes available who have been legally designated and inducted into office.

14–30. **Duties.**—The National Board of Elections exercises general supervision of the election and is the final authority on all matters pertaining to the election. It issues interpretations and instructions for the proper execution of the electoral laws. The National Board of Elections hears all complaints that require its decision. When such action is indicated, complaints should be investigated initially by the political party concerned and then presented to the National Board of Elections by the responsible head of that political party. Petitions and appeals presented for action by responsible citizens must be made in accordance with the law of the country, and in a manner that will uphold the dignity of the National Board of Elections.

14–31. **Secretary of the National Board of Elections.**—This office is generally filled by a commissioned officer of the United States naval or military forces. He is charged with keeping the record of the minutes of all meetings of the National Board of Elections. He prepares all correspondence emanating from the National Board of Elections, and maintains a record of all incoming and outgoing correspondence of that Board. Since the National Board of Elections does not have a communication system of its own, the communication system available to the Electoral Mission is employed to handle the communications of the Board. The Secretariat of the National Board of Elections should be entirely apart from the offices of the Electoral Mission, but should be conveniently located with relation to the latter in order that the necessary contact among the various staff members may be readily maintained. The Secretary has an assistant secretary, and one or more stenographers for clerical assistance. When the offices of the Electoral Mission and the National Board of Elections are separated by several city blocks, it is desirable that the Secretary of the National Board of Elections have his office at the headquarters of the Electoral Mission. The Assistant Secretary of the National Board of Elections is in charge of the office of the National Board of Elections. Matters requiring the signature of the Secretary of the National Board of Elections are sent to the latter's office at the headquarters of the Electoral Mission. It should be clearly understood that the Secretary of the Electoral Mission does not perform a dual function as Secretary of the National Board of Elections. These are completely separate functions.

14–32. **Complaints, appeals, and petitions.**—*a.* All complaints, appeals, and petitions should be presented through the regular official channels. Subordinate agencies should make every effort to settle such matters without the necessity of forwarding them to the next higher agency for action. To further their own interests, complainants will often attempt to take their complaints to the highest authority that will listen to them. Thus, if a complainant succeeds in presenting a complaint before the President of the National Board of Elections, and the latter refers the complaint to a Chairman of a Departmental Board of Elections for an investigation, the complainant will have a tendency, thereafter, to ignore the Chairmen of the Departmental Boards of Elections. Much correspondence and time will be saved if the higher authority refuses to accept complaints, appeals, and petitions when it is obvious that a subordinate agency can handle the matter.

b. When complaints, appeals, and petitions are received through the mail, direct, by the President of the National Board of Elections, they should be returned to the originator with instructions that they be taken up with the proper subordinate authority in order that they may be received through regular official channels. This will generally mean that the originator will be instructed to submit his complaint, appeal, or petition to a Departmental Board of Elections. In order that a record may be made of all such matters, oral complaints, appeals, and decisions should not be accepted. The originator should be instructed that all such matters must be submitted in the form of a written document before action can be taken.

c. Members of the staff of the headquarters of the Electoral Mission, who are personally acquainted with the leading politicians, must be particularly careful to prevent such acquaintanceship from being imposed upon by the politicians. It is to be expected that the latter will attempt to register complaints with the members of the staff of the headquarters of the Electoral Mission with the expressed desire that such staff member make a personal investigation. If such complaints are received, they normally will have to be referred to a Chairman of a Departmental Board of Elections to investigate, and such request should go through the President of the National Board of Elections. This will tend to overburden the Secretariat of that Board, and violates the principles laid down in paragraph 14-32, above. When staff members of the Electoral Mission receive such requests, they should tactfully, but firmly, refuse to accept the complaint and should suggest that the complaint be taken directly to the Chairman of the Departmental Board of Elections concerned. The same principle applies to the acceptance of complaints, appeals, and

NATIONAL BOARD OF ELECTIONS

petitions by the Chairman of a Departmental Board of Elections, when it is obvious that the matter is not to be handled originally by a Chairman of a Cantonal (District) Board of Elections. A strict compliance with the instructions contained in this paragraph will simplify many of the problems arising in connection with the electoral laws and electoral procedure.

14–33. **Assembly.**—The National Board of Elections should be furnished a place for holding its sessions. The location should be such as to furnish complete privacy. The space furnished should be in keeping with the dignity of the high office of the National Board of Elections, and should be free from the curiosity of the general public. The main room for the holding of sessions should be sufficiently large to accommodate a limited number of spectators. Provision should be included for the maintenance of complete privacy during secret sessions. In order that the Secretariat of the Board may occupy offices convenient to the Board while in session, it will often be most convenient to use a private residence of the better class as headquarters of the National Board of Elections. In some cases, the National Board of Elections may be more conveniently located in a commercial building such as an office building. In the latter case, the maintenance of privacy will be more difficult. The headquarters of the National Board of Elections should be within one or two blocks of the headquarters of the Electoral Mission, when such is practicable. The two offices should never be in the same block or under the same roof, in order to prevent interference with the proper performance of duty by both groups due to the close intermingling likely to ensue.

Section V

REGISTRATION AND VOTING

14–34. **Registration.**—*a.* The National Board of Elections designates the day or days on which voters may register. The rules covering the process of registration are issued by the proper authority. Three successive Sundays and two intervening Wednesdays will generally be found sufficient for registration days. The designation of five registration days will encourage the greatest possible number of voters to register and will permit them to do so with the least inconvenience to themselves.

b. Cantonal (District) Boards of Election are organized some time prior to the first date set for registration. The registration of voters is conducted by these Cantonal (District) Boards. In order that the Cantonal (District) Boards of Elections may hear or dispose of any challenges made during registration of voters, a day is set aside for this purpose. It will generally be found convenient to designate a date about a week after the last registration date for the hearing and disposition of challenges.

c. The average voter will judge the efficiency and fairness of the election supervision by the procedure and methods employed during the registration period. The impressions received by the average citizen at this time will determine, in a large measure, the amount and kind of criticism that the Electoral Mission will receive. The creation of a favorable impression of fairness and impartiality will assist the Electoral Mission in carrying out its mission of holding a "free and fair" election, by encouraging a larger proportion of the electorate to vote.

14–35. **Voting.**—*a.* A study of the registration reports by Departmental Boards of Elections will indicate whether any changes are necessary in the designation of Cantons (Districts). It may be found desirable to combine some voting booths, and others may be moved or closed entirely. In some instances, additional voting booths may be needed in sections having poor roads or trails, and in

REGISTRATION AND VOTING

sections where there has been a large increase in population since the last elections.

b. The ballots are prepared and supplied in a form in keeping with the nature of the election and the intellectual attainments of the inhabitants. In some countries, the political parties have a distinguishing color. For example, the color of one political party may be green, and the other red. By the employment of a green or red circle on the ballot, a voter who cannot read, and is also not color blind, is enabled to place an (X) in the colored circle representing the party of his political belief. In some countries, the political parties are identified by certain symbols. For example, in one country, one political party may have for its symbol a rooster and the other may use a bull. The exact form of ballot to be employed should be determined by a study of the customs and methods followed in the country concerned, after consultation with the best local counsel available.

c. In order to prevent multiple voting, it is generally convenient to require each voter to dip one finger in a fluid stain of a secret formula immediately after depositing his ballot. The fluid should be of a type that cannot be removed by ordinary processes available to the inhabitants, and should wear off after the elapse of several days. The color should be such that it will show clearly in contrast with the color of the individual. The formula of the fluid is kept secret to prevent the distribution of neutralizing formulas by persons bent on illegal practices. Since some opposition to the use of marking fluid may be encountered, it is well to have the Chief Executive of the country, leading candidates, and other prominent citizens photographed while dipping their fingers into the fluid. Wide publicity is given the demonstrations, together with the favorable comments from such prominent citizens concerning the requirement.

d. If the registration has been carefully conducted, and disposition has been made of all challenges prior to the day of election, the voting will be expedited, and the work of the Cantonal (District) Boards, subsequent to the day of election, will also be lessened. When all the business pertaining to electoral procedure has been completed by the Cantonal (District) Boards of Election, the members of the Cantonal (District) Boards of Election proceed to the departmental capital with the ballots and records. Each Departmental Board of Elections hears all challenges and complaints of each Cantonal (District) Board in its department. When the Departmental Board has heard and settled all challenges and complaints, the members of the Cantonal (District) Boards of Elections are released from further electoral duty. The

Chairmen of the Departmental Boards of Election then report in person with their complete electoral reports to the National Board of Elections. Serious reports and challenges from any department are heard by the National Board of Elections in the presence of the Chairman of the Departmental Board of Elections concerned. The ruling of the National Board of Elections is final in each case.

14–36. **Final reports.**—*a.* The National Board of Elections submits a complete report of the elections to the Chief Executive of the foreign country after receiving the reports of all the Departmental Boards of Election. After the Chief Executive has received this report, the Electoral Mission is released of its electoral duty by proper United States authority.

b. Upon completion of their duties, the personnel of the Electoral Mission may be required to submit reports of their particular activities. Cantonal (District) Chairmen may be required to describe their cantons (districts), the living conditions encountered, and other matters of interest. Departmental Chairmen may be required to describe the operation of the electoral law as they observed its operation in their departments, together with any recommendations they may wish to make for the conduct of future elections.

c. The Chairman of the Electoral Mission submits a detailed and comprehensive report to the State Department covering the history of the Electoral Mission. The report includes criticisms and recommendations of a constructive nature, and all information likely to be of assistance to future electoral missions.

O

SMALL WARS MANUAL
UNITED STATES MARINE CORPS
1940

✦

CHAPTER XV

WITHDRAWAL

RESTRICTED

UNITED STATES
GOVERNMENT PRINTING OFFICE
WASHINGTON : 1940

TABLE OF CONTENTS

The Small Wars Manual, U. S. Marine Corps, 1940, is published in 15 chapters as follows:

III

SMALL WARS MANUAL

UNITED STATES MARINE CORPS

Chapter XV

WITHDRAWAL

V

INTRODUCTION

15–1. **General.**—In accordance with national policy, it is to be expected that small wars operations will not be conducted with a view to the permanent acquisition of any foreign territory. A force engaged in small wars operations may expect to be withdrawn from foreign territory as soon as its mission is accomplished. In some instances. changes in national foreign policy may lead to the abrupt termination of small wars operations within a given theater. Since eventual withdrawal is certain, it is a governing factor in troop assignment and field operations.

15–2. **Factors to be considered.**—Voluntary and involuntary withdrawal from contact with a hostile force is a tactical operation, the basic principles of which are applicable to any type of warfare. When a withdrawal from a foreign territory is ordered. the mission will usually be to withdraw, leaving the local government secure in its ability to execute all of its functions satisfactorily. Policies, decisions. plans, and alternate plans should be decided upon well in advance of the time of execution. Local governmental functions should be returned to the control of the local authorities as early in the campaign as conditions warrant, in order that it may not be necessary to turn over all such functions at one time. As soon as an approximate date for withdrawal has been decided, the commander of the United States forces makes recommendations to higher authority relative to the methods to accomplish the withdrawal and requests decisions on all matters pertaining to the operation that he is not empowered to decide himself. Decisions are requested on all matters requiring coordination by a higher authority.

15–3. **Phases of withdrawal.**—For convenience of analysis, withdrawal is divided into two phases; the withdrawal from active military operations and the final withdrawal. The withdrawal from active military operations commences when elements of the United States forces initiate the restoration to the local authorities of any governmental authority or responsibility that has been assumed during

1

INTRODUCTION

the course of the campaign. This phase of the operation terminates when the local government is in complete control of the theater of operations. This phase may merge into the final withdrawal phase since the final withdrawal may be proceeding concurrently with the last stages of withdrawal from active military operations. When the United States forces have transferred all governmental authority and responsibility to the local government and no further military operations are contemplated except in case of grave emergency, the final withdrawal phase may be properly considered to have started. This phase ends when all United States forces have been evacuated from the foreign country.

SECTION II

WITHDRAWAL FROM ACTIVE MILITARY OPERATIONS

15-4. Concentration.—During the initial military operations of a small war campaign, the commander of the United States forces is usually free to dispose his forces in accordance with the tactical situation, subject to general directives received from higher authority. When the tactical situation permits, troops may be withdrawn from outlying areas and concentrated at points that will enable them to support native forces if such support is required. When order is restored and the proficiency of the native troops or police is such that there is no further need for United States forces in close support, the latter may be concentrated in a locality or localities where they will be available in case of emergency. This concentration may be gradual and extend over a considerable period of time. It may be hastened or retarded by international or local political considerations. Concentration areas are selected only after a careful estimate of the situation. In a withdrawal that is unhampered by combat operations, logistic considerations will usually be a controlling factor in the selection of such areas. The main or final concentration area or areas will normally be at a seaport. When the final concentration area or areas lie inland, a line of communication thereto must be secured.

15-5. Rights retained.—As the U. S. forces will be relinquishing all, or at least a part, of their military, territorial, and administration functions, consideration must be given to the matter of rights and powers that are to be retained by the United States forces for reasons of policy or security. The right to use all communication facilities is retained in order that the supply and evacuation of United States forces may be readily carried out until the final withdrawal is completed. Means of communication include the unrestricted use of roads, railroads, waterways, as well as telephone and radio facilities. A definite written agreement with the proper authorities is usually made to cover the retention of rights and privileges pertaining to communication facilities. Military control is retained over areas actually used for military purposes, such as camp sites, airfields, and supply bases. Control is also retained over sufficient adjacent terrain to provide for their defense. The right to operate military planes throughout the theater of operations is also retained. Com-

WITHDRAWAL FROM ACTIVE MILITARY OPERATIONS

plete jurisdiction over all members of the United States forces, even if serving in some capacity with the local government, should be retained. Under no circumstances should members of the United States forces be subject to trial by the courts of the foreign country. The detention of members of the United States forces by native police or military authority should be permitted only in the gravest emergency, and then only for the protection of life and property until the offender can be turned over to United States military or naval authority. In all small-war operations, a definite policy relative to the joint functioning, the extent of duties, and the mutual relations among local police forces, local military forces, and United States forces should be enunciated as early as practicable. Modification of the details of the policies agreed upon are transmitted to all concerned from time to time as the withdrawal progresses.

15-6. **Procedure.**—Normally, the first step in the withdrawal from active military operations will be the concentration of troops at some suitable location or locations in each military territorial subdivision. The final step will probably be the withdrawal of troops from these territorial subdivisions to a final concentration point or points. No area is evacuated until adequate local agencies have assumed the responsibility for the maintenance of law and order. Usually, the initial withdrawal of troops is from the more tranquil or remote areas. If conditions permit, troops are assembled at the most advantageous locations in the territorial subdivisions and are then withdrawn by battalions or regiments. In anticipation of withdrawal, and for other cogent reasons, it is desirable to release the United States forces from all routine constabulary and police duties as early in the campaign as the situation permits. Such procedure does not preclude participation in joint combat operations, since the United States forces continue to act as support for the native military forces. Unforeseen developments in the military situation may necessitate active combat operations by United States forces in order to maintain the morale and prestige of the native military forces that find themselves hard-pressed. Tactical units of the United States forces may be evacuated from the theater of operations prior to the date of final withdrawal. In such cases, due regard should be given to the necessity of retaining certain special units that will be required to function until the last troops are withdrawn. Such special units include air, supply, and communication troops. Administrative plans and the logistics pertaining to the evacuation should be formulated well in advance of the actual troop movement.

SECTION III

FINAL WITHDRAWAL

15-7. **General.**—After the withdrawal from active military operations is completed, the United States forces have the status of reserves. Although their active employment is not anticipated, they are held in readiness for active military operations. Their presence is an influential factor in the support of a legally constituted local government. At times, a military commission, legation guard, or other component of the United States forces may remain in the country after the final withdrawal of the major portion of the troops. In such cases, they remain as a result of diplomatic exchanges between the Government of the United States and the country concerned. The final withdrawal is not thereby affected. The plan for the final withdrawal is submitted to higher authority for approval well in advance of the contemplated operation. The final authority empowered to approve, modify, or disapprove all or part of the plan for final withdrawal may be the senior officer present or the Secretary of the State, War, or Navy Department. In some cases, the approval of the Chief Executive of the United States may be required. The necessity for the approval of any or all of these authorities will depend upon the location of the theater of operations, the provisions of the agreement for the withdrawal, or other factors, dependent upon the type of operation. Many important questions requiring action by a higher authority will usually present

FINAL WITHDRAWAL

themselves. Those that can be foreseen are ordinarily submitted for decision well in advance of the date of final withdrawal.

15-8. **Plans and orders.**—Initial plans for withdrawal are usually tentative. Due to unforeseen developments in the military situation, they are subject to changes that may be imposed by directives and instructions from higher authority as they become necessary. As a consequence, initial orders for the withdrawal are issued in fragmentary form in order to allow the maximum preparation period. A formal written operation order with appropriate annexes, confirming fragmentary orders and embodying all pertinent instructions for the withdrawal, is issued well in advance of the final troop movement. All agencies and units of the United States forces should be given ample time to provide for every detail pertaining to the withdrawal. The formulation of a comprehensive plan for the final withdrawal is dependent upon securing definite information relative to the date of withdrawal, the ships available for transportation, the schedules of such ships, the naval support available, naval operations affecting the withdrawal, the destination and final organization of the United States forces involved in the withdrawal, and the policies and decisions of higher authority relative to the political and military features of the situation. The necessary orders for the execution of the withdrawal are issued when the military requirements of the situation have been determined and all plans for the final stages of the withdrawal have been perfected in cooperation with the naval forces involved in the evacuation. The administrative details will ordinarily require the issue of voluminous instructions. Only the essential general instructions pertaining to administrative matters are incorporated in the body of the operation order. Other administrative instructions are issued in the form of annexes to the operation order.

15-9. **Executive staff duties.**—Although the duties and responsibilities of the Executive Staff are essentially the same during the withdrawal period as at any other time during a small wars operation, the members of the staff will usually find that their attention is focused on certain definite phases of their duties that assume relatively greater importance during this period. The Chief of Staff may be designated as liaison officer for the purpose of assuring close cooperation with the local agencies of the State Department and other United States naval and military forces involved in the withdrawal. The Executive Staff concerns itself primarily with those decisions and policies announced by the commander of the United States forces from time to time during the preparation period. These include the following: the date troops are to be withdrawn from each concentration area; the dates on which major troop units are to be transferred to the continental United States; protective measures to be employed; assembly positions, when designated; routes of movement to be used by each unit; composition and strength of the last unit to clear the country; property and supplies to be returned to the continental United States; the disposition of property and supplies not to be returned to the United States; procedure to be followed during negotiations for final settlement of all claims against the United States as a result of military operations; reserve supplies and ammunition to be held available until the date of departure; date of evacuation of ineffectives, noncombatants, dependents, household effects, and excess baggage; and the ceremonies to be conducted upon the return of forts, barracks, or other property to the custody of the country concerned.

15–10. **First section.**—The first section is charged with the preparation of details relative to the evacuation of the members of the United States forces and their families. Advance information as to tentative assignments to new stations of all personnel is required in order that dependents and personal effects may be properly routed. Since the evacuation of dependents will also require the transportation of the personal effects and household goods, this movement should be executed well in advance of the final troop movement. Arrangement is made for the discontinuance of United States postal service ashore and the transfer of such activities to naval vessels within the theater of operations. Steps are taken to obviate the continued arrival of United States mail for members of the command after the withdrawal has been effected by a notification to all postmasters at forwarding ports in the continental United States of the date on which to discontinue forwarding mail. Plans are agreed upon relative to the relationship of military police with members of the United States forces, and with local military and police forces. Preparations are made for the discontinuance of welfare and post exchange activities. The quantity of welfare and post exchange supplies brought into the country during the preparation for withdrawal is limited to the minimum necessary for current needs. Measures are inaugurated to assure that all financial obligations, individual and organizational, of any kind whatsoever, except those pertaining to the Quartermaster and Paymaster Departments, are liquidated prior to departure. This procedure should be comprehensive, thorough, and timely in order to eliminate the submission of claims by individuals against members

7

of the United States forces or against the United States Government after the final withdrawal has been effected. The procurement and retention of releases and signed receipts from firms and individuals having financial dealings with the United States forces or with individual members of the United States forces will eliminate the filing of delayed claims for reimbursement. The first section prepares the administrative annexes with which that section is concerned.

15–11. **Second section.**—The second section continues its normal activities, paying particular attention to the reaction of the populace to the contemplated withdrawal. Steps are taken to uncover the activities of agitators or others who may attempt to interfere with the withdrawal. Appropriate action is taken to see that maps and monographs of the country are up-to-date and as accurate as circumstances will permit. Information thus collected may be of inestimable future value.

15–12. **Third section.**—The third section continues its normal functions. A continuous study of the situation is conducted in order that the section may be prepared to make recommendations for changes in the plan when the circumstances demand. The military situation may change suddenly, and tentative plans must be prepared for the renewal of military operations, should such action become necessary. A thorough study of the ammunition requirements should be made with due regard to the military situation, as well as to the fact that it is desirable to keep such supplies at the minimum necessary for estimated requirements.

15–13. **Fourth section.**—The fourth section is ordinarily confronted with a huge mass of detail prior to and during the withdrawal. The fourth section prepares and distributes administrative instructions and administrative orders to cover the details of the period of withdrawal. Provision is made for the disposition of all supplies, including ammunition, motor transportation, and animals. Steps are taken to dispose of real estate, shelter, and other facilities. In all cases, signed releases are obtained from lessors. Arrangement is made to dispose of all the utilities maintained by the United States forces. In order that the movement of excess supplies may be expedited, a schedule is prepared covering the tonnage to be shipped daily. A continuous study of the situation must be conducted in order that prospective or emergency changes may be met with adequate supply arrangements, and in order to insure the proper disposition of supplies and property. In general, all excess supplies that are to be returned to the continental United States are moved to the United States as early as possible. Care must be taken, however, not to reduce the amount of supplies below the estimated requirements. The equipment for all organizations should be reduced to that authorized for expeditionary units in existing tables of organization, in advance of the date of withdrawal. The normal reserve of supplies carried by the force may be used before drawing on the supplies in the base depot or depots. The depots at embarkation points change from receiving and distributing centers to collecting and shipping centers. They should be kept supplied with outbound freight to insure that all available ship space is utilized. Arrangements are made for the transportation of dependents and troops by land and water, including the operation of ports of embarkation. Available government transportation is employed to the fullest possible extent in effecting the withdrawal of both personnel and matériel. Orders are issued relative to highway circulation and the control of traffic. Decision is made as to the priority of the expenditure of funds. The evacuation of hospital patients to the continental United States should be carried out with a view to reducing the number remaining in field hospitals to a minimum as the final day approaches.

15-14. **Special staff duties.**—The disposition of extra equipment and supplies is an important duty of all special staff officers. Definite arrangements are necessary to prevent the loss of supplies as a result of carrying excess stocks. However, a proper amount of supplies must be maintained until the date of withdrawal.

15-15. **Air officer.**—All active planes are maintained in commission as long as practicable, and are flown or shipped to the continental United States. Since adverse flying weather may delay the departure of the planes by air, the date of departure for planes being flown to the continental United States will normally be placed somewhat in advance of the date of departure of the last troop units. Consideration should be given to the fact that supply and maintenance facilities during the period immediately preceding the date of withdrawal will be somewhat limited. Air operations should be reduced to the minimum required by the military situation in order that all planes may be in proper mechanical condition prior to their departure from the country.

15-16. **Engineer officer.**—The engineer officer is in charge of the dismantling of portable construction that is to be returned to the continental United States. The construction of cranes and other weight-lifting machinery may be required to move and load heavy

FINAL WITHDRAWAL

material. If existing dock facilities are deficient, preparations are made to reinforce and enlarge such facilities or to construct floating docks if required. The cooperation of naval forces at the port or ports of embarkation will be of considerable assistance.

15-17. **Communications officer.**—The communications officer is responsible for the maintenance of communications until the last headquarters is closed. Plans are made for the disposition of communication equipment. Field radio sets are employed from the time permanent stations are closed until the arrival of naval vessels. Arrangements are made with naval vessels to take over radio communication on a date just prior to the final date of withdrawal. Provision is made for the operation of a message center and messenger service until the headquarters of the United States forces is closed. Telephone service is continued at important stations by means of field sets or permanent installations until the final date of withdrawal.

15-18. **Surgeon.**—Field hospitals are maintained to meet maximum requirements until the naval forces present are able to furnish hospitalization. Patients are evacuated from field hospitals as soon as ships are available. Plans are made for the evacuation of casualties to the embarkation point or points in case of casualties during movement to the latter. Special sanitary measures are adopted, provided it is found necessary to concentrate civilian dependents and additional troops in concentration areas prior to embarkation. Provision is made for the assignment of additional medical personnel to units evacuated on ships that are not equipped as transports.

15-19. **Quartermaster.**—*a.* The quartermaster is charged with the preparation for shipment of equipment and supplies that are to be transported to the continental United States. In all cases, excess supplies are shipped at the earliest practicable date. Household effects and baggage should be shipped prior to the date of final withdrawal. The supplies and equipment are, however, maintained in proper quantity until the date of departure in order to care for current needs. Arrangements are made for the operation of quartermaster utilities, storage, and repair facilities to the capacity required until the naval forces afloat can take over such functions. The transportation of supplies and troops by land and water is a function of the quartermaster. Loading plans are completed in detail in order that the movement of troops and supplies may be carried out without confusion or delay.

b. All unserviceable property of every kind should be surveyed and sold in order to avoid congestion and rush in the final days of the withdrawal. Unserviceable property should be disposed of, except such items as cooking ranges, mess equipment, and other similar articles that are required for use up to the date of withdrawal. Property that is serviceable, but so worn that it is not worth transportation to the United States, should be surveyed and sold. Careful consideration should be given to all installations of refrigerating equipment and like items with a view to local disposition of heavy items of such nature whose condition does not fully warrant return to the continental United States for future military use. Steel cots, worn mosquito nets, mattresses, sheets, old office furniture, and other articles of this nature should be surveyed and sold.

c. Stocks of clothing and subsistence stores, as well as other items of normal supply, should be reduced to a minimum prior to evacuation by making requisition for replacements only when absolutely necessary. Supplies of this nature remaining on hand at the time of withdrawal should be returned to the United States. All weapons, ordnance stores, ammunition, and other classes of stores that are serviceable should be returned to the continental United States.

d. Motor transportation and motor transport equipment should be carefully inspected and all that is not considered serviceable for future use should be disposed of by sale. All radio equipment that is serviceable should be returned to the continental United States. Particular care should be taken to see that such equipment is carefully packed under the direct supervision of the communications officer, in order to avoid damage in shipment. Such equipment should be returned complete with all spares and accessories available.

e. The withdrawal period will require the appropriation of more funds than would ordinarily be required. Requirements must be foreseen, and arrangements made for the procurement and allotment of the extra funds well in advance of the date of withdrawal.

f. All contracts for supplies are canceled, final payment is made, and receipts obtained. Proper releases are obtained from lessors of all property in order to avoid subsequent claims for damage. Buildings belonging to the United States forces located on land belonging to the foreign government and those constructed on leased land should be sold.

○